Dominik Robert Naake

Simulation of damage mechanisms in weave reinforced materials based on multiscale modeling

Karlsruher Schriftenreihe Fahrzeugsystemtechnik
Band 79

Herausgeber
FAST **Institut für Fahrzeugsystemtechnik**
Prof. Dr. rer. nat. Frank Gauterin
Prof. Dr.-Ing. Marcus Geimer
Prof. Dr.-Ing. Peter Gratzfeld
Prof. Dr.-Ing. Frank Henning

Das Institut für Fahrzeugsystemtechnik besteht aus den Teilinstituten Bahnsystemtechnik, Fahrzeugtechnik, Leichtbautechnologie und Mobile Arbeitsmaschinen.

Eine Übersicht aller bisher in dieser Schriftenreihe erschienenen Bände finden Sie am Ende des Buchs.

Simulation of damage mechanisms in weave reinforced materials based on multiscale modeling

by
Dominik Robert Naake

KIT Scientific Publishing

Karlsruher Institut für Technologie
Institut für Fahrzeugsystemtechnik

Simulation of damage mechanisms in weave reinforced materials based on multiscale modeling

Zur Erlangung des akademischen Grades eines Doktors der Ingenieurwissenschaften von der KIT-Fakultät für Maschinenbau des Karlsruher Instituts für Technologie (KIT) genehmigte Dissertation

von M.Sc. Dominik R. Naake

Tag der mündlichen Prüfung: 15. April 2019
Hauptreferent: Prof. Dr.-Ing. Frank Henning
Korreferenten: Prof. Prof. hon. Dr. Tim A. Osswald
Dr.-Ing. Luise Kärger (KIT Associate Fellow)

Impressum

KIT Scientific Publishing

Karlsruher Institut für Technologie (KIT)
KIT Scientific Publishing
Straße am Forum 2
D-76131 Karlsruhe

KIT Scientific Publishing is a registered trademark of Karlsruhe Institute of Technology.
Reprint using the book cover is not allowed.

www.ksp.kit.edu

Print on Demand 2020 – Gedruckt auf FSC-zertifiziertem Papier

ISSN 1869-6058
ISBN 978-3-7315-1005-5
DOI 10.5445/KSP/1000104598

Acknowledgments

The work presented in this thesis was carried out between 2014 and 2018 at the Robert Bosch GmbH in the Corporate Sector for Research and Advance Engineering and the Institute of Vehicle System Technology (FAST), Department of Lightweight Technology (LBT) at the Karlsruhe Institute of Technology (KIT). Here, I would like to say a few words of appreciation to the many people who have contributed in countless ways to the success of this work.

I extend my deepest gratitude to Prof. Dr.-Ing. Frank Henning. Even though I was externally employed by Bosch, he supported my research topic continuously and integrated me into the Institute as best as possible. I would like to thank Dr.-Ing. Luise Kärger for her willingness to participate as the co-examiner for my doctoral thesis and for her support throughout this project. The discussions with her made my stays at the institute in Karlsruhe very fruitful periods. I would like to thank Prof. Prof. Dr. Tim Osswald for his interest in my work, for the co-examination of this thesis, and for his effort he spent to review it.

I am sincerely grateful for my advisor Dr.-Ing. Fabian Welschinger. Thanks to countless discussions, his support and motivation, I could develop my ideas and achieve the best quality in my work. My gratitude goes to Prof. Dr.-Ing. Markus André for his support, especially at the beginning of my work. I would like to thank my group leader Dr.-Ing. Kurt Hornberger and department head Dr.-Ing. Martin Giersbeck for the constant support and the freedom granted to pursue my work so independently. Furthermore, I want to thank all colleagues at Bosch and at the FAST for the constant friendly atmosphere and the many discussions on- and offtopic. Special thanks go to Riccardo Pietrogrande, Patrick Hessman, Siegfried Galkin, and Dominik Dörr for their great support during this adventure.

I especially want to thank my friends Franz, Enrico, Veronica, Riccardo, Giada, and Diana, without whose support I would not have been able to complete this project. Special thanks go to my beloved fiancée Federica Dadamo for always being on my side, her patience, and her endless support. Finally, I want to express my deepest gratitude towards my parents, Eleonore and Andreas, and my brother Thomas. Wherever my way has led me, I could always trust on their unconditional love and support. Their constant encouragement has given me the strength to tackle such a project and it will always be remembered.

Stuttgart, December 2019 *Dominik Naake*

Abstract

This thesis deals with the multiscale investigation of a weave reinforced composite material with thermoplastic polypropylene matrix. Three different subscales are investigated: the basic constituents (fibers and matrix), unidirectionally reinforced microstructures, and the full composite of woven fiber structures embedded in a thermoplastic matrix. The aim of the present work is the introduction of a closed multiscale chain to predict the material behavior of the macroscopic composite with respect to deformation and failure. First, the experimental evaluation of the different scales of the present material system is discussed. The basic idea throughout this thesis is to use fitted data as input for microstructure level simulations only on the lowest scale, i.e. for matrix and fibers. The parameters for the higher-scale models are calculated using simulations and a suitable methodology for scale bridging. Moduli of elasticity, plastic properties and material strengths as well as failure behavior are evaluated on the basis of simple specimens, taking into account different loading rates. For the simulation of the deformation behavior of the material on different scales it is of great importance to consider the local orientation and the relative position of the reinforcing fibers. For this reason, the geometry of the structures is first measured. The emphasis of the mechanical tests lies on the determination of the anisotropic influence on the deformation behavior of the composite materials and the comparison of the observed failure types with those from the literature. The measured stress-strain curves are used for the validation of the model responses. The generation of virtual representative volume elements of the micro- and mesostructure based on the results of the geometry measurement and the definition of suitable periodic boundary conditions enables the computational analysis of the constitutive behavior of composite materials. The second part of this thesis defines the theoretical framework for the corresponding material models. Three constitutive models are introduced in the course of this work according to the identified scales. In addition to the prediction of deformation, the focus lies on the consideration of material non-linearities such as plasticity in the matrix material and the damage development in all underlying material systems. In order to ensure thermodynamically consistent behavior, the models provided are derived from the free energy and the dissipation functions. All models presented are defined in the kinematic framework of large strains. The ability of the geometric and constitutive models to predict the deformation and failure behavior of fiber-reinforced composites is demonstrated by means of selected examples.

Kurzfassung

Diese Arbeit beschäftigt sich mit der Multiskalenuntersuchung eines gewebeverstärkten Verbundwerkstoffs mit einer thermoplastischen Polypropylenmatrix. Drei verschiedene Subskalen werden untersucht: die Basiskomponenten, Fasern und Matrix, unidirektional verstärkte Mikrostrukturen und der vollständige Verbund von gewebten Faserstrukturen in thermoplastischer Matrix. Das Ziel der vorliegenden Arbeit ist die Einführung einer geschlossenen Multiskalenkette, um das Materialverhalten des makroskopischen Verbundwerkstoffs in Bezug auf Verformung und Versagen rechnerisch vorherzusagen. Zunächst wird die experimentelle Bewertung der verschiedenen Ebenen des vorliegenden Materialsystems vorgenommen. Die grundlegende Idee dieser Arbeit ist es, gefittete Daten als Input für die Simulationen auf Mikrostrukturebene lediglich auf der untersten Skala, das heißt für Matrix und Fasern zuverwenden. Die Parameter für die Modelle höherer Ebenen werden mit Hilfe von Simulationen und einer geeigneten Methodik zum Skalenübergang rechnerisch ermittelt. Unter Berücksichtigung unterschiedlicher Belastungsraten werden Elastizitätsmoduli, plastische Eigenschaften und Materialfestigkeiten sowie das Versagensverhalten anhand von Probekörpern ausgewertet. Für die Simulation des Verformungsverhaltens des Materials auf unterschiedlichen Skalen ist es von großer Bedeutung, die lokale Orientierung und die relative Lage der Verstärkungsfasern zueinander zu berücksichtigen. Aus diesem Grund werden die Geometrien der Strukturen zunächst vermessen. Der Schwerpunkt der mechanischen Prüfungen liegt auf der Bestimmung des anisotropen Einflusses auf das Verformungsverhalten der Verbundwerkstoffe und dem Vergleich der beobachteten Versagensarten mit denen aus der Literatur. Die gemessenen Spannungs-Dehnungs-Kurven werden für die Validierung der Modellantworten verwendet. Die Generierung von virtuellen repräsentativen Volumenelementen der Mikro- und Mesostrukturen auf Basis der Ergebnisse aus der Geometrievermessung und die Definition geeigneter periodischer Randbedingungen ermöglicht eine rechnerische Analyse des konstitutiven Verhaltens von Verbundwerkstoffen. Der zweite Teil der vorliegenden Arbeit definiert den theoretischen Rahmen für die verwendeten

Materialmodelle. Drei Konstitutivmodelle werden entsprechend der identifizierten Skalen im Laufe dieser Arbeit eingeführt. Neben der Vorhersage der Verformung liegt der Schwerpunkt auf der Berücksichtigung materieller Nichtlinearitäten wie Plastizität im Matrixmaterial und Schädigungsentwicklung in allen zugrunde liegenden Materialsystemen. Um ein thermodynamisch konsistentes Verhalten zu gewährleisten, werden die bereitgestellten Modelle aus der freien Energie und den Dissipationsfunktionen abgeleitet. Alle vorgestellten Modelle sind im kinematischen Rahmen großer Verformungen definiert. Die Fähigkeit der geometrischen und konstitutiven Modelle das Verformungs- und Versagensverhalten faserverstärkter Verbundwerkstoffe abzubilden, wird anhand von ausgewählten Beispielen aufgezeigt.

Contents

1 Introduction

1.1 Motivation

Requirements and needs for mobility solutions are constantly increasing. In line with these developments, a major part of research and development tasks in the upcoming years will be related to the complex of alternative energies and an efficient usage of available resources. Especially in the automotive context, a political rethinking takes place these days, resulting in stricter emission regulatory and environmental specifications. In order to stay competitive in such a regulatory framework, it is of immense importance to drive research topics of resource efficient mobility. In this context, lightweight design of structural parts plays an important role. The usage of fibrous composites has an enormous potential to support the overcoming of the mentioned issues. Here, textile composites gain an increasingly important position in structural engineering, offering a reasonable ratio of performance and weight reduction over classical structural materials when designed correctly. Although continuous glass and carbon fiber composites mostly based on a thermoset matrix have been extensively used in mobility related areas like aeronautic, automotive, and transport applications, their potential regarding high-volume production, repairs and recycle abilities and reliable designs is not yet fully exploited. Composites based on thermoplastic matrix systems show significant advantages in this regard, enabling considerably shorter production cycle times, an easier handling during processing and an enhancement with regard to repairs and recycling possibilities due to their different chemical character. In contrast to these positive effects, the complex microstructure of fibrous composites in general makes it very difficult to accurately predict deformation and damage behavior. Especially the succession of failure processes and the identification of responsible triggers have not been entirely clarified yet. In addition, failure in woven fabric reinforced composites takes place on different scales, where single phases with non-linear behavior interact. Consequently, these materials are often used by means of exaggerated safety factors, based on empirical data. A purely experimental approach to characterize failure in continuous fiber composites results in high costs in terms of resources and time, necessary because of the high testing effort due to anisotropic nature of such materials. To overcome this drawback, numerical methods are used, supplying a valid support. However, available numerical techniques regarding the overall behavior of woven fabric composites including damage formulations have not yet reached a sufficiently matured state. Multiscale analysis can help to provide a better understanding of the interactions of the constituents on designated geometrical levels, allowing the description of failure processes and finally enable the formulation of macroscopic constitutive equations for woven fabric reinforced composites. The advantage of such an approach is the transfer of local material effects occurring on different length scales onto the macroscopic component level.

1.2 Objectives and overview

The work at hand deals with the experimental investigation and consequent modeling of the failure behavior under quasi-static loading of *organo sheets*. The latter are fully impregnated and consolidated semi-finished products, where a thermoplastic material is used as basic constituent. Here, polypropylene serves as a coating matrix and embedded glass fibers form the inter-woven reinforcement structure. The results of an optical analysis of the given material system are depicted in Fig. 1.1, allowing a direct identification of intrinsic hierarchical levels in a top-down manner. Hereby, *global* and *local* entities can be defined on each scale, describing the interfaces of the adjacent geometrical levels:

- **Macroscopic level**: The three-dimensional form defines the *global* behavior of the composite part. As a direct derivation of geometry by means of *local* parameters of the material, the distribution of the local reinforcements, fiber volume fraction, reinforcement orientation, etc., follow. These entities finally describe the local composite stiffness, which corresponds to the *global* level on the mesoscopic level.
- **Mesoscopic level**: This level describes the superordinate reinforcement structure within the composite, defining the weave pattern, preferred fiber directions and volume fractions of tows, shaped during processing.

On a *local* level, two phases can be distinguished, areas of pure matrix and impregnated yarns. While the matrix is assumed to be isotropic, the impregnated yarns yield an anisotropic material behavior due to the principal directions of the aligned fibers. Here, *local* entities such as fiber direction and volume fraction define the overall stiffness of the impregnated yarn. *Local* variables can be determined as the *global* behavior of the microscopic level.

- **Microscopic level**: The arrangement of fibers within the thermoplastic matrix is described yielding the overall (*global*) material behavior of impregnated yarns. *Local* properties are defined by the material behavior of the basic constituents, fibers and matrix.

By inverting this analysis, the used multiscale approach that is pursued in a two-track (experimental and numerical) manner is directly deduced. This structure is also reflected in the structure of this work. Throughout the course of this work a closed modeling chain over all scales is pursued, accompanied by numerical calculations, using the finite element code Abaqus. By applying suitable homogenized material formulations (cf. colored arrows in Fig. 1.2), the given methodology allows the up-ward transfer of local failure mechanisms occurring on the depicted scales. To this end, the definition of a newly proposed macroscopic material model for fabric reinforced thermoplastics taking into

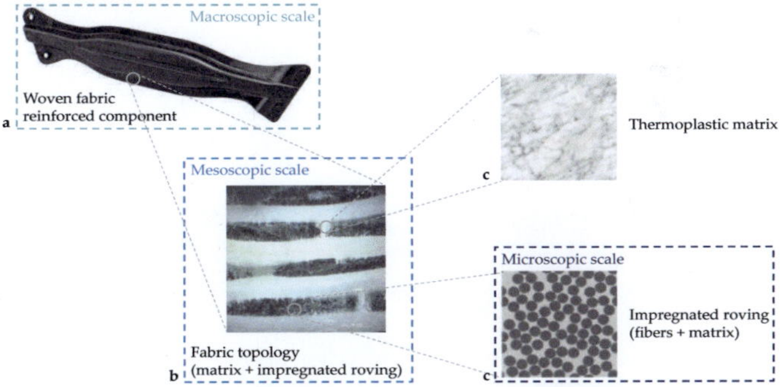

Figure 1.1: Optical analysis of a woven fabric reinforced component made of *organic sheets*. Distinct hierarchical levels in the multiscale structure can be identified as: (a) Macroscopic level, described in the dimensions of the component, (b) Mesoscopic level, resolving the woven structure of the material, and (c) Microscopic level, on which the basic constituents are visible.

account local material orientations and relevant failure mechanisms is presented. An overview of the structure of the presented thesis is given below.

Chapter 1 outlines the major research topics as well as goals and structure of the present thesis.

Chapter 2 is devoted to the introduction of the fundamental continuum mechanical formulations necessary for the definition of material models (cf. Section 2.1). Furthermore, the basic equations used during a multiscale analysis (cf. Section 2.2) are presented. Hereby, both analytical and numerical homogenization techniques are considered. Section 2.3 introduces major geometrical features of the fabric reinforced subcomposites necessary for the virtual assessment of the failure behavior associated with the mesoscopic structure and illustrates the resulting physical failure behavior in the single rovings and fabric reinforced composites. Available modeling techniques for all relevant geometrical scales are shortly reviewed in Section 2.4.

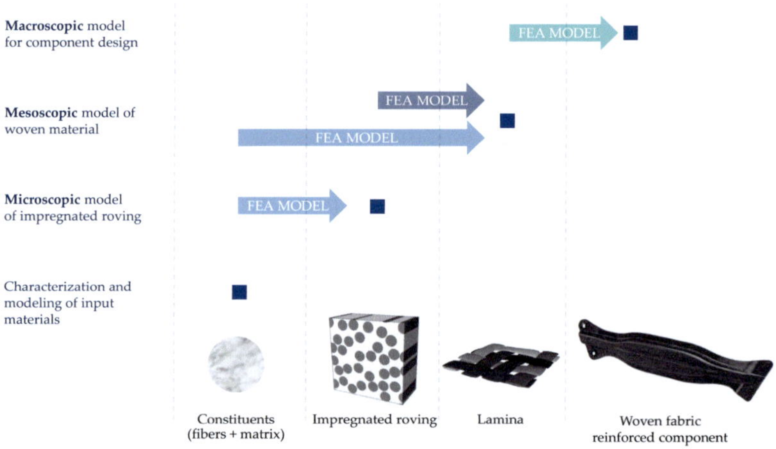

Figure 1.2: Proposed multiscale modeling approach guided by the optical analysis from Fig. 1.1. Definition of the distinct four scales of interest: constituent, microscopic, mesoscopic, and macroscopic level. Scale-bridging, transferring information onto a superordinate scale, is achieved by the definition of homogenized material models used within the finite element code Abaqus (cf. colored arrows, labeled with *Finite Element Analysis MODEL*).

Chapter 3 introduces the materials used for characterization and validation of the developed material model and presents the experimental results. Section 3.1 gives an overview over strain and stress measures used for the presentation of the results. A solid parallel validation of the proposed modeling techniques necessitates adequate experimental characterization of the underlying damage and failure processes on the identified scales and hierarchical levels. Consistent with the stages of the multiscale analysis presented above, three material levels, neat polypropylene matrix (*PP*, cf. Section 3.2), unidirectionally glass fiber-reinforced composite (*PP/GF UD*, cf. Section 3.3), and woven fabric reinforced polypropylene (*PP/GF twill weave*, cf. Section 3.4) have been tested under quasi-static loading.

Referring to Fig. 1.2, the transfer from a lower to an upper scale (*scale-bridging*) is accomplished by using suitable material models. These represent the material behavior of the micro- and meso-structures in such a way, that all relevant material effects in terms of damage are respected.

The development of the resulting three material models is presented in Chapter 4. Section 4.1 focuses on the definition of a rate-dependent elastoplastic material model including a thermodynamically consistent damage formulation of the matrix material for isothermal conditions. After an initial definition of additive plasticity using the small strain framework, the model is extended towards a finite formulation using a formulation in the logarithmic strain space. Incorporating the predominant occurring failure mechanisms, Section 4.2 introduces a fully three-dimensional continuum damage model for impregnated rovings (*UD model*), representing parallel fiber arrangement embedded in a thermoplastic matrix. In order to consider large rotations of the reinforcement structures in the mesoscopic fabric reinforced structure undergoing external loads, the presented material model is defined in a finite strain framework. Suitable criteria for damage onset and a thermodynamically consistent derivation of damage evolution complete the constitutive model.

Section 4.3 focuses on the composite formulation of a constitutive model taking into account the combination of the aforementioned constitutive equations. For the present material system, the preferred directions coincide with the principal directions of the reinforcements which change continuously during loading due to material deformation. Respecting these kinematic conditions by the introduction of an intermediate configuration, separate damage formulations for matrix and rovings are included from the models introduced before. The present work therefore provides a straightforward formulation of a macroscopic material model incorporating plastic and anisotropic damage effects for fabric reinforced materials based on a combination of analytical and numerical homogenization techniques. This procedure allows the transfer of dominant material effects onto the macroscopic scale and the consideration of varying approximated arrangement of reinforcement structures. Remarks on the algorithmic treatment and implementation of the constitutive equations for all scales are presented accordingly.

Chapter 5 deals with the numerical application of the presented material models to realistic geometries. The results of numerical calculations obtained using virtual micro- and mesoscopic structures support the interpretation of experimental results and enhance a further understanding of occurring material effects. An additional objective of such investigations is the assessment of the usability of virtual characterization as a substitute for expensive experimental testing of plastic composites. A major key to retrieve resilient results is the correct reproduction of the microstructural conditions at hand. Section 5.1 is devoted to the introduction of microstructure generation of unidirectionally and woven composites, based on the results presented in Chapter 3 with regard to the geometry of the composite. An extension of an available algorithm for the generation of statistically representative volume elements (*SRVE*) is presented, giving the possibility to respect variable fiber diameters as well as near-field clustering of groups of fibers. For the generation of the virtual woven unit cells an existing software package (*TexGen*) was used and embedded in the preprocessing framework.

Section 5.2 demonstrates the results of virtual characterization calculations, using the generated unidirectional microstructures. Hereby, the focus lies on the calculation of the elastic bodies and the demonstration of plastic and damage behavior of the composite using selected examples. Where possible, the obtained results are compared to experimental results. Following the strategy of the implemented multiscale analysis, the main objective of the presented calculations on this scale is the determination of a failure envelope, that is used to parametrize the homogenized UD model discussed in Section 4.2.

Using the UD model for the calculation of geometrical mesoscopic models of the impregnated fabric structure (*woven unit cell*) in Section 5.3, enables a up-transfer of the subordinate damage effects. On the basis of selected loading cases, the kinematical and damage related behavior of woven fabric reinforced composites is investigated. A comparison of numerically obtained homogenized and available experimental results is given, with respect to stiffness and the non-linear behavior of the stress-strain response.

The features of the newly proposed macroscopic material model are demonstrated in Section 5.4. The model behavior with respect to the change of material orientation under deformation as well as the representation of damage evolution is reviewed in the course of exemplary virtual tests.

The thesis concludes in Chapter 6, summarizing the undertaken modeling steps included in the proposed full-featured multiscale approach. Discussing the obtained results, propositions for improvement of the used methodology and remarks on possible future work are given.

2 Fundamentals and state of the art

The objectives of this chapter are to introduce the fundamental formulations of continuum mechanics (cf. Section 2.1) and basic equations necessary for a multiscale analysis (cf. Section 2.2) using analytical and numerical homogenization techniques. Furthermore, Section 2.3 summarizes the major geometrical features of the woven subcomposites and illustrates the resulting physical failure behavior in the single rovings and multi-layer woven composites. Finally, Section 2.4 reviews relevant available modeling activities on different scales.

2.1 Fundamentals of continuum mechanics

The subsequent section introduces the basic equations of continuum mechanics that are needed to describe arbitrary material response. Throughout this work all quantities are expressed in terms of Cartesian coordinates. This section is by no means complete, for a more detailed representation the interested reader is referred to the work of HOLZAPFEL [169] among others.

2.1.1 Kinematics

In an EUCLIDean space $\mathbb{R}^3$, an orthonormal base with its base vectors $\boldsymbol{e}_i$ and origin O is defined. Within this domain a continuous body $\mathcal{B}$ containing any point of this body $P \in \mathcal{B}$ can be described. The setting of $\mathcal{B} \in \mathbb{R}^3$ is generally called a configuration $\chi(P, t)$, whereby $\mathcal{B}_\mathrm{s} = \chi(\mathcal{B}, t)$ and $\boldsymbol{x} = \chi(P, t)$. $\mathcal{B}_\mathrm{s}$ is the occupied region by $\mathcal{B}$ at time t, and $\boldsymbol{x}$ gives the spatial position of P within $\mathbb{R}^3$ as functions of time t. The initial undeformed configuration $\chi_0 = \chi(t_0)$ is chosen as reference configuration with $\mathcal{B}_\mathrm{m} = \chi_0(\mathcal{B}, t_0)$ and $\boldsymbol{X} = \chi_0(P, t_0)$ and serves as the material (LANGRANGian) description of the body with material coordinates $\boldsymbol{X}$. Deformed configurations are called spatial

(EULERian) descriptions with spatial coordinates $\boldsymbol{x}$. The uniquely invertible map function $\boldsymbol{\varphi}$ maps from reference to any spatial configuration by

$$\boldsymbol{x} = \boldsymbol{\varphi}\left(\boldsymbol{X}, t\right) \quad \text{and} \quad \boldsymbol{X} = \boldsymbol{\varphi}^{-1}\left(\boldsymbol{x}, t\right) \tag{2.1}$$

All entities introduced so far are depicted in Fig. 2.1. The motion of a continuum is the change of the position of any material point P in time. With known coordinates $\boldsymbol{X}$ and $\boldsymbol{x}$ of the undeformed and deformed material points the displacement field reads

$$\boldsymbol{u} = \boldsymbol{x}\left(\boldsymbol{X}, t\right) - \boldsymbol{X} = \boldsymbol{x} - \boldsymbol{X}\left(\boldsymbol{x}, t\right). \tag{2.2}$$

The velocity field $\boldsymbol{v}$ is obtained by the time derivative of the motion

$$\boldsymbol{v}\left(\boldsymbol{X}, t\right) = \frac{\mathrm{d}\boldsymbol{\varphi}\left(\boldsymbol{X}, t\right)}{\mathrm{d}t} = \boldsymbol{v}\left(\boldsymbol{\varphi}^{-1}\left(\boldsymbol{x}, t\right), t\right) = \boldsymbol{v}\left(\boldsymbol{x}, t\right). \tag{2.3}$$

In continuum mechanics the deformation gradient $\boldsymbol{F}$ as well as its inverse $\boldsymbol{F}^{-1}$ play the most important role. They are defined by

$$\boldsymbol{F}\left(\boldsymbol{X}, t\right) = \frac{\partial \boldsymbol{x}\left(\boldsymbol{X}, t\right)}{\partial \boldsymbol{X}} \quad \text{and} \quad \boldsymbol{F}^{-1}\left(\boldsymbol{x}, t\right) = \frac{\partial \boldsymbol{X}\left(\boldsymbol{x}, t\right)}{\partial \boldsymbol{x}} \tag{2.4}$$

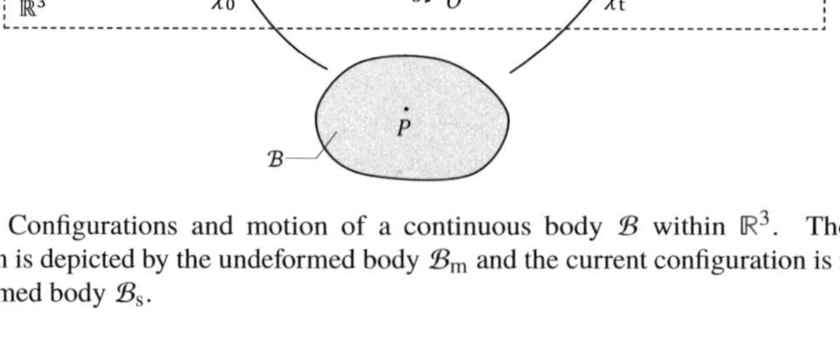

Figure 2.1: Configurations and motion of a continuous body $\mathcal{B}$ within $\mathbb{R}^3$. The reference configuration is depicted by the undeformed body $\mathcal{B}_\mathrm{m}$ and the current configuration is represented by the deformed body $\mathcal{B}_\mathrm{s}$.

and are used to describe kinematic entities. The deformation gradient maps line elements and is used to derive strain measures. In general, $\boldsymbol{F}$ and $\boldsymbol{F}^{-1}$ are non-symmetric second-order tensors that relate two different configurations and therefore are designated as two-point tensors. Due to the existence of the inverse, $J = \det(\boldsymbol{F}) \neq 0$ must hold true for the Jacobian determinant. The displacement gradients in material and spatial configuration can be introduced analogously by

$$\nabla_{\boldsymbol{X}} \boldsymbol{u} = \boldsymbol{F}(\boldsymbol{X}, t) - \mathbb{1} \quad \text{and} \quad \nabla_{\boldsymbol{x}} \boldsymbol{u} = \mathbb{1} - \boldsymbol{F}^{-1}(\boldsymbol{x}, t) . \tag{2.5}$$

At this point, three frequently used decompositions of the deformation gradient

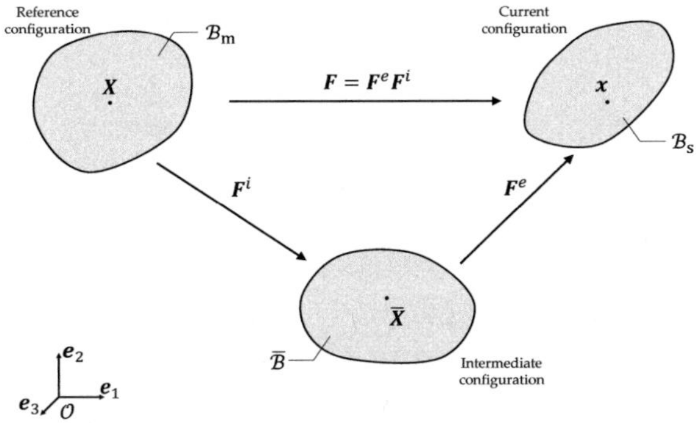

Figure 2.2: Intermediate configuration $\bar{\Omega}$ and the corresponding decomposition of the deformation gradient $\boldsymbol{F} = \boldsymbol{F}^e \boldsymbol{F}^i$ in an elastic and in-elastic part within $\mathbb{R}^3$.

shall be mentioned. The first is the geometrical decomposition in volumetric and isochoric deformations

$$\boldsymbol{F}(\boldsymbol{X}, t) = \boldsymbol{F}_{\text{vol}}(\boldsymbol{X}, t)\, \boldsymbol{F}_{\text{iso}}(\boldsymbol{X}, t) . \tag{2.6}$$

The second consists of an elastic and an in-elastic part, according to

$$\boldsymbol{F}(\boldsymbol{X}, t) = \boldsymbol{F}^e(\boldsymbol{X}, t)\, \boldsymbol{F}^i(\boldsymbol{X}, t) , \tag{2.7}$$

as depicted in Fig. 2.2. The polar decomposition splits the deformation gradient according to

$$\boldsymbol{F} = \boldsymbol{R}\boldsymbol{U} = \boldsymbol{V}\boldsymbol{R}, \tag{2.8}$$

where $\boldsymbol{R}$ is an orthogonal rotation tensor and $\boldsymbol{U}$ and $\boldsymbol{V}$ are the right and left symmetric stretch tensors, corresponding to material and spatial configuration respectively. By squaring the stretch tensor, the definition of the right and left CAUCHY-GREEN tensors follow

$$\boldsymbol{C} = \boldsymbol{U}^2 = \boldsymbol{F}^{\mathrm{T}}\boldsymbol{F} \quad \text{and} \quad \boldsymbol{b} = \boldsymbol{V}^2 = \boldsymbol{F}\boldsymbol{F}^{\mathrm{T}} \tag{2.9}$$

being again material and spatial tensors respectively. Note that $\boldsymbol{R}$ describes pure rigid body rotations, hence, all information about volume change is contained by $\boldsymbol{U}$ and $\boldsymbol{V}$. Therefore, we can rewrite the determinant of $\boldsymbol{F}$ simply by

$$J = \det[\boldsymbol{U}] = \det[\boldsymbol{V}] \quad \text{and} \quad J^2 = \det[\boldsymbol{C}] = \det[\boldsymbol{b}]\,. \tag{2.10}$$

The introduction of general, symmetric, positive definite and local deformation measures can be derived directly from the stretch tensors. For the material configuration the general SETH-HILL strain measures $\boldsymbol{\lambda}$ are often found. They can be written as

$$\boldsymbol{\lambda} = \begin{cases} \frac{1}{m}\left[\boldsymbol{U}^m - \mathbb{1}\right], & \text{if } m \neq 0 \\ \ln\left[\boldsymbol{U}\right], & \text{if } m = 0 \end{cases} \tag{2.11}$$

with $m \in \mathbb{R}$. Analogously strain measures corresponding to the spatial configuration can be found. In the following, two cases which will be used throughout this work are introduced in more detail. For $m = 2$ in Eq. 2.11 the GREEN-LAGRANGE strain measure is obtained

$$\boldsymbol{E}(\boldsymbol{X}, t) = \frac{1}{2}\left[\boldsymbol{C}(\boldsymbol{X}, t) - \mathbb{1}\right], \tag{2.12}$$

whereas $m = 0$ yields the HENCKY or logarithmic strain measure

$$\boldsymbol{\varepsilon}(\boldsymbol{X}, t) = \ln\left[\boldsymbol{U}(\boldsymbol{X}, t)\right] = \frac{1}{2}\ln\left[\boldsymbol{C}(\boldsymbol{X}, t)\right]. \tag{2.13}$$

Taking into account the decompositions in Eqs. 2.6 and 2.7, the additive deviatoric-volumetric split as well as the elastic-inelastic split can be recovered

$$\boldsymbol{\varepsilon} = \boldsymbol{\varepsilon}_{vol} + \boldsymbol{\varepsilon}_{iso} \quad \text{and} \quad \boldsymbol{\varepsilon} = \boldsymbol{\varepsilon}^e + \boldsymbol{\varepsilon}^i, \tag{2.14}$$

which have special significance throughout the course of this work.

Remark. For small deformation gradients, a simplified representation of the motion can be found (cf. DE SOUZA NETO et al. [296]). In general, the CAUCHY-GREEN deformation tensor can be rewritten in terms of the displacement gradients (cf. Eq. 2.5) as

$$\begin{aligned} \boldsymbol{C} &= \mathbb{1} + \nabla_{\boldsymbol{X}} \boldsymbol{u} + (\nabla_{\boldsymbol{X}} \boldsymbol{u})^{\mathrm{T}} + (\nabla_{\boldsymbol{X}} \boldsymbol{u})^{\mathrm{T}} \nabla_{\boldsymbol{X}} \boldsymbol{u}, \\ \boldsymbol{b} &= \mathbb{1} + \nabla_{\boldsymbol{x}} \boldsymbol{u} + (\nabla_{\boldsymbol{x}} \boldsymbol{u})^{\mathrm{T}} + \nabla_{\boldsymbol{x}} \boldsymbol{u} \, (\nabla_{\boldsymbol{x}} \boldsymbol{u})^{\mathrm{T}} . \end{aligned} \tag{2.15}$$

For considerably small displacement gradients $\nabla \boldsymbol{u} \approx \nabla_{\boldsymbol{X}} \boldsymbol{u} \approx \nabla_{\boldsymbol{x}} \boldsymbol{u}$, the quadratic terms and the distinction between material and spatial configuration vanish ($\boldsymbol{C} \approx \boldsymbol{b} \approx \mathbb{1} + \nabla \boldsymbol{u} + (\nabla \boldsymbol{u})^{\mathrm{T}}$) and the infinitesimal or nominal strain can be written as

$$\boldsymbol{\epsilon} = \frac{1}{2} \left[\nabla \boldsymbol{u} + (\nabla \boldsymbol{u})^{\mathrm{T}} \right] = \nabla^{s} \boldsymbol{u} \tag{2.16}$$

with ∇^s being the symmetric gradient. Approximating the deformation by the infinitesimal strain measure, the GREEN-LAGRANGE and logarithmic strains are similar to the nominal representation ($\boldsymbol{E} \approx \boldsymbol{\varepsilon} \approx \boldsymbol{\epsilon}$). Furthermore it can be shown that the available decompositions for the logarithmic strains apply in a similar manner also for infinitesimal strains.

With the abbreviation for the total time derivative $(\dot{\cdot}) := \mathrm{d}(\cdot)/\mathrm{d}t$, the rate form of the deformation gradient can be introduced by the gradient of the material velocity field

$$\dot{\boldsymbol{F}} (\boldsymbol{X}, t) = \frac{\mathrm{d}\boldsymbol{F}}{\mathrm{d}t} = \nabla_{\boldsymbol{X}} \dot{\boldsymbol{\varphi}} (\boldsymbol{X}, t) = \nabla_{\boldsymbol{X}} \boldsymbol{v} (\boldsymbol{X}, t) . \tag{2.17}$$

The spatial counterpart is defined analogously as

$$\boldsymbol{l} (\boldsymbol{x}, t) = \nabla_{\boldsymbol{x}} \boldsymbol{v} (\boldsymbol{x}, t) = \dot{\boldsymbol{F}} \boldsymbol{F}^{-1} = \boldsymbol{d} (\boldsymbol{x}, t) + \boldsymbol{w} (\boldsymbol{x}, t) , \tag{2.18}$$

where $\boldsymbol{d} (\boldsymbol{x}, t) = \frac{1}{2} \left(\boldsymbol{l} + \boldsymbol{l}^{\mathrm{T}} \right)$ and $\boldsymbol{w} (\boldsymbol{x}, t) = \frac{1}{2} \left(\boldsymbol{l} - \boldsymbol{l}^{\mathrm{T}} \right)$ with the symmetric part $\boldsymbol{d} = \boldsymbol{d}^{\mathrm{T}}$ as the rate of deformation tensor and the skew part $\boldsymbol{w} = -\boldsymbol{w}^{\mathrm{T}}$ as the rate of rotation tensor. The independence of the right CAUCHY-GREEN deformation tensor (cf. Eq. 2.9_1) from rotational components becomes visible also in its rate form through

$$\dot{\boldsymbol{C}} = \dot{\boldsymbol{F}}^{\mathrm{T}}\boldsymbol{F} + \boldsymbol{F}^{\mathrm{T}}\dot{\boldsymbol{F}} = 2\boldsymbol{F}^{\mathrm{T}}\boldsymbol{d}\,\boldsymbol{F}. \tag{2.19}$$

The same applies therefore for the GREEN-LAGRANGE strain measure (cf. Eq. 2.12)

$$\dot{\boldsymbol{E}} = \boldsymbol{F}^{\mathrm{T}}\boldsymbol{d}\,\boldsymbol{F}, \tag{2.20}$$

using the standard *pull-back* operation of $\boldsymbol{d}$ in both cases. Inversion of the relation yield the definition of the standard *push-forward* operation, transforming entities from the spatial to the material configuration.

2.1.2 Concept of stresses

Starting from the traction vector $\boldsymbol{t}$ on the surface $\partial\mathcal{B}_s$ of the body $\mathcal{B}_s$ in spatial configuration, the application of the theorem of CAUCHY such that

$$\boldsymbol{t} = \boldsymbol{\sigma}\boldsymbol{n} \tag{2.21}$$

and using the definition of an infinitesimal surface traction force $\boldsymbol{t}\mathrm{d}a$ yields the CAUCHY stress tensor $\boldsymbol{\sigma}$. Analogously, the introduction of the first PIOLA-KIRCHHOFF stress tensor $\boldsymbol{P}$ enables the formulation of the current stress state corresponding to the infinitesimal surface $\mathrm{d}A$ of the body $\mathcal{B}_0$ of the reference configuration. Applying material cuts in accordance to the EULER principle, every cut volume can be seen as a body $\mathcal{B}_\mathrm{s}$ and hence the stress tensor $\boldsymbol{\sigma}$ can be defined in any material point $\boldsymbol{x} \in \mathcal{B}_\mathrm{s}$ as depicted in Fig. 2.3.

Equation 2.22 summarizes frequently used stress measures in literature.

$$\begin{aligned}
&\boldsymbol{\sigma} && \text{CAUCHY stress tensor}\\
\boldsymbol{\tau} &= J\boldsymbol{\sigma} && \text{KIRCHHOFF stress tensor}\\
\boldsymbol{P} &= J\boldsymbol{F}^{-1}\boldsymbol{\sigma} && \text{First PIOLA-KIRCHHOFF stress tensor}\\
\boldsymbol{S} &= J\boldsymbol{F}^{-1}\boldsymbol{\sigma}\boldsymbol{F}^{-1} && \text{Second PIOLA-KIRCHHOFF stress tensor}
\end{aligned} \tag{2.22}$$

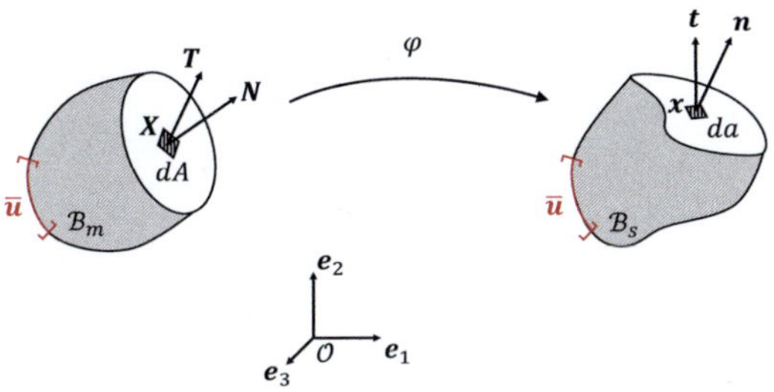

Figure 2.3: Material $\boldsymbol{T}$ and spatial traction vectors $\boldsymbol{t}$ of continuum body in the cutting plane. $\bar{\boldsymbol{u}}$ denotes the displacement boundary conditions of $\mathcal{B}_{\mathrm{m}}$ and $\mathcal{B}_{\mathrm{s}}$, respectively.

By performing standard *push-forward* and *pull-back* operations the description of stresses in the referring configuration can again be changed. The same applies for the strain measures accordingly.

2.1.3 Physical balance principles

In this section a short overview over the necessary balancing equations is given. For a more detailed review the reader is referred to the extensive works of HOLZAPFEL [169], TRUESDELL [388], TRUESDELL & NOLL [389], ERINGEN [106], DOYLE & ERICKSEN [100], GREEN & ADKINS [130], and others. The subsequent equations must be satisfied at all material points within the volume at all times. For the modeling activities throughout this work, the overview here is restricted to a purely mechanical context. Four basic groups of equations regarding a spatial subregion $\mathcal{P}_{\mathrm{s}} \subset \mathcal{B}_{\mathrm{s}}$ ($\mathcal{P}_{\mathrm{m}} \subset \mathcal{B}_{\mathrm{m}}$ for the material framework) can be identified.

1. **Conservation of mass**: Neglecting transport into a subregion $\mathcal{P}_{\mathrm{s}}$ and production within $\mathcal{P}_{\mathrm{s}}$, the now constant mass can be described by a single scalar value by

$$\frac{\mathrm{d}}{\mathrm{d}t} m = 0 \quad \text{with} \quad m = \int_{\mathcal{P}_{\mathrm{s}}} \mathrm{d}m = \int_{\mathcal{P}_{\mathrm{s}}} \rho \mathrm{d}v = \int_{\mathcal{P}_{\mathrm{m}}} \rho_0 \mathrm{d}V \tag{2.23}$$

in terms of the current density ρ and the reference density $\rho_0 = J\rho$.

2. **Balance of linear and angular momentum**: The balance of linear momentum postulates the equality of the temporal change of the linear momentum $\boldsymbol{I}$ of $\mathcal{B}_\mathrm{s}$ and external force $\boldsymbol{F}_\mathrm{ext}$

$$\frac{\mathrm{d}}{\mathrm{d}t}\boldsymbol{I} = \boldsymbol{F}_\mathrm{ext}, \tag{2.24}$$

with the definitions of the linear momentum

$$\boldsymbol{I} = \int_{\mathcal{P}_\mathrm{s}} \dot{\boldsymbol{x}}\mathrm{d}m = \int_{\mathcal{P}_\mathrm{s}} \rho\boldsymbol{v}\mathrm{d}v \tag{2.25}$$

and the resulting external force

$$\boldsymbol{F}_\mathrm{ext} = \int_{\mathcal{P}_\mathrm{s}} \rho\boldsymbol{b}\mathrm{d}v + \int_{\partial\mathcal{P}_\mathrm{s}} \boldsymbol{t}\mathrm{d}a \tag{2.26}$$

with $\boldsymbol{b}$ as a prescribed acceleration field. The surface integral can be rewritten as volume integration, applying the Gauss divergence theorem

$$\int_{\partial\mathcal{P}_\mathrm{s}} \boldsymbol{t}\mathrm{d}a = \int_{\partial\mathcal{P}_\mathrm{s}} \boldsymbol{\sigma}\boldsymbol{n}\mathrm{d}a = \int_{\mathcal{P}_\mathrm{s}} \nabla_{\boldsymbol{x}} \cdot \boldsymbol{\sigma}\mathrm{d}v. \tag{2.27}$$

Localization theorem finally yields the strong form of the balance of linear momentum

$$\rho\dot{\boldsymbol{v}} = \nabla_{\boldsymbol{x}} \cdot \boldsymbol{\sigma} + \rho\boldsymbol{b}. \tag{2.28}$$

In a similar manner the balance of angular momentum can be formulated. It states the equilibrium of the temporal change of the angular momentum $\boldsymbol{D}^0$ with respect to an arbitrary point (usually the origin O of a fixed coordinate system) and externally applied torque $\boldsymbol{M}^0_\mathrm{ext}$

$$\frac{\mathrm{d}}{\mathrm{d}t}\boldsymbol{D}^0 = \boldsymbol{M}^0_\mathrm{ext} \tag{2.29}$$

with

$$\boldsymbol{D}^0 = \int_{\mathcal{P}_\mathrm{s}} \boldsymbol{x} \times \dot{\boldsymbol{x}}\mathrm{d}m = \int_{\mathcal{P}_\mathrm{s}} \boldsymbol{x} \times \rho\boldsymbol{v}\mathrm{d}v \tag{2.30}$$

being the angular moment. The sum of the applied moments resulting from the body forces $\rho\boldsymbol{b}$ and the surface tractions $\boldsymbol{t}$ yield the torque

$$\boldsymbol{M}^0_{\text{ext}} = \int_{\mathcal{P}_\text{s}} \boldsymbol{x} \times \rho\boldsymbol{b}\mathrm{d}v + \int_{\partial\mathcal{P}_\text{s}} \boldsymbol{x} \times \boldsymbol{t}\mathrm{d}a. \tag{2.31}$$

Without showing the derivations, a direct conclusion is the symmetry of the stresses $\boldsymbol{\sigma} = \boldsymbol{\sigma}^\text{T}$.

3. **First law of thermodynamics - Balance of energy**: The first law of thermodynamics describes the equality of the change of total energy $\mathcal{E}_\text{tot}$ of a distinct subregion $\mathcal{P}_\text{s}$ and the sum of external mechanical and thermal power ($\mathcal{P}_\text{ext}$ and $\mathcal{Q}_\text{ext}$)

$$\frac{\mathrm{d}}{\mathrm{d}t}\mathcal{E}_\text{tot} = \mathcal{P}_\text{ext} + \mathcal{Q}_\text{ext}. \tag{2.32}$$

With regard to the context of purely mechanical problems, the specific definition of the thermal power is neglected at this point and $\mathcal{Q}_\text{ext}$ vanishes. The mechanical power can be defined in spatial configuration by

$$\mathcal{P}_\text{ext} = \int_{\mathcal{P}_\text{s}} \rho\boldsymbol{b} \cdot \boldsymbol{v}\,\mathrm{d}v + \int_{\partial\mathcal{P}_\text{s}} \boldsymbol{t} \cdot \boldsymbol{v}\,\mathrm{d}a. \tag{2.33}$$

The time derivative of the total energy being $\dot{\mathcal{E}}_\text{tot} = \dot{\mathcal{K}} + \dot{\mathcal{U}}$, consists of the kinetic contribution

$$\dot{\mathcal{K}} = \frac{\mathrm{d}}{\mathrm{d}t}\int_{\mathcal{P}_\text{s}} \frac{1}{2}\dot{\boldsymbol{x}} \cdot \dot{\boldsymbol{x}}\,\mathrm{d}m = \frac{\mathrm{d}}{\mathrm{d}t}\int_{\mathcal{P}_\text{s}} \frac{1}{2}\rho\boldsymbol{v} \cdot \boldsymbol{v}\,\mathrm{d}v \tag{2.34}$$

and a part regarding internal energy storage mechanisms

$$\dot{\mathcal{U}} = \frac{\mathrm{d}}{\mathrm{d}t}\int_{\mathcal{P}_\text{s}} \rho u\,\mathrm{d}v, \tag{2.35}$$

based on the specific internal energy density per unit volume u. From Eq. 2.32 the balance of kinetic energy can be derived

$$\dot{\mathcal{K}} = \mathcal{P}_\text{ext} - \mathcal{S} \tag{2.36}$$

where $\mathcal{S}$ describes the stress power evaluated in the current configuration

$$\mathcal{S} = \int_{\mathcal{P}_\mathrm{s}} \boldsymbol{\sigma} : \boldsymbol{d} \, \mathrm{d}v. \tag{2.37}$$

Hereby, $\boldsymbol{d}$ denotes the strain rate tensor. Alternative notations in a material configuration reads

$$\mathcal{S} = \int_{\mathcal{P}_\mathrm{m}} w_\mathrm{int} \, \mathrm{d}V \quad \text{with} \quad w_\mathrm{int} = J\boldsymbol{\sigma} : \boldsymbol{d} = \boldsymbol{P} : \dot{\boldsymbol{F}} = \boldsymbol{S} : \dot{\boldsymbol{E}} \tag{2.38}$$

giving the rate of internal mechanical work per unit reference volume and defining work conjugated pairs. The balance of kinetic energy yields directly the formulation for the balance of internal energy,

$$\frac{\mathrm{d}}{\mathrm{d}t}\mathcal{U} = \mathcal{S} \tag{2.39}$$

and the according local strong form in the spatial configuration can be stated as

$$\rho \dot{u} = \boldsymbol{\sigma} : \boldsymbol{d}. \tag{2.40}$$

4. **Second Law of thermodynamics - Entropy inequality Principle**: When a material returns in its initial state upon unloading then it behaves reversibly. However, energy is usually *dissipated* in some kind, e.g. through plastic or damage evolution. In order to quantify the extend of this energy transfer, the state variable of entropy $\mathcal{G}(t)$ is introduced. It is considered as a measure for the orderliness of an investigated system. In general, the entropy inequality condition states a positive entropy production at all times

$$\Gamma(t) = \frac{\mathrm{d}}{\mathrm{d}t}\mathcal{G}(t) - \mathcal{Q} \geq 0 \tag{2.41}$$

with

$$\Gamma = \int_{\mathcal{P}_\mathrm{s}} \rho\gamma \, \mathrm{d}v \quad \text{and} \quad \mathcal{G} = \int_{\mathcal{P}_\mathrm{s}} \rho\eta \, \mathrm{d}v, \tag{2.42}$$

where γ is the local entropy production per unit mass and η the local entropy per unit mass. Q is the rate of entropy input in the system and is governed by non-mechanical entities and therefore neglected here. The above relation in Eq. 2.41 is called the second law of thermodynamics, defining the direction of energy flow and ensuring the irreversibility of thermodynamic processes. Coming back to the example mentioned above, reversible processes are characterized by $\Gamma(t) = 0$, but are handled as an idealized special case. The mechanical dissipation per unit volume is introduced as

$$\mathcal{D} = \gamma\theta \geq 0. \tag{2.43}$$

Neglecting convective contributions, the CLAUSIUS-PLANK inequality for internal dissipation is identified as

$$\rho\mathcal{D}_{\text{int}} = \boldsymbol{\sigma} : \boldsymbol{d} - \rho\dot{\Psi} \geq 0, \tag{2.44}$$

with the HELMHOLTZ free energy $\Psi = u - \theta\eta$ where θ denotes the absolute temperature.

2.2 Multiscale modeling of fiber composites

Figure 1.1 shows the different levels of micro-structures in an organo sheet. The qualitative and quantitative mechanical behavior on the macroscopic scale of organo sheets are a direct consequence of the behavior of materials on the subordinate scales, their microscopical setup, and corresponding interactions. In many cases during the design process it is not necessary to know the exact stress and strain states within the microstructure, but it is sufficient to describe the overall behavior of the material adequately by means of macroscopic and effective substitute models. The mapping of microscopic effects onto a macroscopic description is commonly referred to as *homogenization*. This contrasts with the procedure of *localization* where distinct macroscopic measures are distributed onto the microstructures (cf. SUQUET [367]).

2.2.1 Scale bridging techniques

Multiscale models and modeling of the micro- or meso structures are popular approaches in order to assess the elastic and inelastic behavior of a material. General methods are well described in FISH [113] and SPAHN [360] among others. The relation between a mesoscale and a microscale is described by multiscale models, using homogenization laws. At the microscale, constitutive models are implemented and the resulting stress and strain fields are transferred to the upper scale via transformation field tensors (cf. CHACBOCHE et al. [67], FISH & YU [114], FISH et al. [116], and VOYIADJIS & DELIKTAS [407]) or by solving a numerical problem (e.g. finite element problem, cf. CAR et al. [50] and OLLER et al. [301]). In order to minimize computational effort, a certain degree of periodicity in microstructures is assumed. Throughout this work the microstructures of the material under consideration is taken into account by representative volume elements (*RVE*). Only a few decades ago, such a computational approach was not directly accessible due to the absence of sufficiently performant computers. Therefore, starting in the framework of linear elasticity, a number of analytical or semi-analytical approximation schemes based on RVEs and mean-field homogenization were developed early on, providing efficient algorithms to estimate the overall stiffness and other properties. Key models are the *self-consistent-method* by HILL [160] and HILL [161], its extension towards the *Halpin-Tsai equations* (cf. HALPIN & CARDOS [146]) or procedures based on the *Eshelby method* by ESHELBY [107–109], such as the *Mori-Tanaka method* introduced by MORI & TANAKA [280] and many others. For a more extensive review of existing homogenization schemes, the interested reader is referred to the works of TUCKER & LIANG [394] and KLUSEMANN & SVENDSEN [194].

2.2.1.1 Average stress and strain

Carefully note, that the given models yield reasonable results only in defined limits of volume fractions of the single phase α and furthermore special attention has to be given on the symmetry properties of resulting stiffness or compliance tensors for the case of elastic properties (cf. BENVENISTE [31], BENVENISTE et al. [32]). The main goal and strategy of this modeling type is to substitute and approximate a heterogeneous material by a homogenized equivalent. As a main assumption, *statistical heterogeneity* is stated, i.e. the homogenized material has the same (or at least comparable) overall properties as the heterogeneous material. To this end, the following equation is fulfilled

$$\bar{\sigma}\left(\bar{X}\right) = \bar{\mathbb{C}}\left(\bar{X}\right) : \bar{\varepsilon}\left(\bar{X}\right) \tag{2.45}$$

defining $\bar{\mathbb{C}}\left(\bar{X}\right)$ as the homogenized macroscopic (effective) stiffness tensor and

$$\bar{\sigma}\left(\bar{X}\right) = \frac{1}{V}\int_{\mathcal{B}_\mathrm{m}} \sigma\left(X\right)\mathrm{d}V \quad \text{and} \quad \bar{\varepsilon} = \frac{1}{V}\int_{\mathcal{B}_\mathrm{m}} \varepsilon\left(X\right)\mathrm{d}V. \tag{2.46}$$

$\bar{X}$ is the position vector of a macroscopic point within the material. Note that $\bar{(\cdot)} = \langle(\cdot)\rangle$ indicates the ensemble average and therefore a homogenized macroscopic entity and $X \in \mathcal{B}_\mathrm{m}$ denotes the position vector of any available material point in the representative averaging volume V on the microscopic scale. Figure 2.4 depicts the relation between macro and micro scale. For

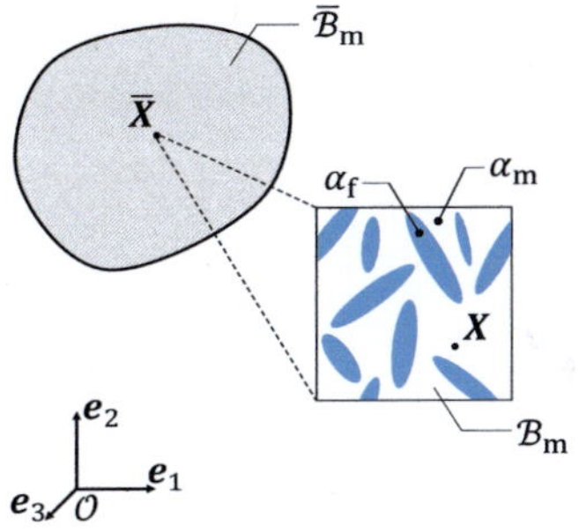

Figure 2.4: Relation between macro and micro scale. The material behavior of a macroscopic material point at a distinct position $\bar{X} \in \bar{\mathcal{B}}_\mathrm{m}$ is defined by its microstructure. Single phases are not distinguishable on macroscopic level. The representative averaging volume $\mathcal{B}_\mathrm{m}$ consists locally of different phases $\alpha = \alpha\left(X\right)$, where α_m corresponds to matrix material and α_f represents the fibrous inclusions. Material points within the microstructure are denoted by X.

improved clarity, $\bar{X}$ and X are not noted from now on. Formulating Eq. 2.46 in words, $\bar{\sigma}$ is defined as the volume average stress of the point-wise stress σ over the volume V. V has to be chosen big enough in order to guarantee statistical homogeneity of the results. Assuming a two-phase composite, consisting of matrix and fibers, V can be decomposed by phase differentiation according to

$$V = V_{\mathrm{m}} + V_{\mathrm{f}}, \tag{2.47}$$

which leads further to the decomposition

$$\bar{\boldsymbol{\sigma}} = v_{\mathrm{m}}\bar{\boldsymbol{\sigma}}_{m} + v_{\mathrm{f}}\bar{\boldsymbol{\sigma}}_{\mathrm{f}} \quad \text{and} \quad \bar{\boldsymbol{\varepsilon}} = v_{\mathrm{m}}\bar{\boldsymbol{\varepsilon}}_{\mathrm{m}} + v_{\mathrm{f}}\bar{\boldsymbol{\varepsilon}}_{\mathrm{f}} \tag{2.48}$$

with $v_{\alpha}=V_{\alpha}/V$ being the volume fractions for fibers (α=f) and matrix (α=m), respectively.

Analogously, according to HILL [159] and HASHIN [154], the homogenized stiffness tensor $\bar{\mathbb{C}}$ can be calculated through the potential energy formulation

$$\bar{U} = \frac{1}{2V}\int_{V} \boldsymbol{\sigma} : \bar{\boldsymbol{\varepsilon}}\mathrm{d}V = \frac{1}{2}\bar{\boldsymbol{\varepsilon}} : \bar{\mathbb{C}} : \bar{\boldsymbol{\varepsilon}}. \tag{2.49}$$

2.2.1.2 Energy-based boundaries of elastic properties

The potential energy defined in Eq. 2.49 gives rise to the introduction of two classical and rigorous bounds of elastic properties, the VOIGT and REUSS bounds (cf. HILL [158]). For the first it is assumed that all phases α see the exact same uniform strain, $\bar{\boldsymbol{\varepsilon}}^{*}$. Applying the principle of minimum energy, the following inequality is found

$$\bar{U} \leq U^{*} = \frac{1}{2V}\int_{V} \bar{\boldsymbol{\varepsilon}}^{*} : \mathbb{C} : \bar{\boldsymbol{\varepsilon}}^{*}\mathrm{d}V = \frac{1}{2V}\bar{\boldsymbol{\varepsilon}}^{*} : \int_{V} \mathbb{C}\mathrm{d}V : \bar{\boldsymbol{\varepsilon}}^{*}, \tag{2.50}$$

where $\bar{U}$ is the true energy and U^{*} the approximated one. Comparing coefficients with those of Eq. 2.49 yields the estimation for the VOIGT bound of the elastic stiffness tensor (cf. VOIGT [404] and WILLIS [418])

$$\bar{\mathbb{C}} \leq \int_{V} \mathbb{C}\mathrm{d}V = \sum_{\alpha} v_{\alpha}\mathbb{C}_{\alpha} = \bar{\mathbb{C}}_{\mathrm{V}}, \tag{2.51}$$

which describes an upper bound of physically consistent elastic material properties. The lower bound, described by the REUSS bound, is therefore derived analogously by assuming uniform stresses $\bar{\boldsymbol{\sigma}}^*$ over all phases and approximating the overall complementary energy

$$\bar{U}_c = \frac{1}{2V} \int_V \boldsymbol{\sigma} : \mathbb{C}^{-1} : \boldsymbol{\sigma} \mathrm{d}V, \tag{2.52}$$

yielding

$$\bar{U}_c \leq U_c^* = \frac{1}{2V} \int_V \bar{\boldsymbol{\sigma}}^* : \mathbb{C}^{-1} : \boldsymbol{\sigma}^* \mathrm{d}V = \frac{1}{2V} \bar{\boldsymbol{\sigma}}^* : \int_V \mathbb{C}^{-1} \mathrm{d}V : \bar{\boldsymbol{\sigma}}^* \tag{2.53}$$

and finally

$$\bar{\mathbb{C}} \geq \left[\int_V \mathbb{C}^{-1} \mathrm{d}V \right]^{-1} = \left[\sum_\alpha v_\alpha \mathbb{C}_\alpha^{-1} \right]^{-1} = \bar{\mathbb{C}}_{\mathrm{R}}. \tag{2.54}$$

For a more detailed review on the boundary derivation it is referred to the work of Wu & McCullough [421]. The above bounds provide stiffness tensors in the range

$$\bar{\mathbb{C}}_{\mathrm{R}} \leq \bar{\mathbb{C}} \leq \bar{\mathbb{C}}_{\mathrm{V}}. \tag{2.55}$$

For the case of isotropic phases, an isotropic homogenized stiffness will yield. Taking into account the geometry of the inclusions, Hashin & Shtrikman [152, 153] proposed a set of narrower bounds.

2.2.1.3 Mori-Tanaka model as an example for analytical homogenization

In what follows, the Mori-Tanaka scheme (cf. Mori & Tanaka [280]) is introduced, that allows a straight forward implementation and respects the microstructure in more detail. The basis for the derivations below is the concept of *localization*, i.e. the mapping of macroscopic states onto the microstructure. According to Hill [159], this is done by fourth-order strain- and stress-concentration tensors, $\mathbb{A}$ and $\mathbb{B}$, such that

$$\bar{\boldsymbol{\varepsilon}}_{\mathrm{f}} = \mathbb{A} : \bar{\boldsymbol{\varepsilon}} \quad \text{and} \quad \bar{\boldsymbol{\sigma}}_{\mathrm{f}} = \mathbb{B} : \bar{\boldsymbol{\sigma}}. \tag{2.56}$$

In general, the overall stiffness tensor can then be rewritten as

$$\bar{\mathbb{C}} = \mathbb{C}_{\mathrm{m}} + \sum_{\alpha} v_{\alpha} \left(\mathbb{C}_{\alpha} - \mathbb{C}_{\mathrm{m}}\right) : \mathbb{A}_{\alpha}. \tag{2.57}$$

An equivalent formulation for the compliance tensor can be formulated using the stress localization tensor $\mathbb{B}$. Equation 2.57 builds the basis for nearly all available analytical homogenization schemes, differing in a varying specification of the localization tensors. The above introduced Voigt and Reuss bounds of physically meaningful stiffness values are for instance recovered by defining $\mathbb{A} = \mathbb{B} = \mathbb{I}$. The fundamental basis of more elaborated models is the ESHELBY's *equivalent inclusion* [107, 109], introducing the Eshelby tensor $\mathbb{E}$. This theory is assumed to be known at this point, for a detailed derivation it is referred to the works of MURA [282], TAYA et al. [377–380], and TANDON & WENG [373]. Applying the solution of Eshelby, it is possible to find the stiffness of a composite with ellipsoidal fibers at *dilute* concentrations. The resulting localization tensor can be identified as

$$\mathbb{A}^{\mathrm{Eshelby}} = \left[\mathbb{I} + \mathbb{E} : \mathbb{S}_{\mathrm{m}} : \left(\mathbb{C}_{\mathrm{f}} - \mathbb{C}_{\mathrm{m}}\right)\right]^{-1}. \tag{2.58}$$

Assuming for a dilute composite the average strain being identical to the applied strain ($\bar{\boldsymbol{\varepsilon}} = \boldsymbol{\varepsilon}^{A}$) yields the strain state in the fibers

$$\bar{\boldsymbol{\varepsilon}}_{\mathrm{f}} = \mathbb{A}^{\mathrm{Eshelby}} : \bar{\boldsymbol{\varepsilon}}. \tag{2.59}$$

In contrast, the Mori-Tanaka scheme assumes that each particle of the composite is subjected to a far-field strain equal to the average strain in the matrix,

$$\bar{\boldsymbol{\varepsilon}}_{\mathrm{f}} = \mathbb{A}^{\mathrm{Eshelby}} \bar{\boldsymbol{\varepsilon}}_{\mathrm{m}}. \tag{2.60}$$

The basic equation for the implementation of a two-phase Mori-Tanaka model reads

$$\mathbb{A}^{\mathrm{MT}} = \mathbb{A}^{\mathrm{Eshelby}} \left[\left(1 - v_{\mathrm{f}}\right) \mathbb{I} + c_{\mathrm{f}} \mathbb{A}^{\mathrm{Eshelby}}\right]^{-1}, \tag{2.61}$$

allowing an explicit computation of the composites stiffness tensors, according to Eq. 2.57. A very comprehensive explanation of the Mori-Tanaka scheme is provided by BENVENISTE [31].

2.2.2 Numerical homogenization

The justification of homogenization of continua is that constitutive equations for basic constituents and geometrical circumstances on a microscopical scale are well, or at least better, understood compared to the macro-scale directly. As described before, analytical homogenization schemes take into account the geometry and the behavior of the single phases, however making assumptions regarding their interactions. As a powerful alternative, the methodology of numerical homogenization is commonly used, directly taking into account the microstructure and using elaborated constitutive models for the constituents. In practice, approaches based on *representative volume elements* (RVE) are widespread and have attracted increasing scientific attention during the last years. Respecting some requirements concerning the case-dependent size of the RVE (cf. TRIAS [387]), these allow the computation of micro-stress and -strain fields within a complex microstructure enabling for example examination of micro-crack initiation, evolution and coalescence at distinct positions in the material in the context of a failure assessment. In literature, the corresponding numerical methods are often referred to as *computational homogenization* methods, and basic works were published by SUQUET [367], SANCHEZ-PALENCIA [358], HILL [159], WILLIS [418], GUEDES & KIKUCHI [137], TERADA & KIKUCHI [382], MATSUI et al. [255], GHOSH et al. [123–125], MIEHE & KOCH [271], SMIT et al. [357], KOUZNETSOVA et al. [197–199], GEERS et al. [121,122], MICHEL et al. [263,264], YUAN & FISH [429], just to name a few. For the special case of fully periodic media, the RVE can be further reduced to what is called a *unit cell* (UC), which by continuously repetition yields the superordinate structure (s. Fig. 2.6b and c as examples for textile unit cells). Equations 2.45 to 2.49 are also valid in the context of numerical homogenization without any further restriction but of course in a discrete way. The choice of correct boundary conditions has a major influence on the plausibility of the results obtained by computational homogenization. With $\boldsymbol{\sigma}$ and $\boldsymbol{u}$ being the active stress and displacement fields, the macroscopical work $\bar{\boldsymbol{\sigma}} : \bar{\boldsymbol{\varepsilon}}$ and the average of the microscopic work $\langle \boldsymbol{\sigma} : \boldsymbol{\varepsilon}(\boldsymbol{u}) \rangle$ must be identical, giving the definition of the macro-homogeneity condition, also known as the *Hill condition*

$$\bar{\boldsymbol{\sigma}} : \bar{\boldsymbol{\varepsilon}} = \langle \boldsymbol{\sigma} : \boldsymbol{\varepsilon} \rangle. \tag{2.62}$$

SUQUET [367] called this equation the *equality of virtual work* between the micro and macro scales.

Similar to the formulation of localization tensors for single phases in Eq. 2.56, the computation of microscopic tensor fields based on macroscopic fields is possible by the introduction of a fourth-order localization tensor that yields

$$\boldsymbol{\varepsilon} = \mathbb{P}^{\varepsilon} : \bar{\boldsymbol{\varepsilon}} \quad \text{and} \quad \boldsymbol{\sigma} = \mathbb{P}^{\sigma} : \bar{\boldsymbol{\sigma}}. \tag{2.63}$$

Note, that the microscopic tensor field can be reformulated, according to

$$\boldsymbol{\sigma} = \mathbb{P}^{\sigma} : \bar{\boldsymbol{\sigma}} = \mathbb{I} : \bar{\boldsymbol{\sigma}} + \tilde{\mathbb{P}}^{\sigma} : \bar{\boldsymbol{\sigma}} \quad \text{with} \quad \tilde{\mathbb{P}}^{\sigma} = \mathbb{P}^{\sigma} - \mathbb{I}, \tag{2.64}$$

identifying a constant part $\bar{\boldsymbol{\sigma}}$ and a fluctuating part $\tilde{\boldsymbol{\sigma}} = \tilde{\mathbb{P}}^{\sigma} : \bar{\boldsymbol{\sigma}}$. The latter represent residual stresses with vanishing average $\langle \tilde{\boldsymbol{\sigma}} \rangle = \mathbf{0}$. The same applies for the micro-field of the strain,

$$\boldsymbol{\varepsilon} = \mathbb{P}^{\varepsilon} : \bar{\boldsymbol{\varepsilon}} = \mathbb{I} : \bar{\boldsymbol{\varepsilon}} + \tilde{\mathbb{P}}^{\varepsilon} : \bar{\boldsymbol{\varepsilon}} \quad \text{with} \quad \tilde{\mathbb{P}}^{\varepsilon} = \mathbb{P}^{\varepsilon} - \mathbb{I}, \tag{2.65}$$

and $\langle \tilde{\boldsymbol{\varepsilon}} \rangle = \mathbf{0}$. Inserting the decomposition of the tensor fields in constant and fluctuating terms in Hill's condition and some simplification steps yield the reduced form

$$\bar{\boldsymbol{\sigma}} : \bar{\boldsymbol{\varepsilon}} = \bar{\boldsymbol{\sigma}} : \bar{\boldsymbol{\varepsilon}} + \langle \tilde{\boldsymbol{\sigma}} : \tilde{\boldsymbol{\varepsilon}} \rangle \tag{2.66}$$

which is only fulfilled in the case that the last summand is equal to zero. This condition allows the formulation of three kinds of boundary conditions, valid on $\partial \mathcal{B}_{\mathrm{m}}$,

1. The approach of anti-periodic stress boundary conditions

$$\boldsymbol{\sigma} \left(\boldsymbol{X}, \bar{\boldsymbol{X}} \right) \boldsymbol{n} \left(\boldsymbol{X} \right) = \bar{\boldsymbol{\sigma}} \left(\bar{\boldsymbol{X}} \right) \boldsymbol{n} \left(\boldsymbol{X} \right) \tag{2.67}$$

2. Assumption of linear displacement field

$$\boldsymbol{u} \left(\boldsymbol{X}, \bar{\boldsymbol{X}} \right) = \bar{\boldsymbol{\varepsilon}} \left(\bar{\boldsymbol{X}} \right) \boldsymbol{X} \tag{2.68}$$

3. A coupled formulation of the above mentioned solution for the anti-periodic stresses from Eq. 2.67 and

$$u(X) = \bar{\varepsilon}\left(\bar{X}\right) X + \tilde{u}(X), \tag{2.69}$$

in terms of the periodic function $\tilde{u}$.

Throughout this work, these *periodic boundary conditions (PBC)*, were chosen and implemented for the numerical simulation of the representative microstructures (cf. Chapter 5). For RVEs or UCs with rectangular cuboidal shape, the set of equations that allow the application for a 3D representative volume element are provided by BARBERO [25, 26]. MELRO [260] states a reduced set of equations needed for the simulation of one-layer woven structures, where only in-plane periodicity is assumed. Both sets of equations are introduced into the simulation as node-wise kinematic constraints for the opposing nodes on the boundaries in the form of additional equations.

2.3 Failure of weave reinforced structures

The microstructure of woven fabric reinforced composites is the decisive factor when it comes to the derivation of macroscopic mechanical properties such as stiffness and strength. Especially, effects related to damage progression depend on the local geometrical topology on the micro- and mesoscopic level of the composite. The following section gives a short introduction on the geometrical topology and the technical terminology for the description of woven composites. A summarizing glossary of important terms for textile applications can be found in PASTORE [309]. An extent description of the variety of fiber reinforced composites as well as the corresponding processing techniques can be found in HENNING [156], NEITZEL [295], LONG [239] and TEN HOMPEL [381]. Furthermore, a general description of relevant damage evolution in unidirectional and woven composites is addressed.

2.3.1 Topology of woven fiber-reinforced plastics

The general definition of textile reinforced composites states that these fiber reinforced composites are characterized by more than one preferred direction.

On constituent level, fibers (or *filaments*) and a matrix system define the behavior of the composite. The fibers, as the basic unit of the textile material, are converted into laminated tapes, yarns or *direct formed fabrics* (cf. Bogdanovic & Pastore [38]). For the present case special attention is given on the yarn, since they can be used to set up a variety of fabric structures. Either carbon, glass, ceramic or polymeric fibers are used within the assembly of the yarn, which can then directly be processed into the chosen fabric structure. The terminologies of *yarns*, *rovings* and *tows* are used in an equivalent manner throughout the present work. Regarding unidirectional or multi-directional layups, several enhancements can be provided by combining interwoven tows. Hereby, each tow consists of several thousand (ideally parallel) fibers. For example, depending on the exact microstructure, a reduction of the tendency of delamination during loading and a better handling during processing are obtained. On the microscopic level woven and more advanced textile composites, such as three-dimensional woven, braided, knitted or stitched textiles, exhibit highly complex reinforcement geometries. By varying the assembling pattern, a vast number of possible spatial arrangements of the curved yarns can be achieved. To this end, the designer is able to tailor the material for a specific requirement, apart from the choice of material systems. Usually, these kinds of composites are manufactured on modern textile machinery. In an industrial environment, fabrics with a high width to thickness ratio are named *two-dimensional* fabrics. These fabrics are usually defined by a thickness that is smaller than three yarn diameters, but widths of the order of magnitude of meters. This definition allows the distinction of three-dimensional fabrics with a much greater inherent thickness ratio. According to the material investigated in this work, the focus of this introduction lies on woven textile composites. Bogdanovic & Pastore [38] give a historical summary of the development of textile composites and more details on three-dimensional textile composites. Clarke & Morales [74] and Ko & Pastore [196] provide a comparative study of fabrics for composites. Figure 2.5 shows three different types of fabric composites where the differences in the complexity of reinforcement orientation is visible. A plain weave (Fig. 2.5a), as an example for a woven textile is produced by initially orthogonal or triaxial interlacing of yarns. Figure 2.5b shows a braided fabric which is set up by the intertwining of three or more strands of yarns. Finally, Fig. 2.5c represents a knitted fiber architecture. It is formed by the interlooping of yarns along the vertical direction and is therefore identified as a *warp knit*. In an analogous manner weft knitted fabrics can be manufactured where the interloop runs horizontally.

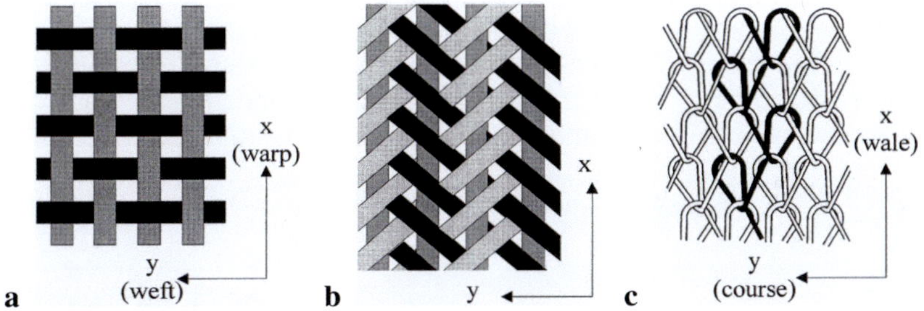

Figure 2.5: Three exemplary types of typical fabric composites: (a) Plain weave, (b) triaxially braided fabric, and (c) warp knitted fabric. Images were taken from GOMMERS et al. [127].

In industry, woven fabrics are used most frequently. The most widely used examples regarding such structures are plain, twill and satin (n-harness weaves) patterns as depicted in Fig. 2.6. Here, the woven structures are set up by two

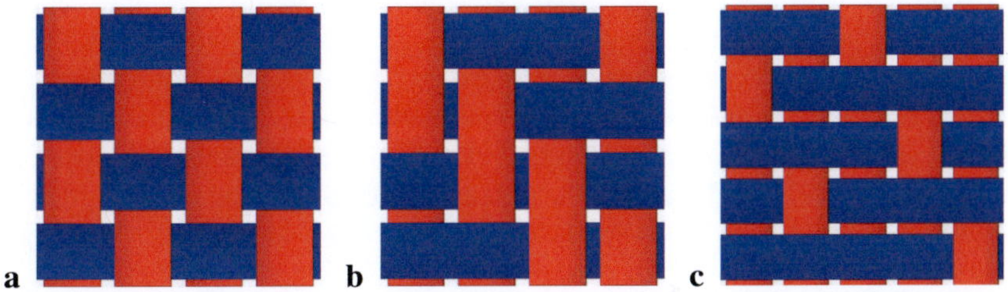

Figure 2.6: Three types of typical two-dimensional weave patterns in schematic form: (a) Plain weave, (b) Twill weave, and (c) 5-harness satin weave. Weave patterns were created using the software TEXGEN [350].

tows (*weft* and *warp* yarns) that are assumed to be identical in their mechanical properties. However, the way of spatial arrangement differs for each type which causes differences in the weaves appearance as well as the mechanical behavior. Looking at the patterns in more detail, either inclined or straight passages can be considered. Regarding plain weaves, all yarns are inclined over their complete length, since they are floating constantly from bottom to top and vice versa. In contrast, yarns in satin weaves show longer linear in-plane segments before they float to the opposite side of the weave. Hereby, the length of the linear parts passing over or under other yarns is dependent on n. Depending on the number and size of linear parts, handling, manufacturing, and mechanics related properties are effected. The area where yarns change sides within the weave is of special interest and is labeled as *crimp* areas (cf. Fig. 2.7). The geometric characteristics of the crimp areas influence the extent of the out-of-plane fiber orientation, the volume fraction, the thickness, and therefore the overall mechanical performance of the textile composite. The dimensions within the

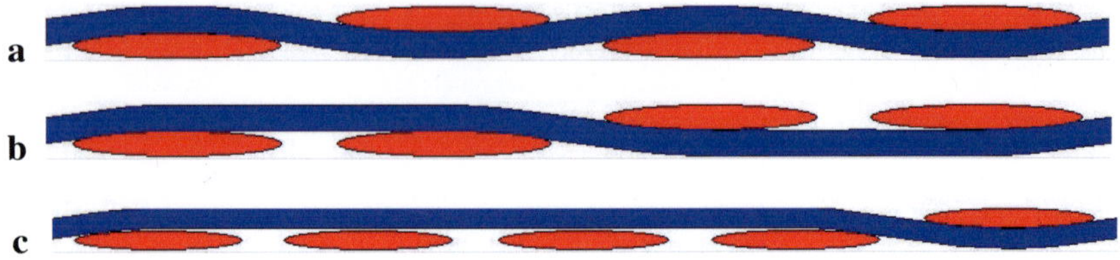

Figure 2.7: Side view cut of (a) a plain weave, (b) a 2x2 twill weave, and (c) a 5-harness satin weave as shown in Fig. 2.6. The *crimp* area, where the yarn floats from top to bottom side of the textile is clearly visible. The side views were created using the software TexGen [350].

created textile composite are of particular interest regarding the mechanical behavior. Due to the loose connection of single filaments within a dry yarn, the individual fibers can move to a great extent against each other. While under longitudinal tension they tend towards a circular cross-section minimizing both cross-sectional lengths and longitudinal curvature, transverse compressive load as applied during the thermoplastic processing leads to a narrow lenticular or elliptical shape of the cross-sections (cf. Bogdanovic & Pastore [38], Tang & Whitcomb [375], Zhou & Yang [433], Xu et al. [426]). Hereby the minor axis is oriented along the direction of applied pressure. For the later

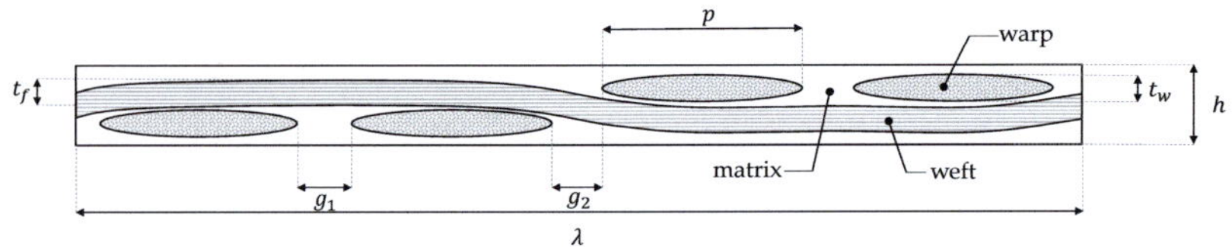

Figure 2.8: Schematic display of a one-layer twill weave showing relevant dimensions. The corresponding descriptions are given in Table 2.1.

modeling of the mesoscopic level of the woven composite in a geometrical model it is imperative to respect the dimensions of the real specimen in order to obtain sufficiently accurate results (cf. Lomov et al. [231], Green et al. [132]). As shown in Fig. 2.8, the essential dimensions of an idealized one-layer twill weave are the tow wavelength (λ), weave height including the matrix domain

(h), tow width (p), tow heights (t_w and t_f) and the gaps between neighboring tows (g_1 and g_2). Furthermore, weaves are usually described by parameters that describe the volume ratios of fiber and matrix rich areas and are applicable in different ways. ν_{tow} describes the volume fraction of tows in the weave. κ represents the fiber volume fraction in the single tow. With this the the total fiber volume fraction in the weave (ν_{fw}) can be calculated according to

$$\nu_{fw} = \kappa\nu_{tow}. \tag{2.70}$$

The relevant dimensions are summarized in Table 2.1. Beside the topology of

Table 2.1: Parameters used to define the dimensions within a one-layer twill weave.

Type	Dimension	Description
Geometry	λ	Tow wavelength
	h	Weave mat thickness
	p	Tow width
	t_w	Tow thickness (warp)
	t_f	Tow thickness (weft)
	g_1	Distance of gap between neighboring yarns
	g_2	Distance of gap between yarns in crimp area
Volume fractions	ν_{tow}	Tow volume fraction
	κ	Fiber volume fraction in the tow
	ν_{fw}	Fiber volume fraction in the weave

single weave layers, stacking plays a significant role when it comes to the mechanical properties of the synthesis of the composite. During the manufacturing process the impregnated layers undergo high pressures in thickness direction, yielding deformed yarn shapes and paths. Measurable local variations of the section and fiber volume fractions are the result (cf. Olave et al. [300], Karahan et al. [188]). Furthermore, random shifts and nesting between the layers occur, which favor the formation of complex contact regions between the yarns (cf. Olave et al. [300], Chen et al. [70], Lomov et al. [238]). The relative position of yarns within the stack and the degree of nesting has a pronounced influence on the local stress and strain distributions, and thus, on damage initiation and propagation (cf. DAGGUMATI et al. [85], LE PAGE et al. [212], JOHN et al. [181] and DOITRAND et al. [99]).

2.3.2 Failure behavior of unidirectional fiber composites

For the successful modeling of damage propagation in woven fabric reinforced composites a sound understanding and description of the related physical phenomena in the microstructure of different scales and hierarchical levels is mandatory. In the previous sections it has been mentioned that woven composites consist of interwoven yarns which by themselves can be described as subcomposites again. The damage phenomena observed in textile composites consist partly but not exclusively of processes occurring in the curved but ideally unidirectional reinforced yarns. In the following, an overview of the physical failure patterns occurring in unidirectional is given.

The correct description of damage progression in continuous fiber reinforced plastic composites is an ongoing topic of research. For the design of parts of such materials PUCK [319] provides an established methodology investigating the structure in a layer-wise manner. Therefore it is necessary to understand and describe the micro and macro processes occurring within the layer. Due to the material symmetry of unidirectional composites, PUCK introduces two types of macroscopic failure occurring between the fibers (*inter-fiber failure*) as a matrix dominated failure mode as well as an ultimate *fiber failure*. Both failure types can be subdivided again, taking into account the sign of the corresponding stress state, resulting in four distinct failure modes (*modes I-IV*). Additionally, PUCK describes *delamination* as an inter-laminar failure mechanism, i.e. a separation between two adjacent layers. The evolution of damage depends on the local strain states during loading. Failure in fiber composites is the final result of a combination of very complex damage processes that occur on different length scales within the material. Each failure mode can be seen as a distinct state in the history of its damage progression, where a specified combination of stress components on a designated possible fracture plane (*action plane*) reaches a maximum, allowing the formulation of a set of failure criteria for the different failure modes. The action plane is identified as the plane with the highest risk of failure depending on the active stress state (cf. PUCK [319]). The stresses on an arbitrary action plane, inclined by an angle Θ around the fiber axis (x_1), are depicted in Fig. 2.9.

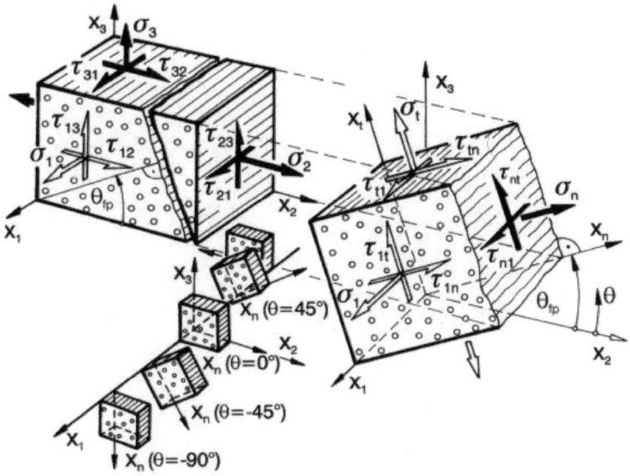

Figure 2.9: Definition of stresses on an arbitrary action plane. $\{x_1, x_2, x_3\}$ correspond to the coordinate system of the lamina with the corresponding stresses $\{\sigma_1, \sigma_2, \sigma_3, \tau_{12}, \tau_{13}, \tau_{23}\}$. The action plane is inclined by the angle Θ around the fiber axis (x_1). The normal stresses $\{\sigma_1, \sigma_n, \sigma_t\}$ as well as shear stresses $\{\tau_{1n}, \tau_{1t}, \tau_{nt}\}$ can be identified. Image was taken from Puck [319].

2.3.2.1 Fiber fracture

The fibers are the actual load-bearing elements, which should be utilized to the limit of strength before other structures are destroyed. Tensile fiber failure (FF^+), is thus the only desired type of failure. Due to their high strength, the fibers are able to withstand very high longitudinal stresses in the composite. In the transverse direction, the strength of the fibers is virtually never achieved, since the yield of the matrix is smaller by several orders of magnitude. In other words, the transverse fracture behavior is not (primarily) influenced by the fibers themselves. In this sense, fiber failure does not mean the breakage of individual fibers, but rather the breaking of a large number of fibers or entire fiber bundles. The load-bearing capacity of the laminate is greatly reduced over a large width. However, while the term fiber breakage appears to be justified in the case of tensile stress (*mode I*), where a "breaking" or "tearing" of the fibers actually occurs, the damage caused by compressive stress (*mode II*) does not necessarily correspond to this term. Here the degradation of the fiber bearing capacity is caused by a lack of elastic support by the surrounding matrix material, which can be seen through a buckling motion (microbuckling, internal buckling, *kinking*) (cf. Fig. 2.10a). Several of such areas can grow

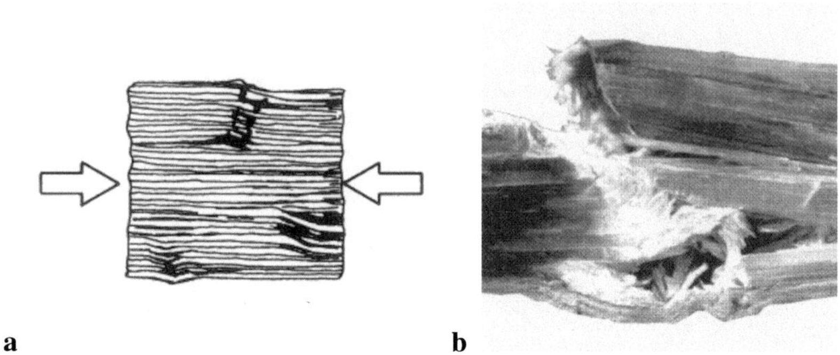

Figure 2.10: (a) Representation of the *kinking* as a precursor to fiber fracture under compression and (b) Compressive fiber fracture in a unidirectional laminate, images taken from PUCK [319].

and merge, yielding finally a compression fracture (cf. Fig. 2.10b). Stochastic effects play a major role in fiber breakage. It can be observed that at about 70-80% of the maximum achievable failure load, the first fractures of both single fibers and fiber bundles occur (cf. PUCK [319]). When fracture occurs, the material cohesion of matrix and fiber is almost completely degenerated due to high energy releases. This leaves a strongly disjointed, brush-like fracture pattern. Degradation models that leave the bond with a residual stiffness or strength across the grain after fiber failure are considered unrealistic according to PUCK [319]. Fiber breakage affects not only the layer in which it was initially formed. It often has a devastating effect on the neighboring layers. In this way, it is possible to trigger the delamination processes in layers that are already on the edge of fiber breakage more easily. This renders failure detection a difficult task and the resolution of damage progression during the breaking process becomes almost impossible.

2.3.2.2 Inter-fiber fracture

In the case of transverse loading, the stresses acting perpendicular to the fibers are transmitted through both matrix and reinforcements. Inter-fiber fracture (*IFF*) describes the crack opening taking place in the plane perpendicular to the fibers. Regarding the damage effects, the cracks grow in the matrix-rich areas, while the initiation occurs mostly in the boundary layer of fiber and matrix (*interface*). Depending on the sign of the normal stresses in the action plane, different forms of inter-fiber failure occur.

Normal tensional load on the action plane (σ_n>0)

This failure mode (*mode III*) is caused by a transverse tensile stress σ_n^+ and/or transverse shear stress $\tau_{1n/1t}$. The resulting damage patterns are cracks in

a 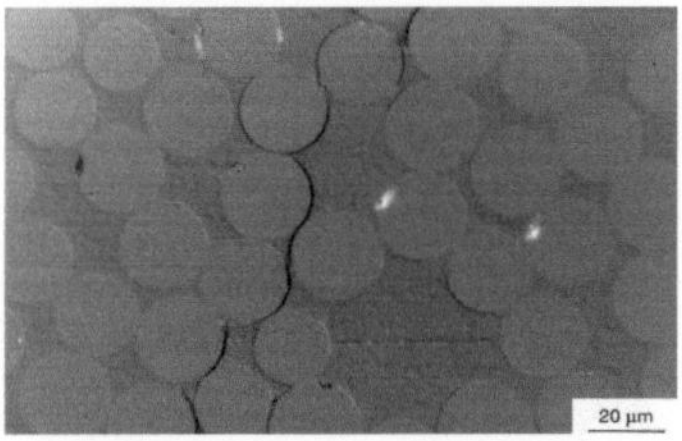 b 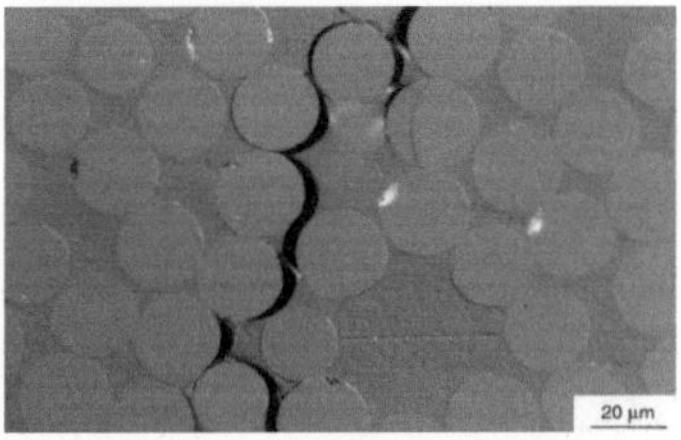

Figure 2.11: Mode III crack under horizontal tensional loading. a) Initial cracking and debonding of the fiber/matrix interface and b) coalescence into micro-cracks, image taken from PRAUD et al. [317].

thickness direction, i.e. on the active plane of the damaging stresses, as shown in Fig. 2.11. The load was applied in horizontal direction. It is observed that the crack opens further the higher the corresponding load σ_n^+ becomes compared to the shear loads $\tau_{1n/1t}$. The cracking observed after the appearance of the first inter-fiber fracture expresses itself through intensive noise emission (cf. DEUSCHLE [90]). The weakening of the material, due to the affected layer being free of force in the cracked areas and only gradually carrying the load again, is noticeable externally as a reduced overall stiffness of the layer. The stiffness will gradually decrease further with increasing crack density. This behavior is used for modeling the post-damage behavior (*post-failure-degradation*). While the matrix material is heavily stressed under such loads, the fibers integrity is hardly affected by inter-fiber fracture, which can be shown by constant longitudinal stiffness and strength as shown by DEUSCHLE [90] and PUCK [319]. The damage processes in the matrix during loading, causing inter-fiber fracture at its end are depicted in Fig. 2.12. Starting from reaching about

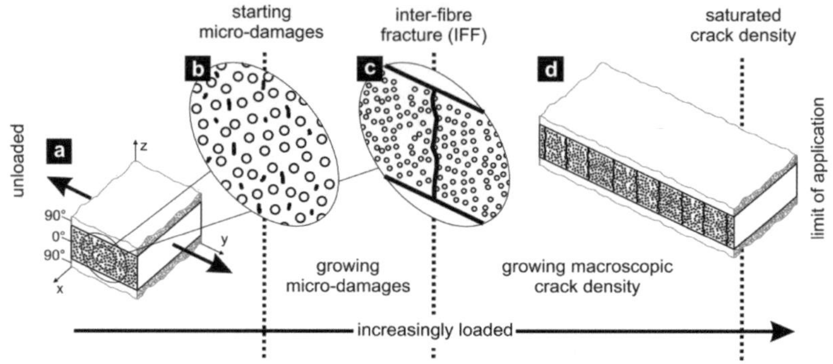

Figure 2.12: Schematic representation of the progressing damage based on increasing σ_n^+ transversal tensional load, image taken and modified from DEUSCHLE [90].

50% of the ultimate load, which causes pure inter-fiber fracture, one can observe beginning micro-damage in the material (cf. Fig. 2.12b, DEUSCHLE [90]). From this threshold, existing micro defects - microscopic cracks in the matrix - begin to grow and new ones are included in the matrix. This effect explains the loss of stiffness, which leads to non-linearities in the stress-strain curve before the ultimate failure exhibiting an anisotropic character. The increase in the number and growth of defects depends on the type of loading. Hereby, strains perpendicular to the fiber (*out-of-plane*) produce and promote more and faster growing defects than *in-plane* strains (cf. DEUSCHLE [90]). If the load on the layer becomes too high, the micro cracks coalesce and macroscopic, visible cracks are formed. If, under certain circumstances, only a single layer had been considered, it would lose its complete load-bearing capacity at this point, since a material separation would occur. However, Puck's theory considers the single layer as a compound. Although the destroyed layer can no longer carry an immediate load, it still contributes to the load-bearing capacity of the entire system by redistributing the load into the neighboring layers. As a result, further load can be applied until other cracks occur. This process will be repeated until a layer saturated with cracks is present. The description of the loss of stiffness due to such crack growth is the subject of post-failure modeling (*post failure degradation*).

Normal compressional load on the action plane (σ_n>0)

The defining characteristic of this type of failure (*mode IV*) is that the resulting cracks do not proceed in the thickness direction, i.e. the common plane of action of the stresses σ_2 and τ_{21}, but in an inclined fracture plane around the fiber axis by an angle Θ . The fracture angle Θ depends on the ratio σ_2/τ_{21} and lies in a range of $\Theta \in [-90°, +90°]$. At *mode III* and at the transition from mode III to IV the angle takes the value of $\Theta = 0°$. It increases with growing ratio until it assumes the approximate value of $\Theta \approx \pm 53°$ at pure compressive stress depending on the chosen fiber/matrix composite and fiber volume fraction. In the fracture plane, now not only an overlay of σ_n^-- and $\tau_{1n/1t}$-stresses, but an additional cross/transverse stress τ_{nt} is active. An important characteristic that distinguishes the intermediate fiber fractures by pressure from those under tension is the fact that the former can no longer be described as harmless. The wedge effect, caused by the oblique breakage in the laminate, causes an explosive force acting in the direction of thickness, resulting in sudden and devastating failure (cf. Fig. 2.13). The resulting fracture angle cannot be

Figure 2.13: Mode IV failure, transverse compression yields an inclined fracture plane (Θ) accompanied by high out-of-plane forces due to the wedge effect, image taken from PUCK [319].

specified a priori for the modes III and IV, but must be calculated. In general, the different fracture modes can only be characterized in the general case by the stress combinations acting on the fracture plane (cf. PUCK [319]).

2.3.2.3 Delamination

The interfaces of single laminae are distinct weak points due to the absence of reinforcements. Delamination is defined as a separation of laminae possible through the application of out-of-plane normal stresses or in-plane shear stresses. These phenomena are caused by σ_3^+, τ_{32}, and τ_{31} in a general 3D stress state, so-called *inter-laminar stresses*. Extensive delamination occurs mostly, if a crack, caused by inter-fiber damage, reaches the laminate interface. Such cracks enhance the formation of delamination due to the concentrated, but inhomogeneous stress states in the crack tip area (cf. DEUSCHLE [90] and PUCK [319]). It is observed that layer separations grow with increasing monotonous load as well as with cyclic load of constant amplitude. The extreme case is the complete detachment of whole unidirectional layers, as described in SCHREIBER [347]. However, delamination is a form of failure occurring in the laminate and not in the lamina and is therefore not respected in the homogenized description of unidirectional composites.

2.3.3 Failure behavior of woven fiber composites

Interwoven composites (especially more advanced textile composites, e.g. 3D woven, braided, knitted, stitched) show improved impact resistance and damage tolerance in comparison to traditional laminates (cf. BOGDANOVIC & PASTORE [38]). They are able to withstand a higher number of matrix cracks and fiber debondings before reaching ultimate failure. In the case of three-dimensional fiber scaffolds, the material does not contain predefined delamination paths due to the absence of planar interfaces. The undulation of the fiber tows yields a rather random distribution of crack orientations within the tow. Thus, there are typically no obvious and clear weak directions. Therefore, the formation of extensive crack propagation which would lead to direct ultimate failure of the structural part is prevented. The internal micro- and mesoscopic topology is hereby the key factor when it comes to damage progression in textile composites.

To this end, the biggest difficulty regarding the understanding of such composites is that the damage processes have to be considered on different scales and hierarchical levels. This circumstance makes the analysis more complicated than it is the case for classical unidirectional laminates. Furthermore, changing the weave architecture has a significant effect on the appearing damage modes within the composite but even specimens that belong to the same architecture showed varying damage features (cf. JOHN et al. [181]). Finally, the in-plane properties of multilayer woven composites are affected by the misalignment of the fibers, resin rich areas due to the weave design, pre-damage in the fiber tows due to the weaving process (cf. FARLEY et al. [112], DING et al. [97], ARENDTS et al. [9], NORMAN [298], BYUN & CHOU [45], GUESS et al. [138]) or pre-damage induced by composite processing, e.g. due to thermal or cure stresses (cf. LOMOV et al. [237]). All these influences on the mechanical performance of a certain material system in combination with a chosen weave architecture necessitate the detailed examination of the damage processes for each textile composite. Looking at existing works dealing with the description of failure in woven fiber composites, it is immediately noticeable, that compared to their laminated unidirectional counterparts much less general literature is available. Publications concentrate on certain weave architectures or material systems as a consequence of the above mentioned reasons.

Important similarities can be found in the methodology of the monitoring and characterization of damage during (*online*) and after (*post-mortem*) testing. LOMOV et al. [237] proposes the usage of a variety of analytical techniques in order to systematically study failure phenomena during tensile deformation, consisting of

1. Online acoustic emission (AE)
2. Online full-field strain mapping (SM)
3. X-ray/CT and ultrasonic C-scan examination (online and post-mortem)
4. Cross-sectioning and optical microscopy (post-mortem).

The several steps of this sequence complement each other in an upward direction in order to identify both *locus* and mode of occurring failure. Acoustic emission analysis as defined by WEVERS & SURGEON [410] detects transient stress waves propagating in a material as a result of fast release of strain energy. It is a widely used method, to gain information about the damage processes, its mode, and location in the specimen without interrupting the test. The damage thresholds can be identified by analyzing the curves of cumulative energy, as described in Truong et al. [391], Lomov et al. [234, 237], Daggumati et al. [84], and Carvelli et al. [57]. This technique becomes interesting when combined with the cluster analysis of AE events as performed by Lomov et al. [231], John et al. [181], El Hage [105], Benmedakhene et al. [30] in glass/epoxy and carbon/epoxy two- and three-dimensional woven composites and by Carvelli et al. [56] for a thermoplastic PPS carbon woven composite. Taking into account not only the acoustic energy thresholds registered by the AE sensors but furthermore other descriptors of AE events such as the frequency of the emitted signal (cf. Gutkin et al. [140], Sause et al. [342], Baker et al. [23], Maillet et al. [243]) allows a precise identification and distinction of the appearing damage modes after calibration. The found clusters correspond to damage events originated from transverse matrix cracks (events with low frequency and low amplitude), local delaminations (low frequencies and high amplitude), and fiber breakage (high frequencies). Together with the technique of strain mapping, where strains on the stress-free surface are measured, the failure locations and types can be identified in a non-destructive manner, helping to identify characteristic strain levels for damage onset or change of damage modes. The specimen tensioned up to a certain critical strain level, identified by the aforementioned techniques, are then to be examined further by the usage of X-ray and ultrasonic C-scanning technologies. While X-ray investigations allow the precise presentation of cracks within the material, C-scan tests, as applied in Truong Chi et al. [391], give a rougher picture of damage in the specimen and are hence more suitable to give an impression of the overall damage extend, periodicity and to observe more widespread damage patterns, related to the underlying textile architecture (cf. Lomov et al. [237]). Both techniques can be used while the specimen is loaded and, in case of a tensional load, the cracks are opened and hence nicely visible. To enhance visibility, contrast agents are often used in the case of X-ray analyses. Cross-sectioning the identified locations of the cracks and their adjacent areas with the help of the recorded X-ray and C-scan data allows further studying the damage behavior on a microscopic level. In some cases further investigation of the fiber-matrix interface is needed, which can be inquired by using a *scanning electron microscope* (SEM). The proposed suite of tests presented here result in a complex and time-consuming test preparation and execution.

As one of the first authors in recent history, JOHN et al. [181] introduce a systematic "taxonomy" of damage processes in three-dimensional woven E-Glass vinyl ester composites using an alternative approach to analyze damage processes in woven composites. By investigating the stiffness during tensile tests in measured stress-strain relationships (cf. Fig. 2.14) and analyzing the step-wise softening behavior of the material, different levels of damage onset were detected. By using solely post-mortem optical analyses,

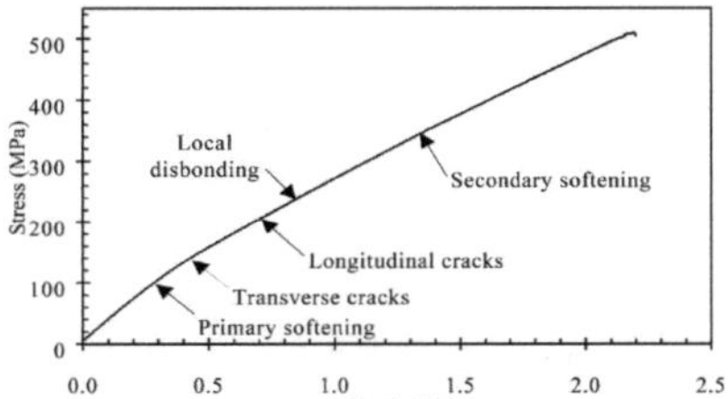

Figure 2.14: Stress-strain relation of a woven composite tested under tensile load in warp direction. The arrows indicate the identified failure modes by analysis of the secant stiffness and optical microscopy. Image taken from JOHN et al. [181].

the essential failure modes appearing under applied load in weft and warp yarns could be identified by the authors. They describe the damage processes as a complex combination of transverse and longitudinal cracks in the fiber tows followed by consecutive debonding on the boundaries of the impregnated yarn-matrix and delamination finally yielding to ultimate failure. Other works confirmed the results with regard to the sequence of damage processes in both three-dimensional woven carbon/epoxy composites (cf. EL HAGE [105]) and also for two-dimensional woven composites as demonstrated in the works of TANG & WHITCOMB [375], KURASHIKI et al. [209], and UETSUJI et al. [397], where the authors used optical analysis techniques for the detection and investigation of damage. CARVELLI et al. [56] and SUGIMOTO et al. [364] used acoustic emission for the identification of characteristic strain levels at which damage initiation of different modes commences. Results of damage investigations of composites with varying fiber architecture indicate that the same basic mechanisms are responsible for the ultimate damage. Using the broad spectrum of possible analysis techniques presented above, LOMOV et al. [231], MASTERS & IFJU [254], and IVANOV et al. [176, 178] focused on the damage evolution in braided composites, and Greve & Pickett [134], Edgren et al. [103], Truong Chi et al. [390,391], Asp et al. [16], Mattson et al. [256], Mikhaluk et al. [278] studied appearing patterns in non-crimp fabrics, all of them with the resulting conclusion that the same sequence of damage event governs the failure behavior of such composites. Carvelli et al. [56] proposed the sequence of damage events appearing

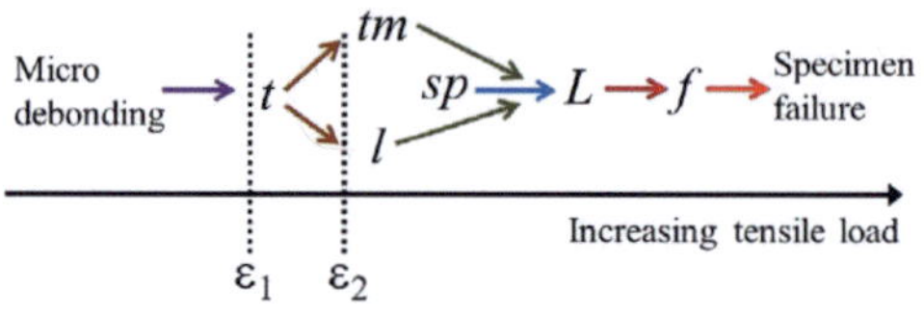

Figure 2.15: Typical sequence of damage events appearing under tensile loading according to Carvelli et al. [56]. *t*: transverse cracking, *tm*: transverse cracking in resin pockets, *l*: local delamination of impregnated yarns, *sp*: splitting in longitudinal yarns, *L*: inter-ply delaminations, *f* fiber failure in longitudinal yarns. Identification of characteristic strain levels where certain damage modes are triggered. ε_1: Onset of *t*-cracks, and ε_2: onset of local delaminations (*l*) and the formation of large transverse cracks (*tm*).

in textile composites under tensional loading in warp direction as depicted schematically in Fig. 2.15. The initial underlying mechanism that triggers the damage evolution in woven composites is the local micro debonding of single fibers of the transverse tow from their surround matrix. Coalescing micro cracks form larger transverse cracks (*t*) as depicted in Fig. 2.16. While initially

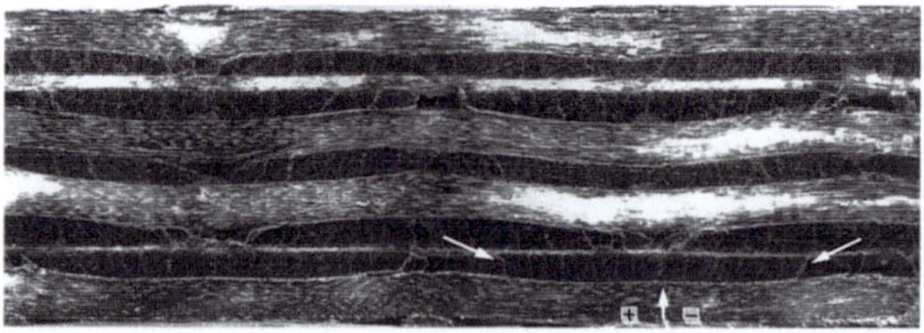

Figure 2.16: Transverse cracks (*t*) in weft yarns (out-of-plane) after a tensile loading in warp direction (horizontal). Image taken from John et al. [181].

only a few cracks exist in the weft tows, which furthermore do not span the whole width of the yarn, the number, size, and length increases upon further loading, until eventually reaching a state of *saturation*. The characteristic strain level ε_1 indicates thereby the onset of *t*-cracks. The experimental characterization of the processes occurring by increasing the load further indicate a parallel development of resulting damage modes. Originating inside the weft yarns, transverse cracks propagate inside the matrix rich areas between the fiber tows (*tm*). Reaching the interface of other, not aligned reinforcement structures, e.g. longitudinal (warp) yarns or sheared weft yarns, the transverse cracks cause the appearance of local delaminations (*l*). The main responsibility for that formation lies in the active inhomogeneous stress states at the crack tip and especially the shear components. Figure 2.17 shows local delamination phenomenon due to *t*-crack grown out of a weft yarn. Other experimental investigations indicate

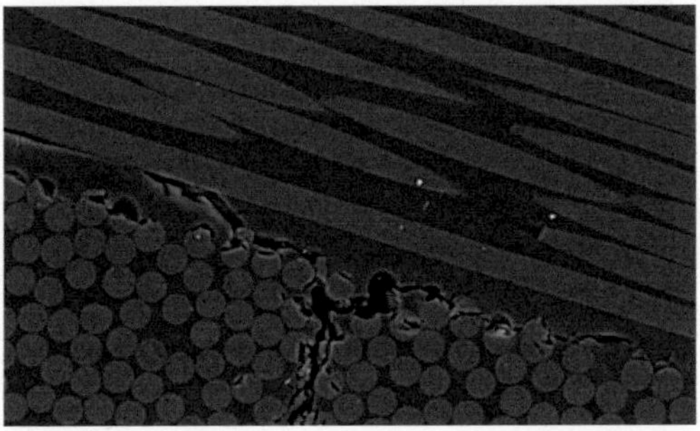

Figure 2.17: Starting local delamination (*l*) initiated by a *t*-crack. Image taken and modified from LOMOV et al. [237].

that local delaminations (*l*) can also appear without being initiated by *t*-cracks (cf. LOMOV et al. [237] and EL HAGE [105]). The formation of *tm*- and *l*-cracks are identified by AE events happening at a characteristic strain level, ε_2. Looking at the warp fiber tows oriented in loading direction, the damage mode of splitting (*sp*) can be identified. The initially undulated longitudinal yarns straighten under tensile load. Especially in the crimp areas this deformation is suppressed by the adjacent reinforcements and surrounding matrix. Furthermore, being subjected to the Poisson contraction, the transversal deformation of warp yarns is constrained, yielding internal transverse stresses. For the case that these stresses exceed the transversal strength, inter-fiber fracture occurs in the longitudinal yarns. Figure 2.18 shows a view of a micro cut of such a damage behavior. As local delaminations (*l*-cracks) grow due to increasing loading, they eventually coalesce to larger delaminated areas (*L*-cracks). These

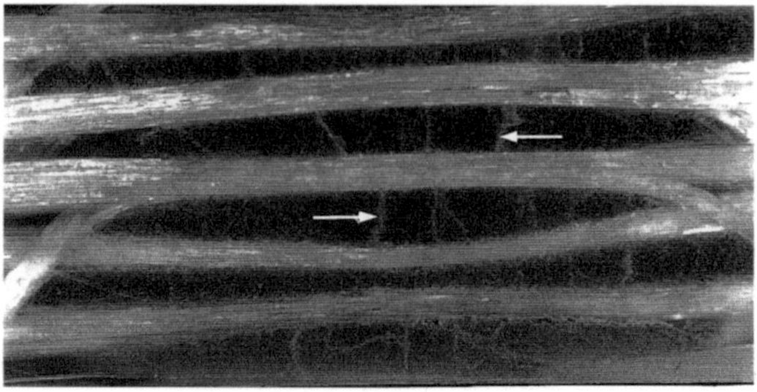

Figure 2.18: Transversal cracks in the longitudinal yarns (warp), oriented perpendicular to the image plane here, due to the damage mode splitting (*sp*) indicated by arrows. Image taken from John et al. [181].

preferably form in the interface between single plies, effecting their separation, and are therefore often described as an inter-ply failure, depending mainly on the inter-laminar fracture toughness of the composite. The ultimate failure of the specimen is defined by fiber failure (*f*). Lomov et al. [237] state, that fiber failure starts at locations of delaminations. Experimental observations show a reduced failure strain in warp direction for textile composites compared to their non-crimped counterparts. Carvelli et al. [56] explain this reduction by the prohibited stress transfer inside the fiber bundles due to the adjacent micro damage. This effect is intensified by the off-axis orientation of the fiber bundle due to the undulation.

2.4 Constitutive modeling approaches on different scales

Increasing attention has been given to composites with textile reinforcement in the recent years with the purpose to close the gap between "high tech" unidirectional laminates and "high volume" composites consisting of short-fibers or particulates (cf. Gommers et al. [127]). While the idea of multiscale analysis of woven textile composites is relatively old, there has been some recent progresses in the modeling of the material behavior on the different scales. The focus of the following section lies on the review of already existing modeling approaches of the thermoplastic constituent (matrix), subcomposites and macroscopic formulation of impregnated weaves. The structure of this presentation is based on the intrinsic hierarchy of the woven composite material at hand.

2.4.1 Plasticity and damage in isotropic matrix material

The last decades have been marked by many developments in the field of describing elastic and inelastic behavior of a large number of different material classes, such as metals, ceramics, plastics, natural materials, or tissues, just to name a few. They cover a wide range of the most diverse applications and disciplines, such as classical stress analysis, soil mechanics in civil engineering, deformation analysis during manufacturing, food processing, mining operations or tissue engineering in biomedical applications (cf. DE SOUZA NETO [296]). The following short review of available models is exclusively carried out in the framework of continuum mechanics. In the context of the presented work, formulations of viscoplasticity and failure in plastics and plastic composites are of primary interest. Looking at the performed material tests of the pure matrix in Section 3.2, those were the predominant phenomenologically observable effects occurring under monotonous loading.

2.4.1.1 Plasticity

Concerning the inelasticities associated to plasticity in a ductile metal, i.e. the remaining of permanent deformation upon unloading during a uniaxial tension test, DE SOUZA NETO [296] summarized the visible effects as follows:

1. The domain of stresses, where no permanent (plastic) strain evolution is occurring and the material response can be assumed to be purely elastic, is called the *elastic domain*. In the easiest isotropic case, it can be characterized by one scalar value, the *yield stress*.
2. If the applied stress exceeds the yield stress, *plastic yielding* or *plastic flow*, thus the evolution of remaining strains upon unloading, takes place.
3. Simultaneously, a change of the yield stress is observed, which is denoted as *hardening* in the following.

These observations are valid analogously for plastic materials. Additionally, most of the observable behavior is dependent on the rate of application of loads and/or the timescale of the problems considered (crash vs. creep or relaxation). Generally, plastic materials show both viscoelastic and viscoplastic behavior. The former shall be neglected in the current study since no significant expression was observed in the experimental results. However, the interested reader is referred to CHABOCHE [64], LEMAÎTRE & CHABOCHE [224], or the textbook of BERGSTRÖM [33], among many others for basic works on this topic. Self-evidently, the respective characteristics of these effects depend strongly on the material to be considered and the micromechanical processes occurring in it, depending on the actual rate, time and temperature. Therefore it is a common approach to investigate the processes on the micro scale, since they are particularly important for the understanding of the material behavior under more complex rate/time dependent loading scenarios, including thermal effects, the influence of hydrostatic pressure or multi-axial loading. To describe the observed behavior qualitatively and quantitatively (either phenomenologically or via micromechanical motivations) is the purpose of the mathematical theory of plasticity using continuum constitutive models (cf. DE SOUZA NETO [296]). Hereby, a multitude of modeling possibilities is given in the literature. In an effort to establish a general model, the basic elements denoted in Box 2.1 have to be defined by the constitutive plasticity model.

Box 2.1: General formalism for a definition of a constitutive plasticity model.

1. Definition of the kinematic framework and elastoplastic strain decomposition.
2. Formulation of the ground-state elasticity.
3. Determination of a suitable yield criterion describing the elastic domain.
4. Construction of the plastic flow rule defining the evolution of inelastic strains.
5. Definition of a hardening law, characterizing the evolution of the elastic domain.

Each of these elements were the topic of research for more than four decades, and efforts are still ongoing. In Section 2.1.1 the major kinematic frameworks have been introduced, distinguishing infinitesimal (cf. Eq. 2.16) and finite deformation (cf. SETH-HILL strain measures, Eq. 2.11). According to MIEHE et al. [269], no canonical kinematic framework has been found so far concerning finite plasticity. At this point, it is referred to NAGHDI [286] and XIAO et al. [425] for a discussion and a critical review of available approaches. Besides hypoelastic rate formulations, which are neglected in this present work, and the decomposition of strain in the logarithmic framework (cf. Eq. 2.14_1), basically two kinematical approaches to finite plasticity are available: (a) Multiplicative decomposition of the deformation gradient into the elastic and inelastic parts $\boldsymbol{F} = \boldsymbol{F}^e \boldsymbol{F}^i$ (cf. Eq. 2.7) referred as the *Kröner-Lee decomposition* (cf. KRÖNER [207], LEE [216], and MANDEL [250]). (b) A framework according to the *Green-Naghdi theory* (cf. GREEN & NAGHDI [131]) that introduces the notion of an additional primitive Lagrangean pastic strain $\boldsymbol{E}^p$. Building on this work, MIEHE [273, 274] developed a third framework based on the *evolving reference plastic metric* $\boldsymbol{G}^p$. This framework is of particular interest, since it allows the transfer from a plasticity model initially defined as a function of logarithmic strains to a finite setting (cf. MIEHE et al. [268], PAPADOPOULOS & LU [306, 307], MIEHE & APELT [275]). The elastic part $\boldsymbol{\varepsilon}^e$ of the Hencky strain $\boldsymbol{\varepsilon}$ (cf. Eq. 2.13) can be defined by

$$\boldsymbol{\varepsilon}^e = \boldsymbol{\varepsilon} - \frac{1}{2} \ln\left(\boldsymbol{G}^p\right) = \boldsymbol{\varepsilon} - \boldsymbol{\varepsilon}^p . \tag{2.71}$$

A comparative study was performed by MIEHE et al. [268] showing the closeness of multiplicative approach with the newly designed additive finite plasticity framework. According to the authors, another advantage is the easy algorithmic implementation of the proposed setting. The basic constitutive equations for geometrically linear theory of elastoplasticity is presented for example in HILL [163], RECKLING [331], BACKHAUS [21], among others. A detailed description of the theory and remark on the numerical implementation, especially in the framework of large deformations, is given by SIMO & HUGHES [352], and MIEHE [272]. Within these works, the concept of *hyperelasticity*, i.e. the free energy function as the starting point for the definition of constitutive laws for stress is also presented. DE SOUZA NETO [296] provides hyperelastic models for both frameworks, including corresponding comprehensive derivations. Examples for non-isotropic free energy functions leading to more complex symmetry classes can be studied in SCHRÖDER [346]. To some extent interconnected, the remaining elements (3-5) of box 2.1 can be used to introduce time dependency in terms of viscoplasticity, using different methods. The following statements apply analogously to both small and large strain frameworks. In the literature, a vast number of yield criteria is available. The formulations for the elastic domain according to TRESCA [385], VON MISES [405], or the pressure dependent criteria by MOHR-COULOMB, described in CHRISTENSEN [73], and DRUCKER-PRAGER [102] are the most classical examples. Graphical depictions of the yield surface in the stress space are provided by DE SOUZA NETO [296].

Focusing on yield behavior of polymers, a serious effort to experimentally characterize the three-dimensional yielding behavior and cast the observations in a general mathematical formulation of an elastic domain has been made over the years. As a result, the data indicates the necessity of pressure dependency and asymmetry of yield strength under tension and compression loading for a sensible yield criterion for plastics (cf. RAGHAVA [328]). Modified von Mises yield criteria are often used for this purpose, as provided by MELDAHL [258], SCHLEICHER [345], STASSI D'ALIA [362], or TSCHOEGL [393] among many others. For a more extensive state-of-art review on available yield criteria, please refer to ALTENBACH et al. [3]. The borders of the elastic domain are described in terms of strength values, which can be formulated as functions of rate/time in order to introduce a viscoplastic formulation, thus directly influencing the formulation of hardening (cf. DE SOUZA NETO [296]). A detailed review on modeling possibilities concerning hardening is neglected at this point of the work and it is referred to DE SOUZA NETO [296]. While the existence of a yield surface is essential for the definition of rate-independent plasticity models, this is not required for some viscoplasticity models. This means, that the model yields plastic deformation whenever stress occurs (cf. LEMAÎTRE & CHABOCHE [223], SKZYPEK [356], and others). Examples for such models are widely used, concerning the analysis of creep (cf. *Norton's creep model*, *Lemaître-Chaboche law*, and others). The most intuitive way to adapt a rate-independent plasticity model towards a viscoplastic formulation is done via the plastic flow rule. In general, the evolution equation for the plastic strain in a small strain framework can be written as

$$\dot{\boldsymbol{\varepsilon}}^p = \dot{\gamma}\boldsymbol{N}, \tag{2.72}$$

where $\dot{\gamma}$ is the *plastic multiplier* and $\boldsymbol{N}$ denotes the direction of plastic flow. The latter is described in general by a flow potential Ψ^p (*non-associative flow rule*). In the case that yield function (Φ) and the potential for plastic flow are chosen to be identical, $\Psi^p \equiv \Phi$, a *associative* (or *associated*) plasticity model is defined. Numerically speaking, for time-independent models an *elastic predictor/plastic corrector algorithm* is usually applied in order to find the solution of elastic and plastic strains, as a function of the plastic multiplier $\dot{\gamma}$ (cf. DE SOUZA NETO [296]) in an iterative manner. Viscoplastic formulations, however, use an explicit function of $\dot{\gamma}$, in order to model how the rate of plastic straining varies with the level of stress. In the literature many approaches for such a definition are available (cf. PERIC [311]), such as the *Bingham model* (cf. DE SOUZA NETO [296]) or the *Perzyna model* provided by PERZYNA [313,314]. CHABOCHE [66] provides another extensive review on plasticity and viscoplasticity models for small strain settings.

As already mentioned, similar developments were undertaken also in the framework of finite strain. In the context of polymeric materials, phenomenological models date back to the works of EYRING [110]. Although the parameters of this model have a physical meaning, their physical significance remains dubious, according to MIEHE et al. [269], BOWDEN [40], and CRIST [80]. Therefore, early on models based on thermally activated *inter-* and *intramolecular* mechanism were developed (cf. MARSCHALL & THOMPSON [252], HAWARD & THACKRAY [155], RIDER & HARGREAVES [335], VINCENT [403], ROBERTSON [337]). The basis for a multitude of actual models is the work of ARGON [10], who introduced the *double-kink model*.

With regard to the extension toward a viscoplastic formulation of (mainly glassy) polymers, many contributions were provided in the last years. Hereby, the essential foundations were laid by works of BOYCE and ARGON among others (BOYCE et al. [41], ARRUDA & BOYCE [11], ARRUDA et al. [13], HASAN & BOYCE [150]). Extensions of these models are provided by WU & VAN DER GIESSEN [422–424], TOMITA & TANAKA [384], GOVAERT et al. [128], ANAND & GURTIN [4], BARDENHAGEN et al. [27], MIEHE et al. [269], and many more. The *eight-chain-model* by ARRUDA & BOYCE [12] and the *non-affine microsphere model* by MIEHE [267] are particularly popular in this respect. Most of the named models are formulated in the Kröner-Lee decomposition. The automatic outcome of this approach is the existence of the hypothetical intermediate plastic configuration (see Fig. 2.2). An important assumption for the upcoming developments in the course of this work, is the suggestion made by BOYCE et al. [42], that the elastic part of the deformation gradient has to be symmetric. A comprehensive review on available viscoplastic models for polymers and their numerical implementation is given in BERGSTRÖM [33].

2.4.1.2 Damage and failure

Failure describes the complete loss of load-carrying capability of the material. Ahead of the ultimate failure event, progressive internal damage processes occur in the material, which correspond to the formation of cracks and cavities at the microscopic level. The theory of *Continuum Damage Mechanics (CDM)* as a branch of continuum solid mechanics, formulates constitutive models, in the best case taking into account these micromechanical effects with the aim of a reliable life time prediction (cf. KACHANOV [184], KRAJČINOVIĆ [204] and LEMAÎTRE & CHABOCHE [223]). The phenomena in the material leading to damage evolution and failure are for most materials essentially different from those characterizing deformation (cf. DE SOUZA NETO [296]). A direct consequence thereof is the introduction of state variables that are connected to damage processes additional to the standard variables for elastoplasticity (e.g. (plastic) strain). Thus, the new state variables represent the density and/or distribution of microscopic effects characterizing directly or indirectly the damage state within the material (cf. DE SOUZA NETO [296]). KACHANOV [184] proposed the first continuum damage model in the literature, by means of a scalar internal variable in order to model creep damage for metals under uniaxial loading scenarios. Despite the lack of any physical meaning of the damage variable, this pioneer work marks the beginning of a series of many important publications. Early on, RABOTNOV [327] interpreted the physical meaning of a scalar damage variable as the reduction of the net area of the loaded material due to microcracking as a suitable measure of internal damage. In the following years and in the context of creep modeling, processes were made based on these original developments regarding the generalization towards three-dimensional descriptions of damage (e.g. LECKIE & HAYHURST [213]). MURAKAMI & OHNO [285] and MURAKAMI [284] made noteworthy developments regarding the anisotropic formulation of damage states, presented therefore by second-order tensors. CHABOCHE [61–63, 65] developed a new phenomenological approach for the description of creep-damage evolution based on thermodynamic considerations and proposed the existence of mechanically equivalent *ficticious undamaged configurations*, also known as *stress/strain equivalence*. This hypothesis is also used in contexts other than creep (e.g. damage due to monotonous loading). Further developments were provided by LEMAÎTRE [219, 221, 222] and GURSON [139] for ductile damage. Anisotropic damage as well as unilateral damage effects due to the asymmetry of the damage effect during tensile and compressive load was provided by LEMAÎTRE et al. [225]. A discussion of the separation of CDM theories into two categories, *micromechanical* and *phenomenological* models, is presented in BASISTA et al. [28]. In their work, SIMO & JU [353, 354] give very comprehensible derivations of generally anisotropic strain- and stress-based damage models, consistent with the theory of Lemaître's work. The usage of this hypothesis generally leads to a non-symmetric tangent operator. To overcome this problem CORDEBOIS & SIDOROFF [77] published an elastoplastic damage model based on the assumption of *energy equivalence* and using second-order tensors to describe the damage state. LECKIE & ONAT [214] and ONAT [302] show in their work, that distribution of voids in metallic material can be mathematically presented best by tensors of an even rank. Apparently unaware of this statements, in literature a number of damage models based on a vectorial definition of the damage state are available (cf. KRAJČINOVIĆ & FONSEKA [203], FONSEKA & KRAJČINOVIĆ [117], or KRAJČINOVIĆ [201], among many others). KRAJČINOVIĆ [202] furthermore introduced a model which is able to distinct between a set of active and passive systems of microcracks. KACHANOV [182] and MITCHELL [279] provide an overview over available vector-based damage models. Very extensive reviews of available modeling techniques concerning damage is given in CHABOCHE [60, 61], LEMAÎTRE & CHABOCHE [223], PEERLINGS [310] and LEMAÎTRE & DESMORAT [226] among others. Here, also formulations for other loading scenarios, such as fatigue, are addressed. Based on the above mentioned concepts, constitutive models of the type of LEMAÎTRE & CHABOCHE [223] make use of the degradation of the elastic moduli, hence using the stiffness as the macroscopical measure of damage.Their derivation based on thermodynamical consistent assumptions for hyperelastic materials is presented very comprehensively in DE SOUZA NETO [296] within the framework of small strains. Hyperelastic damage models in the context of finite strain available in the literature are introduced mainly by the example of the *Mullins effect* in rubber materials. Description of the occurring damage effects as well as (general) derivations for the thermodynamically consistent derivation of the isotropic damage model are given in HOLZAPFEL [169]. Further work concerning this topic is provided by LION [230] and MIEHE & KECK [270].

Representative volume elements are one of the main remedies in order to perform numerical studies of the microstructure of unidirectional fiber reinforced composites with the aim of computing and characterizing macroscopic behavior (i.e. stiffness, plastic behavior, failure and many more). For further reading please refer to the works of MELRO et al. [260], WONGSTO & LI [419], LI & WONGSTO [228] or TRIAS et al. [386]. These investigations showed furthermore, that it is necessary to model the microstructure of UD material by means of *statistically representative volume elements (SRVE)* when the characterization of plasticity and damage evolution is the intended goal. The arrangement of the fibers has a significant influence on the stress and strain distribution on micro level. Works by PYRZ [323, 324], SØRENSEN & TALREJA [361], and BROCKENBROUGH [43] are just a few examples that underline the importance of a random spatial arrangement of the reinforcement. For a very extensive review on available as well as a presentation of an intuitive and easy to implement method to generate virtual unidirectional microstructures for the use in finite element applications, please refer to MELRO [260].

A review of the state of the art concerning the modeling of thermoplastic composites with a unidirectional fiber reinforcement shows that viscoplastic behavior is neglected in a large number of publications for the analytical and numerical characterization of the macroscopic behavior. One partial objective of this work is therefore the derivation of a elasto-viscoplastic damage model for the thermoplastic matrix material. It will be used as a basic constitutive model during the meso-modeling of the subsequent microstructures and, thus, can be used to investigate the viscoplastic portion of deformation in the respective composites virtually.

2.4.2 FE-Modeling approaches of textile composites

This section deals with meso-level analysis of mechanical behavior of impregnated textile composites. The simulation of dry composites as well as forming simulation will not be addressed directly and is out of scope here. A short review of available literature is given, that concentrates on the meso-modeling with the aim of maximum detail level with regard to the reinforcement geometry consequently in a very accurate prediction of stress-strain states, using finite element modeling. These analyses allow further investigation of damage initiation and development. When it comes to modeling the mesoscopic level of woven fiber composites regarding deformation and failure onset/progression, engineers have to face a two-fold problem, apart from current issues of the numerical and conceptual framework of multiscale analysis within an FE environment (*consideration of symmetry, correct formulation of boundary conditions, application of numerical homogenization techniques, pre-/post-processing of the FE model*). It consists of (a) the consideration of geometrical effects in multi-layered woven composites and (b) the necessity of reliable constitutive modeling of the single phases (rovings and embedding matrix). LOMOV et al. [231] presented a wide-ranging review of the implementation of progressive damage modeling for different kinds of textile composites. Furthermore, they propose a "road map" in order to set up a multiscale analysis of textile composites, containing the aforementioned issues with focus on the implementation of a closed process chain, starting at the generation of the meso-structure up to the evaluation of damage initiation criteria and damage development.

2.4.2.1 Virtual geometric modeling of weave reinforced composites

The geometrical modeling of woven composites in a finite element context is very challenging. The interlacing of the yarns yields complex microstructures, which have to be accurately rendered in its virtual counterpart. The main difficulties here is the correct definition of the geometry of the single yarns embedded in the composite. Therefore, neighboring yarns and crimp regions, where yarns with different directions meander to the opposite side of the weave layer, are of special interest. In the literature, a large number of authors deal with the FE-meso modeling of textile fabrics (cf. CARVELLI & POGGI [54], KING [192], ZAKO et al. [430], BARBERO et al. [24], and many others, e.g. [22, 35, 36, 54, 55, 58, 81, 82, 86, 103, 116, 142, 165, 179, 209, 215, 232, 235, 236, 297, 325, 348, 368, 371, 374, 375, 397, 411–413, 416, 420, 431, 433]) with variations in the fiber architecture and resolution of the inner structures. Please see CROOKSTON et al. [81] and LOMOV et al. [231] for a detailed overview of available modeling approaches. Overall, mostly woven composites are addressed. Given a sufficiently accurate geometry representation including an adequate FE-mesh, it was shown that the computation of elastic properties can be performed on a very reliable level. The success of these basic investigations and the promising results obtained, encouraged researchers to approach more complex problems such as the investigation of nesting effects (cf. HALE & VILLA [142], KURASHIKI et al. [209], LE PAGE et al. [212]) and damage progression (cf. NICOLETTO & RIVA [297], EDGREN et al. [103], ZAKO et al. [430], DAGGUMATI et al. [85], or WANG et al. [408]). An important step, yielding to a significant simplification and enhancing model possibilities, was the development of microstructure generators for textile composites. The two most prominent representatives are the software packages *WiseTex* (cf. LOMOV et al. [231] and VERPOEST & LOMOV [402]) and developments from researchers of the University of Nottingham, who provide the open-source solution *TexGen* (cf. SHERBURN [350]). Both software solutions deliver accurate representations of both dry and impregnated textile microstructures. In combination with finite element meshers, e.g. *meshTex* or *TetGen* (cf. LOMOV et al. [231] or SHERBURN [350] respectively), an easy transfer to commercial FE programs like Ansys or Abaqus (cf. SIMULIA [355], CADFEM [47]) is possible, distinguishing between matrix rich areas (*matrix pockets*) and weft/warp tows. Due to given information about the element-wise fiber orientation and fiber volume fraction, it is possible to accurately assign the local stiffness properties of the single phases. Using this technique, the majority of the current works deal with the investigation of the local structural behavior of unit cells using FE simulations. Three-dimensional formulations of periodic boundary conditions are mainly used, exploiting the idea of the unit cell under investigation originating from the center of a laminate (cf. LOMOV et al. [231], WHITCOMB [414], WHITCOMB [415], BOGDANOVICH [37]). However, recent studies show the influence of free surfaces and edges on the local stress behavior of shell-like structures (cf. IVANOV et al. [177], OWENS et al. [305], LOMOV et al. [238]), therefore motivating the usage of in-plane periodic boundary conditions (see a definition in MELRO [260]). The resulting virtual model on the mesoscopic scale contains two phases: (a) the impregnating matrix and (b) the undulated rovings. For the approach taken here, a homogenized substitute model is needed for the latter that takes into account the anisotropic nature of elasticity as well as all relevant damage modes. In the following, an overview of some already known models and the requirements for the formulation of damage within the framework of the CDM that were derived from such is given.

2.4.2.2 Constitutive damage modeling of impregnated rovings

An intuitive way for simulating damage in composites is the theory of fracture mechanics, thus, introducing discrete cracks in the FE model. In general, it is however difficult in a computational sense, to create free boundaries in a complex microstructure. Furthermore, the crack path has to be known and well defined beforehand, which is normally the only case during delamination of yarns from the surrounding matrix, as reported by CAMANHO et al. [49]. Hence, only simplified two-dimensional or abstracted geometries were evaluated using classical fracture mechanics (cf. EDGREN et al. [103], LE PAGE et al. [212]). As a consequence, in the context of continuum damage mechanics, models based on internal damage variables are widely used. The advantage here is the possibility of considering a stiffness degradation without introducing cracks directly in the mesh. It does not necessitate mesh refinement, is therefore computationally simple (cf. LOMOV et al. [231]), and hence an important tool for modeling damage evolution in fiber composites. With regard to textile composites, it is widely used (cf. CARVELLI & POGGI [58], TANG & WHITCOMB [375], WHITCOMB & SRIRENGAN [411], FISH & YU [115], ZAKO et al. [430], HAMELIN & BIGAUD [147] just to name a few). In the majority of available works it is assumed, that the matrix behaves isotropic, impregnated tows locally behave like unidirectional composites, showing an initially transversal isotropy and corresponding failure modes. WANG et al. [408] state the importance of further investigations on the exact failure behavior of tows, especially in crimped areas, based on their virtual assessment of the failure behavior of 2D twill woven composites. Most of the damage models used, follow a local formulation. In other words, the damage initiation and progression depends solely on the deformation of the element where the damage criterion is evaluated. In contrast to that, other formulations like phase-field approaches exist, where damage is an additional field (for example to stress, strain or temperature field) that is solved simultaneously (cf. FISH et al. [116]). These concepts are not taken into consideration throughout this work. In order to relate stresses and experimentally determined material strengths to the onset of failure, stress-based failure criteria are commonly used and a large number of available criteria are available in the literature (see e.g. SODEN et al. [359] or PARIS [308]). As a further development of the *von Mises criterion*, HILL [157] proposed the first anisotropic failure criterion, named the *Tsai-Hill criterion*. Criteria of TSAI & WU [392] (*Tsai-Wu criterion*) and HOFFMAN [168] followed, but all three were unable to distinguish the occurring failure modes. To overcome this weakness, HASHIN [151] published the *Hashin criterion*. Intentionally formulated for fatigue failure, it is formulated depending on tensile or compressive loads in combination with shear and distinguishes furthermore between longitudinal (fiber) and transversal (matrix) direction, therefore corresponding directly to the nature of unidirectional plies. SUN et al. [366] and PUCK et al. [320, 321] take into account a detailed investigation of failure modes, appearing during different loading scenarios. More recently, a criterion was presented by DAVILA et al. [88], named *LaRC03*, corresponding to a plain-stress problem. Extending this formulation towards a general stress state, the criterion called LaRC04 was developed (cf. PINHO et al. [315], see also for a review on failure criteria). Additionally, CATALANOTTI et al. [59] proposed another three-dimensional failure criterion for plastic fiber composites, taking special attention to the effect of ply thickness. The extensive work of HINTON et al. [164] evaluated common damage initiation criteria in their *world wide failure exercise*. The reviewed criteria were then adopted by most researchers towards the usage in textile fabrics and build the basis of the assessment of damage evolution. Hereby, the composite lamina (here, this corresponds exclusively to the impregnated yarns) are assumed to be a homogeneous, albeit anisotropic material. The effect of delamination of plies, is usually addressed by methods of linear-elastic fracture mechanics. Here, mostly the *virtual crack closure technique* (cf. KRUEGER et al. [208]) or cohesive formulations (cf. ALFANO & CRISFIELD [1], ALLIX & CORIGLIANO [2], CAMANHO et al. [48], TURON et al. [395], DE BORST [89], JANSSON & LARSSON [180], LI et al. [229]) are used. MAIMÍ et al. [244–246] published a two-dimensional continuum ply damage model for composite laminates, presenting both the theoretical framework as well as the numerical implementation. To ensure thermodynamical consistency during damage evolution, the *crack band model* of BAŽANT [19] is employed. In the context of standard dissipative material theory, a failure model including an anisotropic damage formulation to describe fiber-reinforced composites was proposed by MATZENMILLER et al. [257]. The corresponding rate-equations for damage evolution are consistent with the laws of thermodynamics. A three-dimensional generalized damage model was presented by GOVINDJEE [129] in the context of anisotropic brittle damage of concrete, shows however similarities to a model suitable also for fibrous composites. The appeal of this model is the usage of thermodynamic forces in order to specify the damage evolution. As a result of the hyperelastic theory, these forces are a direct product of the energy potential, which is hence specifying the whole model. The difficulty is however the lack of physical meaning and the therefore non-intuitive usage of such formulations. Based on their previous works, MAIMÍ et al. [247] published an extended damage model to consider general three-dimensional stress states intending to incorporate damage effects due to delamination. The presented constitutive models account for crack closure and passive damage effects under load reversal. In his work, MELRO [260] used this model to evaluate damage evolution in the yarns of a 5-harness stain weave unit cell micro structure under uni- and biaxial loading in combination with a thermodynamically correct elastoplastic damage formulation for the matrix (cf. MELRO et al. [261]). Elastoplasticity in fiber-reinforced composites is investigated and modeled by CHEN et al. [71]. Taking into account RVEs based on measurements of real geometries, BAHEI-EL-DIN et al. [22] presented a damage progression model for three-dimensional weaves, predicting the overall behavior under general, multi-axial stress states. Special attention was paid on the compressive behavior of three-dimensional braided composites by FANG et al. [111]. Other non-linear numerical material models for the prediction of textile composites under damage were provided by COUSIGNÉ et al. [78] and MARTÍN-SANTOS [253]. More recently, ZHONG et al. [432] published a study, in which the prediction of the damage initiation and the development in a three-dimensional woven composite is addressed. Therefore, the authors provide another continuum damage model, covering the predominant failure modes in fiber and in transverse direction as well as matrix failure separately. Damage onset was predicted by using the Puck failure criteria. WANG et al. [408] present a progressive damage modeling strategy based on a multiscale approach to investigate the damage and failure behavior of tow-dimensional woven composites. An anisotropic damage model based on the *Murakami-Ohno damage tensor* is provided. A slightly different approach was presented by THIEME et al. [383], who developed a probabilistic approach to assess failure in glass fiber reinforced weft-knitted thermoplastics. This summary is not intended to be exhaustive and should serve as an introduction. Please note, that the presented damage model used the framework of geometrically linear formulations. A major lesson that can be drawn from the literature review is that the prediction of damage evolution in fibrous composite materials necessitate the capturing of the anisotropic nature of damage. Consequently, with regard to the mathematical representation of directional damage, an at least second or higher order tensor is needed (see e.g. CAROL [51–53], SIMO & JU [353, 354]). For the case of higher order tensors many damage models for several kind of material classes are available (see e.g. DRAGON & MRÓZ [101], KACHANOV [185], CHABOCHE [62], MURAKAMI & OHNO [285], ORTIZ [303], CHOW & WANG [72], MURAKAMI [283], SUARIS [363], YAZDANI & SCHREYER [428], WEI et al. [409], VALANIS [398], LEMAÎTRE & CHABOCHE [223], HANSEN & SCHREYER [148], BETTEN [34], ODA [299], LUBARDA & KRAJČINOVIĆ [240], VOYIADJIS & PARK [406], among many others). CAROL et al. [52, 53] presented a standardized formulation of important concepts in the context of continuum damage mechanics (e.g. *effective stress and strain* and *energy equivalence*) and provided a very comprehensive description of damage models using second-order damage tensors at small strains, following the theory of MURAKAMI & OHNO [285], CORDEBOIS & SIDOROFF [77], and MURAKAMI [283]. This notation will also be used throughout the present work. MENZEL & STEINMANN [262] provide the formulation of anisotropic continuum damage mechanics at large strains and also deal with the algorithmic treatment.

Literature research has shown a large number of available works when it comes to meso modeling of textile composites. Techniques to reliably model the geometry of meso structures based on real geometrical data exist and are widely used. In terms of damage modeling of such composites, many constitutive models are available in numerous variations and were applied in many studies. This high variety and the concomitant discrepancies in notations makes it difficult to compare the single models. Works with the effort of standardization of formulations have been mentioned in the above review. For the purpose of finite element analysis of the mesostructure of two-dimensional woven composites, which will be carried out throughout this work, a homogenized constitutive model representing the impregnated and undulated rovings is needed. This model has to take into consideration the occurring physical failure modes, their effects on the stiffness tensor and has to be able to predict post-failure behavior. To this end, an anisotropic and thermodynamically consistent damage model based on hyperelasticity will be presented in this thesis. Experimentally observed large rotations of the reinforcement structures are hereby described within the framework of finite strains.

2.4.3 Homogenized material models for woven composites

Taking into account the level of detail and the corresponding intensive computational effort necessary by the methods introduced in Section 2.4.2, a homogenized material formulation is needed in order to calculate the deformation and failure behavior of entire structural parts consisting of layered textile composites. Hereby, the knowledge of both, in-plane and through-thickness properties are essential. To address this problem, reliable three-dimensional homogenization methods implemented in material models for the usage in industrial applications (cf. Hu [170], Mouritz et al. [281], Ayranci & Carey [18]) are needed, respecting all geometrical factors, such as fiber architecture (woven, braided or knitted) as well as properties of the subcomponents (mechanical properties, volume fractions, orientations, etc.).

2.4.3.1 Analytical modeling approaches for the computation of the elastic stiffness tensor for textile composites with a specifically defined topology

In analogy to short fiber composites, first analytical homogenization techniques to model the elastic behavior of textile composites were developed. Two-step homogenization procedures, executed on meso- and macro-levels set up the general scheme of these techniques. The idea here is to first predict the elastic properties of a sub-volume of a subdivided unit cell or RVE, taking into account realistic geometrical conditions, where the reinforcements can be assumed to be piecewise unidirectional lamina with long fibers. By calculating the homogenized stiffness of the sub-volumes at micro-level (usually within the local $\{123\}$ coordinate system, where the 1-direction corresponds always to the direction of the local reinforcement) and transforming them into the global coordinate system $\{XYZ\}$ with respect to the RVE, the second homogenization step can be performed. Applying a particular homogenization technique the results from the single sub-volumes are assembled, yielding a prediction of the overall global stiffness matrix of the complete RVE. Hallal et al. [144] cluster available analytical methods in five categories based on the corresponding homogenization method used at meso- and macro-levels: Classical Laminate Theory (*CLT*), iso-strain assumption, mixed iso-strain/iso-stress assumptions, inclusion methods and the methods of cells (*MOC*).

Classical Laminate Theory

Well known from applications with 2D laminated composites, models based on CLT yield good results in comparison with experimental and FE results. In the studied literature, this technique was mostly used on 2D woven textile composites, supported by the relatively simple fiber architecture, such as plain weaves, twill weaves, satin weaves, etc. The advantage of such structures is, that the RVE geometries can be decomposed into layers along the thickness direction rather easily, enabling a straight forward usage of CLT. Ishikawa & Chou [171–175] provide fundamental and basic work on modeling textile composites using CLT as homogenization technique. Ishikawa & Chou [173] derive two approaches, iso-strain and iso-stress condition, yielding finally upper and lower bounds of the in-plane elastic properties. In the course of the just quoted works, three basic models were introduced, the *mosaic*, the *undulation*, and the *bridging* model. Figure 2.19 shows the used ways of microstructure abstraction. Starting point of their developments was the mosaic model, where the RVE is idealized as a asymmetrical cross-ply laminate as depicted in Fig. 2.19a. Further development lead to a one-dimensional crimp model, the fiber undulation model, that described the undulation of the yarn under loading using a sinusoidal function, whereas transverse yarns were assumed to be straight (cf. Fig. 2.19b). This extension of the mosaic model makes use of the iso-stress assumption. Of particular interest applying this model is the subdivision of the undulated sections of the fabric composite, where the stiffness matrices are calculated locally and then assembled under iso-stress condition using CLT. The authors state that this modeling technique works best for two-dimensional plain weave composites (cf. Ishikawa & Chou [173]). Finally, they also proposed the bridging model, where the unit cell is further decomposed in five sub-regions (cf. Fig. 2.19c), where region III corresponds to the crimp region. All sub-regions are then assembled using iso-stress conditions in series and iso-strain condition in parallel. The authors state, that this model is best used for satin fabrics with long straight regions. The models just introduced were the starting point for further developments, with increasing complexity regarding the abstraction of the microstructure as well as the sequence of homogenization using different assumptions. Yang et al. [427] presented an extension of the fiber undulation model, called fiber inclination model, where the unit cell is assumed to consist of inclined yarns. Using CLT together with the iso-stress condition,

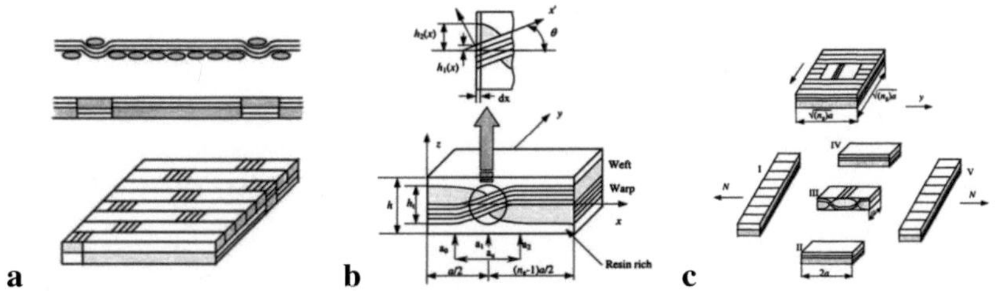

Figure 2.19: Schematic representations of assumptions used by analytical homogenization techniques by ISHIKAWA & CHOU [173]. Approximation of the weave geometry and definition of sub-regions by (a) mosaic model, (b) undulation model, and (c) bridging model. Images taken from HALLAL et al. [144].

the effective elastic properties for a three-dimensional braided composite are derived. In a similar manner, WHITNEY & CHOU [417] provide a model for the prediction of in-plane thermo-elastic properties of a three-dimensional angle interlock composite. Attempting to combine the iso-strain and iso-stress conditions, BYUN et al. [46] introduce a micro-cell model in combination with CLT. Comparison of the model response with experimental results show an improved prediction of the axial Young's modulus as well as in-plane Poisson's ratio, whereas less accurate results for the axial shear stiffnesses are obtained. With the *cross-over model*, RAMAKRISHNA [330] proposed another two-dimensional analytical model for the prediction of the elastic behavior of plain knitted composites. Using CLT together with an iso-strain condition, the calculation of elastic entities in sub-volumes of the curved yarns was carried out in four symmetric sub-regions of the RVE. In the second homogenization step a micromechanical model defined by UEMARA et al. [396] was used to assemble the elastic properties of the yarns and the surrounding resin matrix in order to obtain the stiffness matrix of the RVE.

Iso-strain models

In literature it is reported about all above presented models, that the application of CLT is considerably more difficult when it comes to the modeling of three-dimensional woven, braided or knitted composites, with a more complex yarn structure. In such cases it is more reasonable to adopt a simpler homogenization method. Analytical models using the iso-strain method allow a straightforward and easy calculation of homogenized stiffness properties by only using a discretization of curved or undulated yarns in sub-volumes, with the fiber volume fraction and orientation in each sub-volume and a subsequent averaging over the volume of the yarns and matrix under an iso-strain assumption. Prominent works were published early on by TARNOPOL'SKII et al. [376] and KREGER & MELBARDIS [205], who introduce the analogous formulation of the rule of mixture for textile composites, called the *orientation average model*. Similar to the models using CLT, the microstructure is idealized and decomposed into piecewise unidirectional sub-structures and are then averaged using either iso-stress (compliance average model) or iso-strain (stiffness average model) assumptions, defining the upper and lower bounds of the elastic properties respectively. Modifications were made by COX & DADKHAH [79] for the application on three-dimensional interlock composites. Further developments were presented by HAHN & PANDEY [141], where all yarns were assumed to be undulated and the overall stiffness can be computed in a three-dimensional model. Modifications of this model for the usage on tri-axially braided composites are given by SHOKRIEH & MAZLOOMI [351]. Other examples for the application on different scaffold types are presented by CHANG et al. [69] for the calculation of the stiffness of three-dimensional angle interlock woven ceramic composites. EL HAGE [105] and EL HAGE et al. [104] applied a similar model to predict the elastic properties of orthogonal and angle interlock woven composites. For the former, the authors state a good agreement of the model with experimental data, especially when it is applied on the RVE instead of just a unit cell. HALLAL et al. [145] predict the properties of a three-dimensional angle interlock composite where especially transverse properties are in good agreement with experimental results. Treating three-dimensional braided fabrics, the iso-strain model shows equally good agreement as presented in LI et al. [227]. By its nature, the iso-strain model yields elastic properties in both in-plane as well as out-of-plane directions, which makes it an effective model for almost all advanced textile composites (cf. SHOKRIEH & MAZLOOMI [351], QUEK et al. [326], LI et al. [227] and EL HAGE et al. [104]). However, this simplicity of using only an iso-strain condition is the major drawback concurrently, since it is to be assumed, that such assumption will yield less accurate predictions of ultimate strength and fatigue (cf. HALLAL et al. [144]). It should also be noted that at this time there is not enough experimental data available to confirm the validity of the prediction of out-of-plane properties.

Mixed iso-strain / iso-stress assumptions

With the compliance and stiffness average models being the limiters of physically reasonable results, further developments were focused on finding a model taking into account iso-strain and iso-stress assumptions in a superimposed way. To that end, the intention is to respect the stress-strain fields in the RVE better than it is the case with pure iso-strain models while predicting the mechanical behavior of textile composites. By applying the CLT method and using different assembling approaches in series or parallel the total stiffness of the composite is computed, taking into account the actual geometry of the microstructure, i.e. architecture and yarn cross-sections. NAIK & GANESH [288, 290], NAIK & SHEMBEKAR [292, 293], SHEMBEKAR & NAIK [349] and GANESH & NAIK [118, 119] provide early on work, resulting in a two-dimensional woven fabric model *(2DWF)* predicting the in-plane elastic properties of woven composites. More recently, this methodology was extended to a three-dimensional model in order to compute also the out-of-plane behavior of two- and three-dimensional weaves (cf. NAIK et al. [287], NAIK & KUCHIBHOTLA [291], NAIK & SRIDEVI [289]). The results for 3D orthogonal and angle interlock woven composites were found to be in good agreement with experimental data. Moreover, modifications towards the prediction of thermoelastic properties of textile composites were provided by SANKAR & MARREY [341] by their *selective averaging method (SAM)*. TAN et al. [369–372] developed models in order to address the estimation of thermal expansion coefficients of three-dimensional orthogonal and angle interlock woven fabrics. A model to predict anisotropic elastic stiffness of three-dimensional textiles was presented by POCHIRAJU & CHOU [316]. Here, the macroscopic stiffness is computed using an *effective response comparison (ERC)* technique, based on the decomposition of the RVE in small elements. The authors compared calculated and measured results of angle interlock composites and found good agreement. A very simple and intuitive way of combining iso-strain and iso-stress was the model of KALIDINDI &ABUSAFIEH [186] and KALIDINDI & FRANCO [187]. Introducing a factor α, giving the weight of the different shares of the two single solutions according to

$$\mathbb{C} = \alpha\mathbb{C}_{\bar{\varepsilon}} + (1 - \alpha)\mathbb{S}_{\bar{\sigma}}^{-1}, \tag{2.73}$$

where $\mathbb{C}_{\bar{\varepsilon}}$ and $\mathbb{S}_{\bar{\sigma}}^{-1} = \mathbb{C}_{\bar{\sigma}}$ are the stiffness solutions coming from iso-strain and iso-stress assumptions respectively. However, difficulties in the physical motivation and the finding of a generalized form of such a factor arise.

2.4.3.2 Analytical modeling approaches for the computation of the stiffness tensor for textile composites with an arbitrary topology

The commonality of all approaches so far, was the complex discretization of the RVE, restricting the models to certain kinds of textiles. A slightly different and more flexible approach was pursued by the now following authors. While so far the RVE had to be fully discretized in sub-volumes, which then had to be assembled in parallel or series along their loading directions, according to their relative positions, the following approach is based on the assemblage by components. In other words, first sub-volumes of curved parts of the yarns are assembled and in a second step complete homogenized yarns and matrix are combined to estimate the overall stiffness of the RVE. An example for such a procedure is the model of Vandeurzen et al. [400, 401], who presented an improved *fabric geometrical model*, based on thorough geometrical and mechanical study on two-dimensional woven fabrics. The stiffness of combi-cells, which consist of a pure yarn and a matrix layer, is computed using the complementary variational principle. The assembly is then carried out by applying either iso-strain or iso-stress models. Other examples are given by the mixed iso-strain/iso-stress model of Ruan & Chou [340] for weft-knitted composites, the three-dimensional model proposed by Lee et al. [217] for the calculation of the elastic properties of plain weave composites or the *three stages homogenization method (3SHM)* first used on three-dimensional angle interlock composites (cf. Hallal et al. [145]) and later on braided and knitted composites (cf. Hallal [143]). The nature of the second type of models allow a more general usage and applications on different kinds of architectures.

Inclusion models

To promote further generalization, the *Mori-Tanaka scheme (MT)* was employed by Gommers et al. [127] and applied on many types of textile composites, including woven, braided and knitted scaffolds. Comparing the Mori-Tanaka scheme with other frequently used analytical homogenization schemes, it shows the advantage of delivering an explicit formulation for the effective stiffness tensor, regarding local stresses and strains. The general approach of MT is to approximate the inclusion shape as ellipsoidal, resulting in a constant stress and strain field in each inclusion. In the case of textile composites, the inclusions correspond to curved yarns and can be approximated by a set of ellipsoids with circular or elliptical cross-section in such a way that the fiber orientation distribution is not altered. GOMMERS et al. [127] state that textile composites have to be modeled using a non-trivial fiber orientation distribution, i.e. not aligned nor random, of non-spherical, anisotropic inclusions. The yarns are therefore partitioned into sub-volumes, which are then treated as single inclusions with elliptical cross-sections at an infinite length. Good agreement with experimental and Finite Element results was found by LOMOV et al. [233] for three-dimensionally woven composites. PERIE et al. [312] reports good agreement for 3D angle interlock composites with the remark that the accuracy of the geometrical modeling has a high influence on the validity of inclusion models.

Methods of cells

One of the most recent developments in modeling the elastic behavior is done by PRODROMOU et al. [318], who presented the *method of cells (MOC)* in the context of analytical modeling of woven composites, while analyzing a wide range of two- and three-dimensional woven fabrics. For the application of this model, a full discretization of the microstructure into smaller cells containing yarn and matrix layers is necessary. Therefore, the authors used the software Wisetex. While the stiffness of the yarn layers is computed using *Chamis' micromechanical model* [68], the assembly of the sub-cells and homogenization of block-cells is done by using the complementary energy minimization method. Analyses of results from MOC show an improvement with respect to the MT model for a wide range of two- and three-dimensional woven composites (cf. PRODROMOU et al. [318]). However, the authors state as well that the computation time is significantly higher compared to that of other analytical models, but yet less than Finite Element solutions (cf. PRODROMOU et al. [318]).

For a more detailed review of available homogenization approaches using the different categories, the interested reader is referred to the very extensive and comprehensive reviews of Hallal et al. [144] and Dixit [98]. Furthermore, these authors performed a study, in order to compare the elastic results of particular homogenization techniques and point out substantial differences. They report that some models were designed to predict properties for some specific kind of composite whereas others show a more universal character, whereas only few studies focus on the development of generalized models. Moreover, it can be stated that the modeling of three-dimensional fiber architectures pose a major challenge due to their more complex microstructure. Hallal et al. [144] show in their comparative study that the considered analytical models are quite capable of delivering reliable results and are at the same time easy to use.

2.4.3.3 Damage modeling in homogenized material models for textile composites

With the emphasis of the presented thesis on modeling the failure behavior of textile composites on structural level, the next logical step is to look at damage initiation, progression and ultimate strength. On the macroscopic level rather few works have been published on this matter. In terms of analytical models, Naik [294] presented a study for woven and braided fabric composites where he used the iso-strain assumption together with different failure criteria for yarns and matrix respectively. According to Doitrand et al. [99], macroscopic models that predict damage evolution and failure in textile composites have been published more recently for two- and three-dimensional textile composites respectively (cf. Hochard et al. [166], Hochard et al. [167], Marie & Chaboche [248], Marcin [251], and Rakotoarisoa et al. [329]). With the focus lying on woven composites, the two-dimensional model of Hochard et al. [166] is presented in more detail in the following. To the knowledge of the author, this model is the only one of its kind. Based on the *meso scale* model presented by Aubourg [17], a model to describe the non-linear degradations with progressive damage and inelastic strains was presented. Hereby, a thermodynamic consistent, generalized formulation for the progression of internal variables (d_1, d_2, and d_{12}) associated with the decrease of elastic moduli is used, which are directly related to the brittle fiber fracture in warp and weft direction as well as the decreasing shear stiffness, respectively. The development of d_{12}, as the progressive damage variable for shear, depends on the shear load as well as traction loads, which yield micro-cracks at the fiber/matrix interfaces in the warp and weft yarns and are supposed to be oriented in parallel. The complementary free energy potential of woven plies can therefore be written as

$$\Lambda^D = \frac{1}{2}\left[\frac{\langle\sigma_1\rangle_+^2}{E_1^0\,(1-d_1)} + \frac{\langle\sigma_1\rangle_-^2}{E_1^0} - 2\frac{\nu_{12}^0}{E_1^0}\sigma_1\sigma_2 \right.$$
$$\left. + \frac{\langle\sigma_2\rangle_+^2}{E_2^0\,(1-d_2)} + \frac{\langle\sigma_2\rangle_-^2}{E_2^0} + \frac{\sigma_{12}}{G_{12}^0(1-d_{12})}\right], \tag{2.74}$$

assuming plain stress and small strains. $\langle\cdot\rangle_\pm$ denote the positive or negative Macaulay brackets which distinguish between tensile and compressive loads. In such a way the unilateral feature of damage is included in the model. The derivation of the strain energy with respect to the damage variables yield the thermodynamic driving forces $\boldsymbol{Y}$ associated with the damage variables for tension and shear d_i where $i = 1, 2$ and d_{12} according to

$$Y_{d_i} = \frac{\partial\Lambda^D}{\partial d_i} = \frac{\langle\sigma_i\rangle_+^2}{2E_i^0\,(1-d_i)^2}$$
$$Y_{d_{12}} = \frac{\partial\Lambda^D}{\partial d_{12}} = \frac{\sigma_{12}^2}{2G_{12}^0(1-d_{12})^2}. \tag{2.75}$$

The values of the timely maximum of the thermodynamic forces drive the development of the internal variables. In the underlying model, it was assumed that failure corresponding to d_1 and d_2 is instantaneous in order to represent the brittle characteristics of fiber failure. An equivalent thermodynamic force Y for the development of d_{12} was formulated by

$$Y = \alpha_1 Y_{d_1} + \alpha_2 Y_{d_2} + Y_{d_{12}} \quad \text{with} \quad \underline{Y}(t) = \sup_{\tau\le t}\left(Y(\tau)\right) \tag{2.76}$$

in order to take into account the traction/shear coupling. α_i denote the coupling coefficients which are treated as material parameters. The damage formulation closes with the definition of the damage evolution law

$$d_{12} = \left\langle \frac{\sqrt{\underline{Y}} - \sqrt{Y_0}}{\sqrt{Y_c} - \sqrt{Y_0}} \right\rangle_+, \quad d_1 = 0 \quad \text{and} \quad d_2 = 0 \tag{2.77}$$

if $d_{12} < 1$ and $Y_{d_1} < Y_{1f}$ and $Y_{d_2} < Y_{2f}$ or else

$$d_1 = d_2 = d_{12} = 1. \tag{2.78}$$

Y_0 and Y_c are constant parameters, corresponding to the threshold and critical value of the development of d_{12}. Y_{1f} and Y_{2f} define the thermodynamic forces at which ultimate failure in the warp and weft yarns occur. Besides the development of damage, HOCHARD et al. [166] considered inelastic strains that were observed in $[45]_8$ laminates, as a consequence of slipping phenomena between damaged fiber bundles/matrix interfaces during shear loading. They considered occurring significant inelastic strains only under shear since the stiff fiber bundles prevent the formation in fiber direction. In order to describe these strains, a plastic-hardening model provided by LADEVÈZE & LE DANTEC [211] was used. Taking into account the formulations for effective stress $\tilde{\boldsymbol{\sigma}}$ and strain $\tilde{\boldsymbol{\varepsilon}}$ (cf. LADEVÈZE [210]) the coupling of damage and plasticity was obtained. The elastic domain was modeled by

$$f\left(\tilde{\sigma}_{12}, \varepsilon^p_{\text{acc}}\right) = |\tilde{\sigma}_{12}| - \left(R\left(\varepsilon^p_{\text{acc}}\right) - R_0\right) \tag{2.79}$$

with R_0 being the initial threshold for inelastic strain and $R\left(\varepsilon^p_{\text{acc}}\right)$ describes the hardening as a function of the accumulated plastic strain $\varepsilon^p_{\text{acc}}$ as described by HOCHARD et al. [166]. In the underlying work, a power law was chosen for R and associative flow was assumed. The combination of inelasticity and damage evolution shows good agreement with experimental results for several laminate stackings.

It is important to mention that throughout the complete loading path the woven plies are regarded as homogeneous orthotropic, thus, not taking into account reorientations of reinforcements due to off-axis loadings. Non-linearities in the loading/unloading hysteresis, due to slipping/friction effects on micro level were neglected. Furthermore, it has to be underlined at this point that all presented models concerning the prediction of the behavior of textile composite so far are formulated in the framework of infinitesimal strains.

The lack of an existing three-dimensional macroscopic failure model taking into account the damage mechanisms taking place in the single reinforcement structures as well as in the adjacent matrix and furthermore respects the local material re-orientations due to deformation motivates the formulations carried out in the following chapters. With the suggested approach, the aim is to address the need of the industry for a failure model for textile composites at the structural level.

3 Experimental investigations

This chapter introduces the materials used for characterization and validation of the developed material models and presents the experimental results. Compliant with the suggested multiscale approach, the following distinct scales have been under investigation in particular:

1. Neat polypropylene (*PP*) matrix
2. Unidirectionally glass fiber-reinforced composite (*PP/GF-UD*)
3. Fabric reinforced composite (*organo sheets*)

All mechanical and micro-optical investigations presented in the following were conducted in the in-house laboratory at Robert Bosch GmbH where a standard 200 kN universal testing machine of the manufacturer Zwick was used. For the mechanical testing of the unreinforced matrix, an alternative 20 kN load cell was installed. Longitudinal expansion was captured by an integrated multiXtens extensometer, whereas transversal deformation was measured by a digital image correlation technique (ARAMIS) for some specific cases. The tests were executed by applying different strain rates in a normed environment, i.e. at an ambient temperature of $T = 23°\text{C}$ and a relative humidity of $\phi = 50\%$. In what follows experimental data in forms of stress-strain relationships will be presented. Thereby, recorded raw experimental data will be displayed using gray lines, whereas the colored lines depict an average of the grouped curves up to the minimal fracture strain of respective velocity set. The used strain rates are equally spaced in a logarithmic scaling.

3.1 Experimental strain and stress measures

The determination of mechanical properties for the subsequent use in a material model is only possible by conclusive preceding material testing. Due to very large numbers of possible varieties and combinations of loading states (e.g.

stress states, loading rates, or environmental influences like temperature or humidity) in real parts, the relevant influencing factors have to be identified. For the experimental campaign presented in this work, environmental influences were not considered. Mechanical quasi-static loads taking into account time-dependent effects were investigated only. Therefore, one fundamental aspect of material characterization becomes the assessment of stress and strain response of the material. In order to avoid superposition of different material mechanisms, a first approach concentrates on macroscopically uniaxial loading states. In contrast to multi-axial experiments, these types of tests are easier to conduct on standard testing equipment with regard to the required testing technology. With the stress and strain states being not directly accessible from the specimen, they have to be calculated from measured forces and displacements.

To correlate stress and deformation, it is necessary to define an appropriate strain measure. In Eq. 2.11 the general SETH-HILL strain measurements have been introduced. For the one-dimensional case, they relate the state of current deformation L to a defined initial length L_0. For large deformations or non-linear processes it is recommended to use the GREEN or logarithmic (HENKY) strain measures (cf. GRELLMANN & SEIDLER [133]). The latter was used throughout this work and is explicitly defined by

$$\varepsilon = \int_{L_0}^{L} \frac{1}{L} \mathrm{d}L = \ln\left(\frac{L}{L_0}\right) \in [-\infty, \infty] \, . \tag{3.1}$$

Deformations that appear in transversal direction ($\varepsilon_\perp$) are related to those appearing in the principal loading direction ($\varepsilon_\parallel$) by the Poisson's ratio ν via

$$\nu = -\frac{\varepsilon_\perp}{\varepsilon_\parallel} \tag{3.2}$$

or in terms of relative changes of these quantities

$$\nu = -\frac{\Delta\varepsilon_\perp}{\Delta\varepsilon_\parallel} \, . \tag{3.3}$$

The Poisson's ratio can be calculated in all three spatial directions. For isotropic materials all ratios assume the same values defined in the range of $\nu \in [-1, 0.5]$, where $\nu = 0.5$ stands for incompressible material behavior. For composites with complex microstructure, this domain does not apply in general. Becker [29] presents an overview of standard values for the Poisson's ratio of a selection of pure plastics. For polypropylene, Poisson's ratios of $\nu = [0.3, 0.45]$ are reached (cf. Dassow [87], Schenkel [343]) but are not generally assumed to be constant during deformation. Due to their particular molecular structure, plastics exhibit a rate dependent deformation behavior. In order to describe this time-dependency, the entity of strain rate is introduced here. Analogously to the strain itself, the strain rate is a dimensionless measure to describe the velocity during the deformation process, defined by

$$\dot{\varepsilon} = \frac{\mathrm{d}\varepsilon}{\mathrm{d}t}. \tag{3.4}$$

For a local measurement of the strain rate it can be written alternatively as

$$\dot{\varepsilon} = \frac{v_{\mathrm{T}}(t)}{L(t)}, \tag{3.5}$$

where $v_{\mathrm{T}}(t)$ is the velocity measured by the extensometer. In general, $\dot{\varepsilon}$ can change locally in the course of the experiment, e.g. due to localization effects, and can therefore not be considered as constant. In analogy to Eq. 3.5, a *technical strain rate* can be defined as

$$\dot{\varepsilon}_t = \frac{v_{\mathrm{T}}(t)}{L_0}, \tag{3.6}$$

in terms of the initial free measurement length L_0. The true stresses are defined by

$$\sigma(t) = \frac{F(t)}{A(t)}, \tag{3.7}$$

where F is the measured force and A is the current cross-sectional area of the test sample. If the loading direction and the normal of A coincide, the stress is defined as a normal stress σ otherwise as a shear stress τ. In general, for the computation of the stresses from experimental observations, the current forces are related to the initial cross-sectional area A_0. In this case, the corresponding stress measure is called *technical stress*

$$P(t) = \frac{F(t)}{A_0}. \tag{3.8}$$

For the sake of reducing the experimental effort, the change of cross sectional area is not taken into account. The experimental results are given in terms of the technical stresses.

3.2 Neat polypropylene matrix

3.2.1 Specimen preparation

The unreinforced polymeric polypropylene matrix (PP) builds the basic constituent of all considered composites on upper scales. In order to operate on consistent material systems throughout the scales it is required to use the same materials on all scales, ranging from the pure matrix material to the organo sheet. This concept holds true for the basic polypropylene matrix. In addition to the readily assembled sheet material, neat matrix raw material could be provided by the material supplier in the original composition, i.e. containing additives as used in the final woven composite. The preparation of injection moulded specimens was carried out by SKZ - KFE gGmbH in Würzburg, Germany. Since the matrix material is to be tested under tensile and compressive conditions three specimen geometries were generated. For tensile testing under moderate strain rates the specimen geometry according to DIN EN ISO 3167 [93] is proposed as displayed in Fig. 3.1. For compressive tests, specimens were cut by water

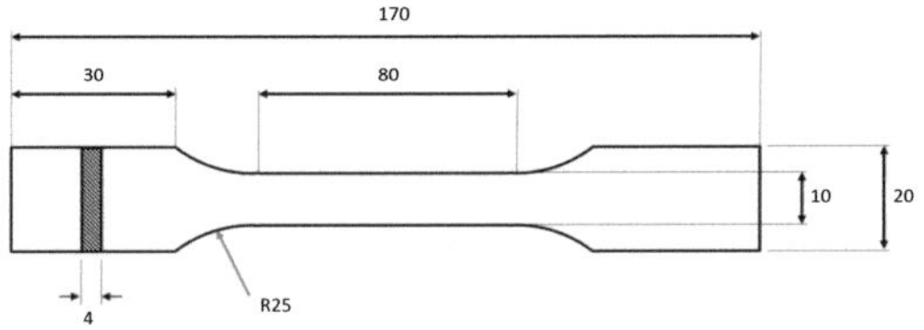

Figure 3.1: Dimensions of the Campus A1 specimen (in mm), cf. DIN EN ISO 3167 [93].

jet from the middle part of the tensile dumbbells in two ways according to DIN EN ISO 604 [96]. Both, tensile and compressive tests are performed to determine the modulus of elasticity and the stress at rupture. The dimensions of the prismatic specimens are summarized in Table 3.1. The measurement of density according to DIN EN ISO 1183-1:2012 [91] of the unreinforced matrix revealed an average value of $\bar{\rho} = 0.907\,\text{g/cm}^3$, which lies in good accordance with literature values for polypropylene. More detailed results are presented in Table 3.2. The glass transition temperature (T_g) was determined to $T_g = 281.7\,\mathrm{K}$ by using dynamic mechanical analysis (DMA). Hence, the amorphous phase of the material is in rubber-like state at testing temperature of $T = 296\,\mathrm{K}$.

Table 3.1: Dimensions of specimen used for compression testing according to the testing standard DIN EN ISO 604 [96] (in mm).

Type	Measurement	Length	Width	Thickness
A	Modulus	50 ± 0.2	10 ± 0.2	4 ± 0.2
B	Strength	10 ± 0.2		

For the subsequent material tests the above geometries were used. The measuring length for tensile specimen was $L_0 = 50\,\mathrm{mm}$ and was measured in the lean centered part of the specimen. Change in length during compressive deformation was recorded by a flat extensometer with a gauge of $L_0 = 20\,\mathrm{mm}$ for specimen type A. In case of a type B specimen, the direct displacement of the machines cross beam was evaluated. In order to characterize the deformation behavior of the thermoplastic polypropylene matrix, quasi-static tensile and compressive tests with and without unloading were performed.

Table 3.2: Density measurements of neat polypropylene matrix. Average value of density $\bar{\rho}$ and the standard deviation s.

ID	Density (g/cm^3)
1	0.90734
2	0.90705
3	0.90708
4	0.90709
5	0.90723
$\bar{\rho}$	0.907158
s	$1.1 \cdot 10^{-4}$

3.2.2 Tensile tests

As shown in Table 3.3, tensile tests were performed at six different position-controlled cross beam velocities. With this test the material stiffness and

Table 3.3: Cross beam velocities and corresponding strain rates for tensile tests of pure matrix (measurement length L_0 = 50 mm).

Cross beam velocity v_T (mm/min)	Strain rate $\dot{\varepsilon}$ (1/s)
1.6	$5.33 \cdot 10^{-4}$
5	$1.67 \cdot 10^{-3}$
16	$5.33 \cdot 10^{-3}$
50	$1.67 \cdot 10^{-2}$
160	$5.33 \cdot 10^{-2}$
500	$1.67 \cdot 10^{-1}$

strength as well as the stress-strain relation at different strain rates are identified. The specimens were loaded monotonously under tension up to failure, characterized by a sudden drop of force and the formation of separation fronts in the material. Contrary to first expectations, the material failed in a very brittle manner without pronounced necking. Figure 3.2 shows a series of matrix specimens exhibiting brittle failure. Figure 3.3 displays the measured stress-strain relations. The detailed information for each velocity set of specimens, the material stiffness E^t_m, the material strength R^t_m, and the fracture strain ε^t_F were determined and are summarized in Figs. A.1 to A.6. The initial stiffness was measured in the strain range of $0.05 \leq \varepsilon \leq 0.25\%$ according to DIN EN ISO 527-1 [94]. Strain at failure ε^t_F is identified as the strain where the tensile material strength, hence the maximum value of stress, is reached. The stress-strain relation in the post-failure regime is not displayed. By comparing the averaged curves, the influence of the external strain rate on the material behavior becomes apparent. With increasing loading rate, the specimens exhibit an increasing initial stiffness and material strength as well as decreasing strain values at fracture. Figure 3.4 displays the correlation of strain rate and material strength. Poisson's ratio was measured to ν_m = 0.42 by using a digital image correlation (DIC) via the optical analysis tool ARAMIS, applying three different loading rates (v_T = [5 mm/min, 50 mm/min, and 500 mm/min]).

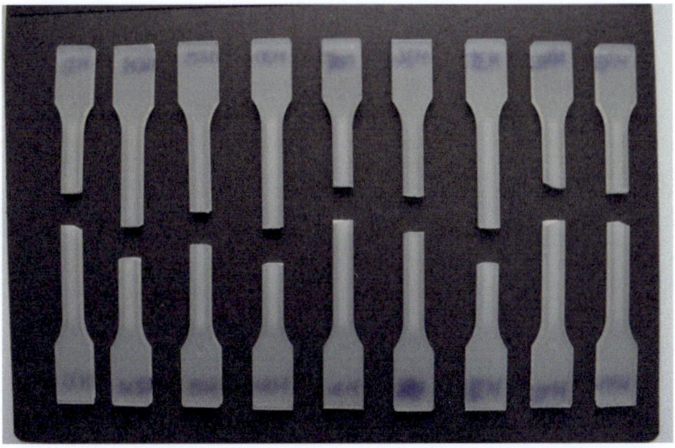

Figure 3.2: Campus A1 matrix specimens after being tested under tension at different loading rates. All specimens exhibit brittle failure.

However, no significant dependency on the loading rate was detected and ν_{m} stayed constant in the course of the quasi-static tests. The optically determined parameters stiffness $E^{\mathrm{t}}_{\mathrm{m}}$, strength $R^{\mathrm{t}}_{\mathrm{m}}$, and failure strain $\varepsilon^{\mathrm{t}}_{\mathrm{F}}$ were in good agreement with the conventionally determined values.

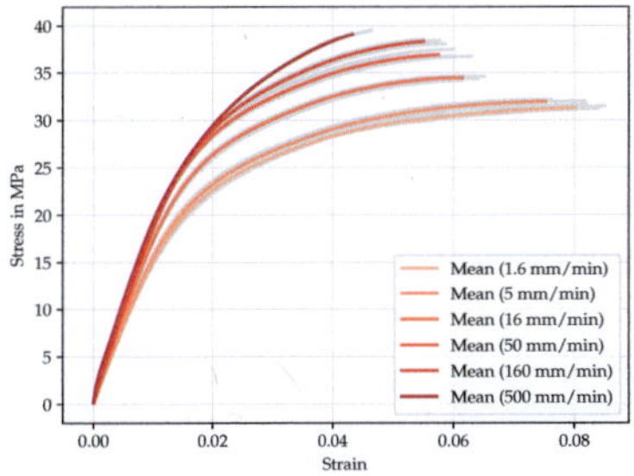

Figure 3.3: Stress-strain relations of quasi-static tension test of pure matrix at different loading rates.

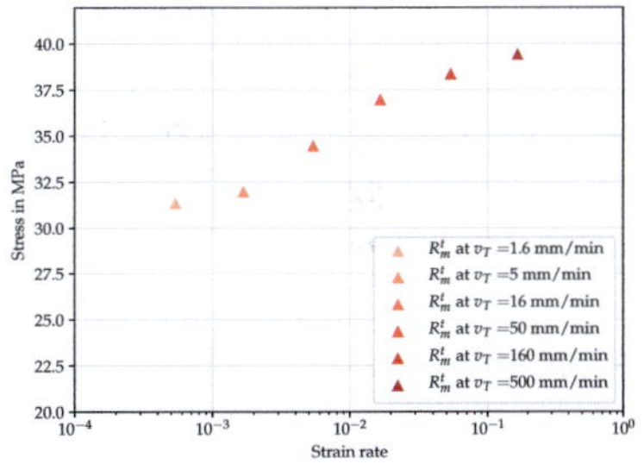

Figure 3.4: Matrix tensile strengths at different strain rates.

3.2.3 Compression tests

Compression tests were performed at four different position-controlled cross beam velocities for two different specimen geometries respectively (cf. Table 3.1). For type A specimens, Table 3.4 shows the corresponding strain rates with a measurement length of $L_0 = 20\,\text{mm}$. The following experiments

Table 3.4: Cross beam velocities and corresponding strain rates for compression tests of pure matrix (measurement length $L_0 = 20\,\text{mm}$).

Cross beam velocity for type A specimen v_T (mm/min)	Strain rate $\dot{\varepsilon}_c$ (1/s)	Cross beam velocity for type B specimen v_T (mm/min)
0.5	$4.16\cdot 10^{-4}$	0.25
5	$4.16\cdot 10^{-3}$	2.5
50	$4.16\cdot 10^{-2}$	25.0
500	$4.16\cdot 10^{-1}$	250.0

identify the material stiffness and strength as well as the stress-strain relation at different loading rates. Therefore, the material was loaded monotonously under uniaxial compression. The upper and lower specimen flanks were covered with grease prior mounting into the testing machine in order to reduce friction effects between specimen and steel blocks. Figure 3.5 shows a schematic stress-strain response of a type B specimen under compression. For all conducted experiments the characteristic local stress minimum in a strain range $\varepsilon > 12\%$ could be observed after exceeding a precursory local stress maximum, accompanied by the formation of shear bands. In the following, the local maximum was used for the definition of the the material strength R^c_m. Strain at failure ε^c_F is identified as the strain where the compressive material strength R^c_m is reached. Figure 3.6 displays the measured stress-strain relations of type B specimens loaded with the defined strain rates to determine the compressive strength. The initial stiffness E_c was measured in the strain range of $0.05\% \leq \varepsilon \leq 0.25\%$ according to DIN EN ISO 604 [96] using specimen type A. The stress-strain pathway beyond the defined failure point is not displayed. A more detailed overview of the measured results is given in Figs. A.7 to A.10. Analogously to the behavior under tension, compressive loading yields a time-dependent

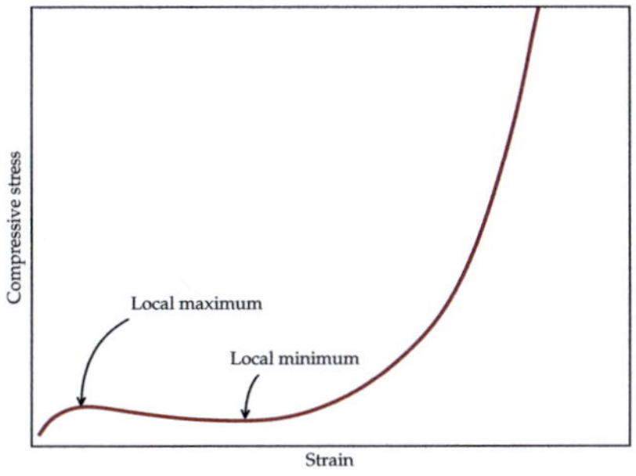

Figure 3.5: Schematic stress-strain relation of quasi-static compression test of pure matrix.

material response. Confronting the averaged curves, the influence of the external strain rate on the material behavior becomes visible. Again, specimens with increasing loading rate show an increasingly higher initial stiffness and higher material strength. The correlation of strain rate and material strength is displayed in more detail in Fig. 3.7. Regarding the stiffness values, a unimodal behavior can be observed. However, in contrast to the before presented results under tensile loading, the value of strain $\varepsilon_{\mathrm{F}}^{\mathrm{c}}$ at reached material strength $R_{\mathrm{m}}^{\mathrm{c}}$ increases with rising loading rates.

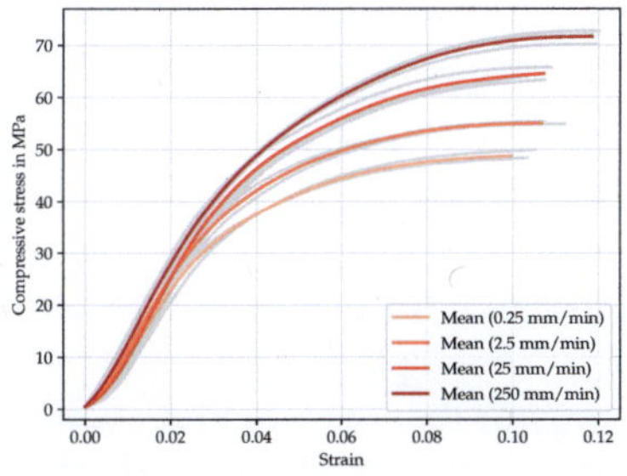

Figure 3.6: Stress-strain relations of quasi-static compression tests of pure matrix at different loading rates.

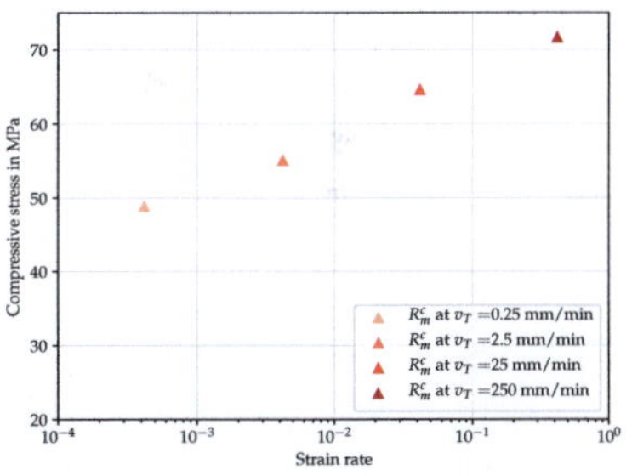

Figure 3.7: Measured matrix compressive strengths at different strain rates.

3.2.4 Tensile loading and unloading experiments

Figures 3.8, 3.9 and 3.12 show the stress-strain relations of the tensile test with 12 loading and unloading cycles for two strain rates. The experiments were conducted up to a maximum strain of $\varepsilon = 5\%$. In addition, the respective quasi-static material response is displayed as an envelope in gray. For each loading rate five specimens were tested. However, to enhance visibility, only one representative curve is displayed here. In a consistent manner to the already presented behavior under tensile loading, increasing loading rates yield higher stresses at the same values of strain. The diagrams show that hysteresis

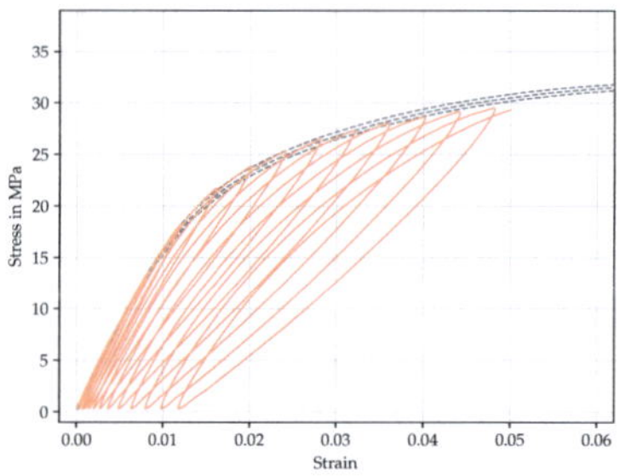

Figure 3.8: Stress-strain relation of the tensile test of pure matrix with 12 loading and unloading cycles at $v_T = 5\,\text{mm/min}$ up to $\varepsilon = 6\%$. The gray curves mark the quasi-static envelopes at the corresponding loading rate.

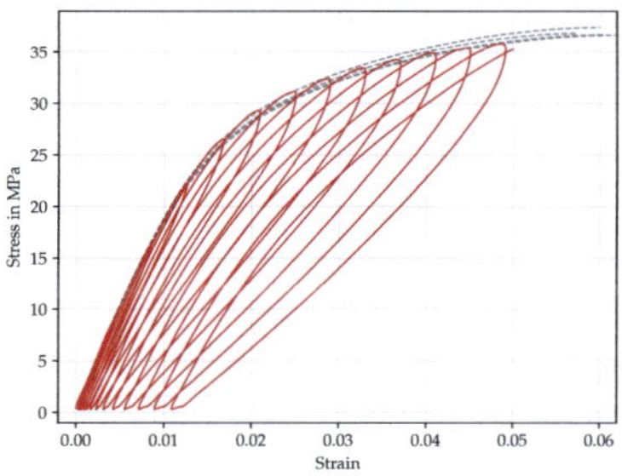

Figure 3.9: Stress-strain relation of the tensile test of pure matrix with 12 loading and unloading cycles at $v_T = 50\,\text{mm/min}$ up to $\varepsilon = 6\%$. The gray curves mark the quasi-static envelopes at the corresponding loading rate.

phenomena appear already at very low strain levels. It is assumed that for small deformations, the residual strain upon unloading is completely recovered if the material is given time for relaxation. This implies that the material shows a viscoelastic behavior first. For higher strains, the residual strains after unloading are only partly recovered in time, hence introducing a viscoplastic component. Brusselle et al. [44] conducted a thorough study to further investigate these effects. To assess the development of material stiffness over loading cycles, the secant between the two extreme values in the half-cycle during unloading was investigated (cf. Figs. 3.10 and 3.11).The full line in Fig. 3.13 shows the calculated slopes of the secants over strain. With increasing deformation, the secant stiffness decreases continuously. The degradation seems to converge

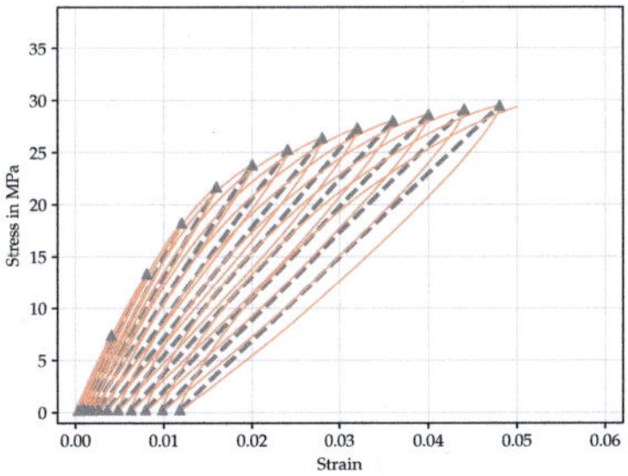

Figure 3.10: Stress-strain relation of the tensile test of pure matrix with 12 loading and unloading cycles at v_T = 5 mm/min up to $\varepsilon = 6\%$. The gray dashed lines depict the decreasing secant stiffness during unloading.

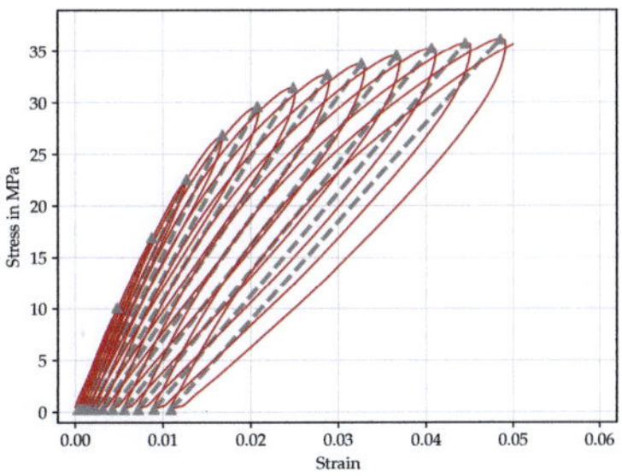

Figure 3.11: Stress-strain relation of the tensile test of pure matrix with 12 loading and unloading cycles at v_T = 50 mm/min up to $\varepsilon = 6\%$. The gray dashed lines depict the decreasing secant stiffness during unloading.

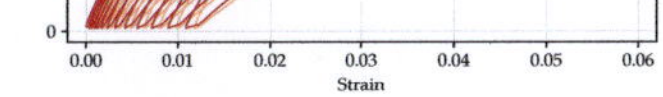

Figure 3.12: Stress-strain relations of cyclic loading and unloading tension tests of pure matrix at different loading rates.

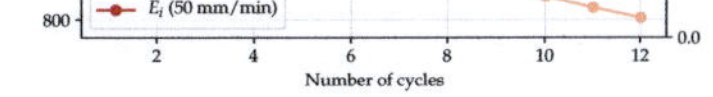

Figure 3.13: Change of secant stiffness over loading cycles. The full lines represent the absolute secant stiffness values, the dashed lines show the data normalized ($f_n = E_i/E_0$) with respect to the initial value.

against a limit value for both strain rates. The dashed lines in Fig. 3.13 represent the normalized secants with respect to their initial value. Both sets of curves are plotted over the number of cycles. Since the two curves nearly coincide, it can be stated that the loading rates have no significant influence on the secant degradation. Figure 3.14 shows a fractured matrix specimen after cyclic loading and unloading. Small cracks in the vicinity of the edges as well as a whitened core are visible in the matrix specimen under transmitted light. It is assumed that the formation of this micro defects is mainly responsible for the deterioration of the material stiffness.

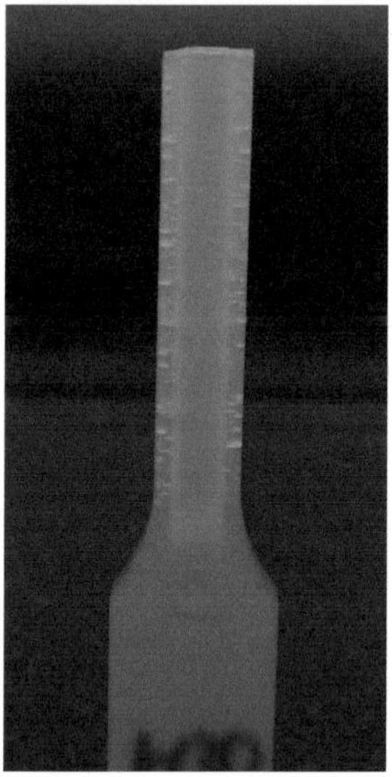

Figure 3.14: Microcracks formed during cyclic loading/unloading in a neat matrix specimen made visible by transmitted light.

3.2.5 Discussion

The preceding section has shown that the basic constituent polypropylene, that is used throughout this work, exhibits a significant time-dependent material behavior. This results in differing stress-strain relations, stiffness, and strengths all depending on the strain-rate. The reasons for this viscoelastoplastic behavior can be affiliated to time-dependent processes occurring in the semi-crystalline molecular structure of polypropylene. Shear yielding is assumed to be the main deformation mechanism of unmodified polypropylene (cf. Kinloch et al. [193], Oswald & Menges [304]). In polypropylene as an example of a semi-crystalline polymer, this effect appears near to crystalline areas, as depicted in fig. 10.21 in Oswald & Menges [304]. Amorphous regions around crystalline zones allow crystals to exhibit positional distortions (e.g. shear and intra-lamellar slipping and rotation). If the deformations inducing this motion are considerably small, these processes are reversible. Otherwise they lead to an irreversible break-up of lamellar aggregates and rearrangement of the polymer chains. In an extreme case both amorphous and crystalline zones are oriented along the loading direction leading to quasi-fibrous stretched characteristics (cf. Arencon et al. [5], Arencon & Velasco [8]). Subjecting the material to small loading rates, polymeric chains can glide, slip, or rotate past each other more easily, allowing higher strains upon failure. With the secondary bounds being primarily subjected to load, the achievable forces are lower than during loading with high loading rates. Here mainly the polymeric backbone is on load, yielding lower stiffness and strengths. Gliding processes on the molecular level become visible by earlier appearance of non-linearity in the stress-strain relations under low loading rates.

The stress-strain relations under cyclic tensile loading show a degradation of secant stiffness. It is assumed here that the degradation itself has several origins. Elasticity in the energy-elastic range in plastic materials relates to intra-molecular secondary bonds. Again, increasing elastic deformation causes a deflection of neighboring atoms from their equilibrium position. In an ideal case, the necessary energy is equivalent to the stored elastic energy in the material that allows fully reversible back-deformation upon unloading. However, deflected atoms result in deteriorated secondary bond forces as well as a lower effective stiffness of the covalent primary bonds (cf. Roesner [339]). Hence, the overall material stiffness reduces. By increasing the deformation, the degradation of the material stiffness is decelerated. This effect is related to reorientation processes of crystalline areas (*spherulites*) in the direction of the active load. In general, this is accompanied by significant plastic shares. Due to the additional plastic contribution at higher strains with $\varepsilon > 3\%$, a decreasing amount of the deformation is to be endured by elasticity. The effect of the disturbance of the equilibrium position of the chains is reduced, since larger parts of the deformation can be described by rotation of spherulites, see e.g. Rösler et al. [338], leading to a diminished change of secant stiffness. Moreover, microscopical investigations have shown a noticeable formation of microcracks. It is assumed that these microvoids are created mainly due to an effect called *stress whitening* as described by Osswald & Menges [304]. Here, microcracks form in the interface between neighboring spherulites. According to Arencon [8], the degradation process is accompanied by the formation of *crazes*. Both processes are irreversible, causing permanent deformation in the polymer. If these microvoids coalesce, macroscopic visible cracks form, finally leading to ultimate fracture. Due to the relatively high percentage of crystalline areas, the damage processes take place in such a high extent that brittle failure through coalescence of microvoids is reached before the typical necking behavior of polypropylene at room temperature appears. The reason for the high percentage of crystalline areas can be found in the manufacturing process of the specimens or in the exact composition of additives in the polymer causing a higher tendency towards crystallinity. A further discussion of these two effects is not part of the present work. However, besides processing-induced structures and the molecular characteristics of polypropylene (e.g. molecular mass), the micro/nanostructural morphology has a particularly strong influence on the fracture behavior (cf. Arencon et al. [6, 7]).

3.3 Unidirectional fiber reinforced polypropylene

The full experimental assessment of all scales addressed within the multiscale approach of organo sheets requires specimens with unidirectional reinforcements. It is important that in the material only one preferred direction exists and no perpendicular structures are present that have falsifying effects. The material provider of the organo sheet does not offer materials complying with such conditions. Therefore, unidirectional glass fiber reinforced tape material (PP/GF UD) of the type CELSTRAN®CFR-TP PP GF70-13-PP with a fiber volume content of $v_f = 45.3\%$ from the material supplier TICONA is used for characterization (cf. B.1 for the data sheet). One major uncertainty that arises is the mismatch of the matrix material of this supplier and the material that is used later in the weaves. Therefore, a consistent usage of the same materials throughout the scales is violated at this stage. The material is delivered as a coiled 0.25 mm-thick tape. Several trimmed layers of these tapes have been stacked manually and consolidated in an industrial press under the influence of temperature. The manufacturing was realized by the Fraunhofer Institute for Chemical Technology in Pfinztal (Fraunhofer ICT), Germany. The compression force was $F_P = 186\,kN$ with a temperature of $T_P = 200°C$ for a time of $t = 200\,s$. The dimensions of the resulting plates was 380 x 380 x d mm^3, where d differs for tensile and compressive specimens. In addition to the unidirectional sheets, plates with a $[0/90]_s$ stacking order were produced. Figure 3.15 shows a

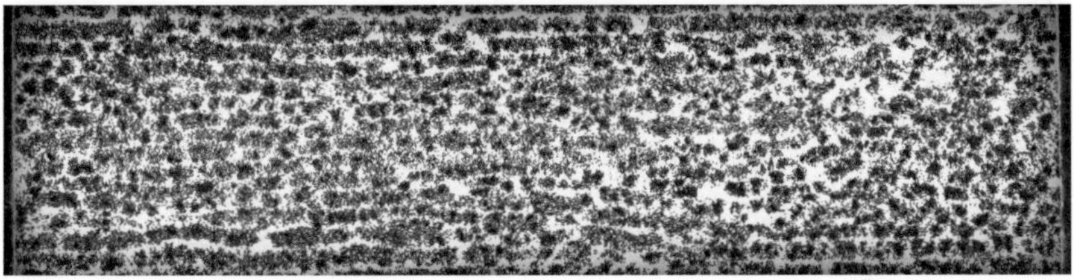

Figure 3.15: Microstructure obtained from a cross-section transverse to fiber direction in unidirectional composites.

dark-field section of the microstructures perpendicular to the fibers. The actual volume fraction of the fibers was determined to $v_f = 43\%$ by using image segmentation on multiple microsections and evaluating the surface areas of matrix and fiber regions. Figure 3.16 depicts a detailed view of the microstruc-

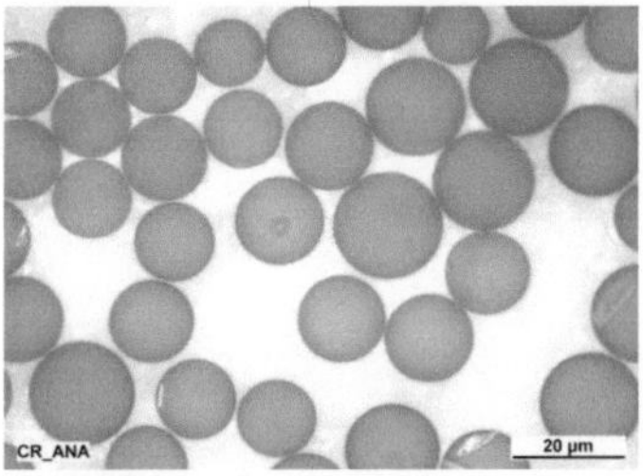

Figure 3.16: Detailed microscopy of a UD specimen showing non-constant fiber diameters.

ture. In addition to the statistically dispersed fibers, a significant distribution of the fiber diameter can be observed. The average diameter was determined to $\bar{d}_f = 13\,\mu m$ with a standard deviation of $s_f = 2.4\,\mu m$. Furthermore, no indications of voids were found and the sections showed an unmitigated impregnation of the fibers. The conducted experiments presented in the following enhance the understanding of the overall anisotropic failure processes occurring in the unidirectional composite during quasi-static loading and further allows a comparison with failure behavior in unidirectionally reinforced plastics as described in literature. Hence, quasi-static tension and compression tests as well as ±45° off-axis (shear) tests under tension were performed at various strain rates. The dimensions of the specimens were chosen according to relevant norms as 250 x 25 x 2 mm^3 for tension (cf. DIN EN ISO 527-4 [95]) and type B2 with 125 x 25 x 10 mm^3 for compression (cf. DIN EN ISO 14126 [92]). For the tensile cases a measurement length $L_0 = 50$ mm was chosen, while the displacement during compression was recorded directly by slim extensometers with a distance of $L_0 = 20$ mm. All specimens were cut out of the corresponding plates by waterjetting taking into consideration the direction of the reinforcements.

3.3.1 Longitudinal tensile tests

Tensile tests in fiber direction were performed at three different position-controlled cross beam velocities. Table 3.5 shows the corresponding strain rates. In Fig. 3.17 the resulting material responses are displayed. After a

Table 3.5: Cross beam velocities and corresponding strain rates for tensile tests of unidirectional specimens (measurement length $L_0 = 50$ mm).

Cross beam velocity v_T (mm/min)	Strain rate $\dot{\varepsilon}$ (1/s)
5	$1.67 \cdot 10^{-3}$
50	$1.67 \cdot 10^{-2}$
500	$1.67 \cdot 10^{-1}$

linear regime the material fails instantaneously for all cases. Figures A.11 to A.13 show more detailed results and summarize the stiffness $E^t_{\parallel}$, material strength $R^t_{\parallel}$, and strain at failure $\varepsilon^t_{\parallel}$. Investigating the confronted stress-strain curves for longitudinal tension tests, small but present viscous effects can be observed. This is due to slight fiber undulations occurring in the specimen, hence, introducing viscous effects dominated by the matrix. The fracture strain lies at about $\varepsilon^t_{\parallel} = 3‰$ and the corresponding stiffness is $R^t_{\parallel} = 900$ MPa, which agrees with the indicated values from the data sheet. Figure 3.18 shows a

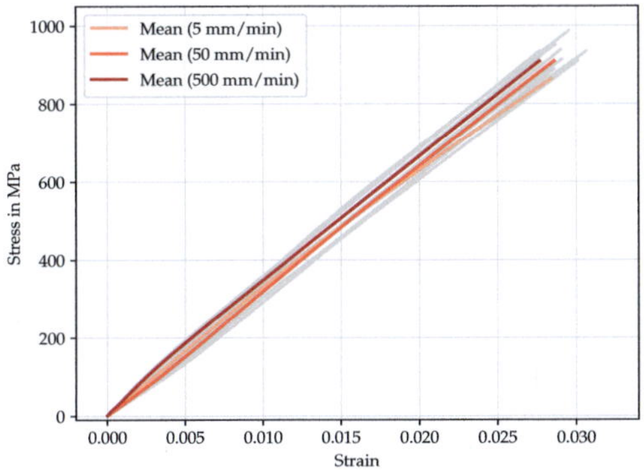

Figure 3.17: Stress-strain relations of quasi-static tensile test in fiber direction of UD specimen at different loading rates.

side view of a UD specimen that failed under longitudinal tension loading. It can be seen that, as soon as failure occurs, the integrity of the material is lost completely. No sharp failure zone can be identified, but the whole specimen gets destroyed to a large extent.

3.3.2 Transversal tension tests

Tensile tests transverse to the fiber direction were performed at three different position-controlled cross beam velocities, see Table 3.5. Figure 3.19 shows the respective material responses. After a degressive stress-strain relation, brittle failure terminates the experiment. In Figs. A.14 to A.16 more detailed results together with the corresponding measured stiffness $E^{\mathrm{t}}_{\perp}$, material strength $R^{\mathrm{t}}_{\perp}$ and strain at failure $\varepsilon^{\mathrm{t}}_{\perp}$ can be found. Confronting the single curves, no significant strain rate effect could be observed regarding initial stiffness and

Figure 3.18: Side view of a UD specimen failed under longitudinal tension.

failure strength. However, the degree of non-linearity seems to be affected by the rate of external loading, since the material fails at larger strains given a slow loading rate ($\varepsilon^{\mathrm{t}}_{\perp}(v_{\mathrm{T}}{=}5\,\mathrm{mm/min}) = 0.65\%$) compared to high loading ($\varepsilon^{\mathrm{t}}_{\perp}(v_{\mathrm{T}}{=}500\,\mathrm{mm/min}) = 0.48\%$). Even though the deformation behavior is matrix dominated in this direction, a brittle behavior like in fiber direction is observed. The strength lies at about $R^{\mathrm{t}}_{\perp} = 22\,\mathrm{MPa}$, which is significantly lower compared to the tensile strengths of the pure matrix specimens as discussed in Section 3.2.2. Figures 3.20 and 3.21 show views of a UD specimen that failed

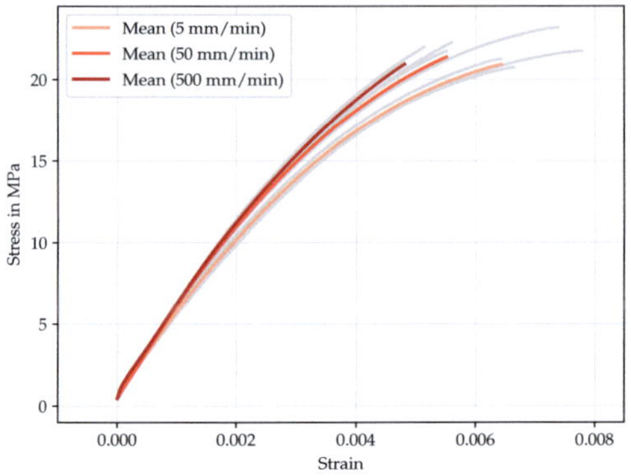

Figure 3.19: Stress-strain relations of quasi-static tensile test in transverse direction of UD specimen at different loading rates.

under transversal tensile loading. A clear fracture plane can be identified that lies perpendicular to the loading direction. Microsections show that the crack grows in the matrix material around the fibers.

3.3.3 Longitudinal compression tests

Compression tests in fiber direction were performed at three different position-controlled cross beam velocities. Table 3.6 states the corresponding strain rates. Figure 3.22 shows the respective material responses. After a linear regime the material fails instantaneously for all cases. More detailed results and a summary of the corresponding measured stiffness $E^{\mathrm{c}}_{\parallel}$, material strength $R^{\mathrm{c}}_{\parallel}$, and strain at failure $\varepsilon^{\mathrm{c}}_{\parallel}$ are given in Figs. A.17 to A.19. Again, no significant strain

Figure 3.20: View of a UD specimen failed under transversal tension. The failure plane lies perpendicular to the loading direction.

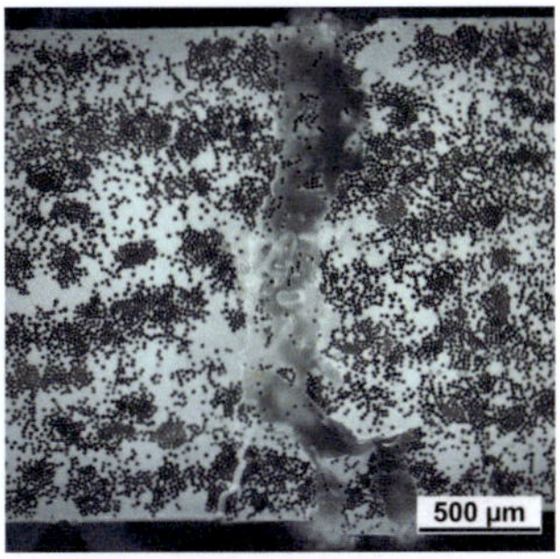

Figure 3.21: Microsection of a UD specimen failed under transversal tension. Load was applied horizontally.

Table 3.6: Cross beam velocities and corresponding strain rates for compression tests of unidirectional specimens (measurement length $L_0 = 20\,\text{mm}$).

Cross beam velocity v_T (mm/min)	Strain rate $\dot{\varepsilon}$ (1/s)
0.5	$4.16\cdot10^{-4}$
5	$4.16\cdot10^{-3}$
50	$4.16\cdot10^{-2}$

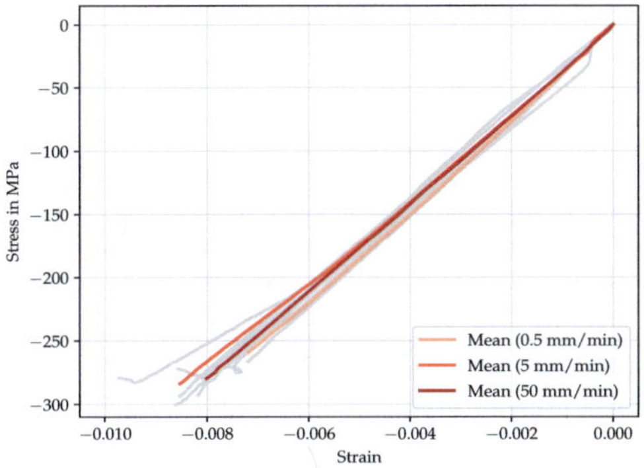

Figure 3.22: Stress-strain relations of quasi-static compression test in longitudinal direction of UD specimen at different loading rates.

rate effect could be observed regarding stiffness, strength, or failure strain when confronting the single curves. The fracture strain lies at about $\varepsilon_{\parallel}^{\mathrm{c}} = 0.8\%$ and the corresponding strength is about $R_{\parallel}^{\mathrm{c}} = 280\,\mathrm{MPa}$. Failure under compression is mainly induced by fiber buckling representing a local material instability. A further investigation on the influence of an initial fiber misalignment and process-induced undulations on the stability failure is not part of the present work.

3.3.4 Transversal compression tests

Compression tests transverse to the fiber direction were performed at three different position-controlled cross beam velocities. For the resulting strain rates see Table 3.6. Figure 3.23 shows the material responses. Figures A.20 to A.22 state more detailed results and summarize the measured stiffness $E_{\perp}^{\mathrm{c}}$, material strength $R_{\perp}^{\mathrm{c}}$, and strain at failure $\varepsilon_{\perp}^{\mathrm{c}}$. As expected, the material

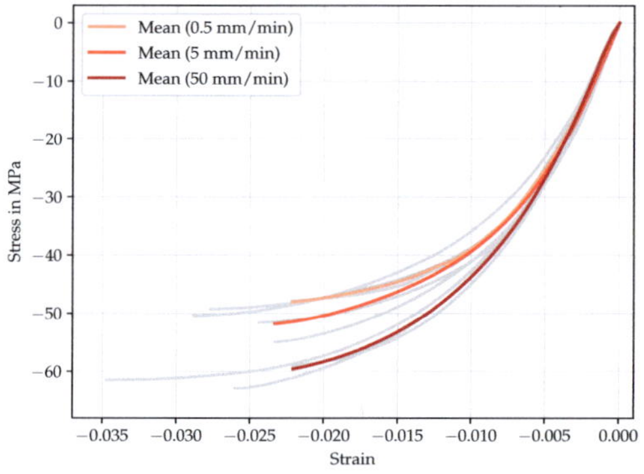

Figure 3.23: Stress-strain relations of quasi-static compression test in transversal direction of UD specimen at different loading rates.

behavior in this material direction is matrix dominated and shows a high degree of non-linearity in the stress-strain relation. Confronting the single curves, no significant strain rate effect could be observed regarding initial stiffness and failure strain. Material strength however, shows a dependency on the strain rate. The material fails at about $\varepsilon_{\perp}^{\mathrm{c}} = 2\%$. Figure 3.24 shows a view of a UD specimen that failed under transversal compressive loading. Looking at the failure image, a shear failure can be seen, expressed by a clearly identifiable fracture plane that is inclined by about $\Theta_{fp} = \pm 50°$ with respect to the loading direction. The microsection shows that the crack grows in the matrix material

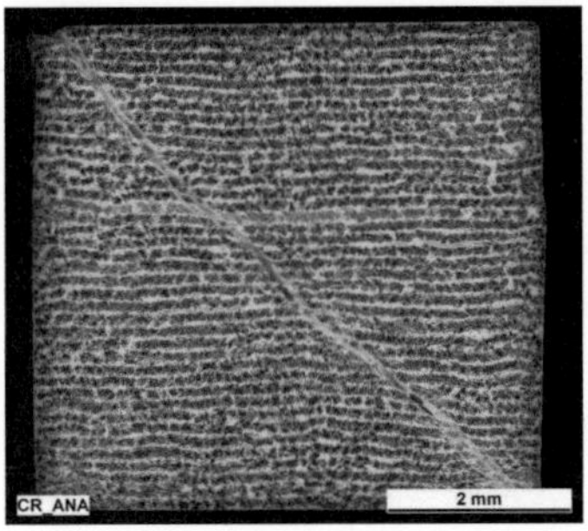

Figure 3.24: Microsection of a specimen failed under transversal compression.

around the fibers. Due to the fracture plane angle that deviates significantly from zero, it is assumed that a critical shear load leads to failure in one distinct plane. This behavior lies in good agreement with the behavior reported in relevant literature, e.g. PUCK [319].

3.3.5 Tensile tests of off-axis specimens

Tensile tests on ±45° off-axis specimens were performed at three different position-controlled cross beam velocities, resulting in the effective strain rates as summarized in Table 3.5. For each strain rate three specimens were tested. Figure 3.25 shows the respective material responses. Due to the relatively high ductility in the thermoplastic matrix during shear loading, the specimens show large deformations and a pronounced non-linear behavior in this loading direction. For more detailed results the reader is referred to Figs. A.23 to A.25, where the single material responses together with the corresponding measured initial stiffness E_s, material strength R_s, and strain at failure ε_s are documented. The reason for reaching such high strain values of more than 25% lies in the ability of the fibers to re-orientate along the loading direction, see Fig. 3.26. This is possible due to the ductility of the surrounding matrix. The motion within the material yields a significant change in cross-sectional area and therefore the true stresses in the material differ from the measured tensions.

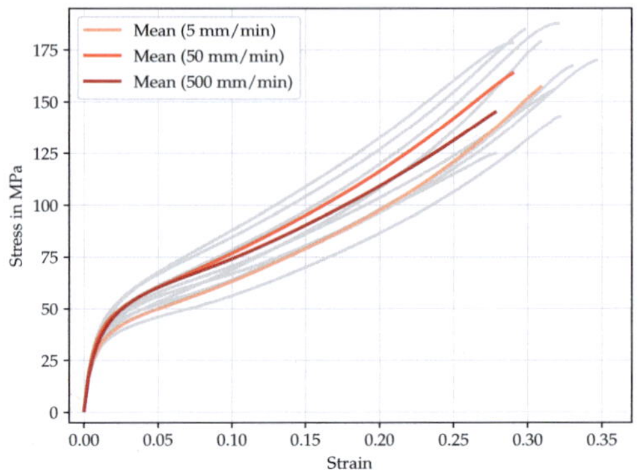

Figure 3.25: Stress-strain relations of quasi-static shear test of a UD ±45° off-axis specimen at different loading rates.

Even though this material direction activates matrix dominant processes, no significant strain rate effect could be observed regarding initial stiffness and failure strength and strain. Figure 3.27 shows the fracture area of a ±45° off-axis UD specimen that failed under tensile loading. Delamination of the single layers is the dominant failure mode, leading to a separation of the specimen.

3.3.6 Discussion

The quasi-static experiments presented here summarize the basic deformation behavior of unidirectionally glass fiber reinforced thermoplastics. It has been shown that those materials show a significant dependency on the loading direction with respect to their fiber direction as well as a pronounced asymmetry

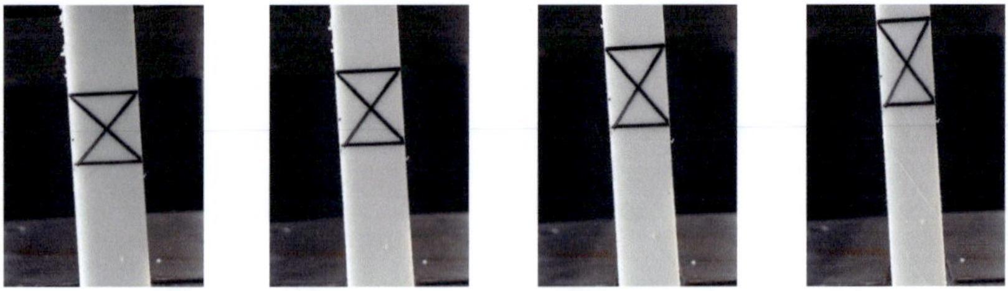

Figure 3.26: Development of fiber angle with increasing strain.

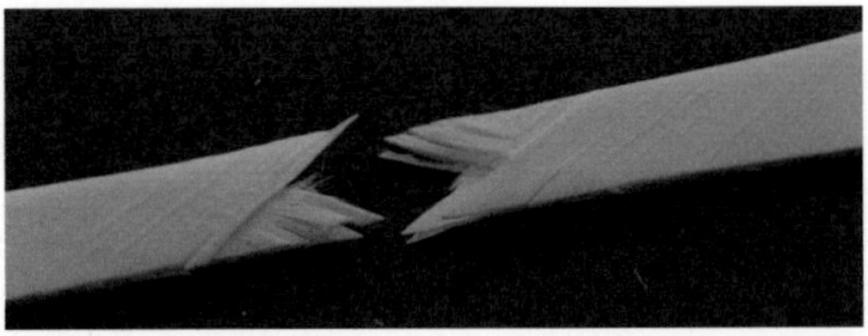

Figure 3.27: Failed ±45° off-axis specimen after tensile test. Close-up view of fracture area.

under tensile and compressive loading. Furthermore, especially directions that are matrix dominated show an influence on the loading rate. The fibers introduce a heterogeneity in the composite with local stress concentrations when the material is deformed. The morphology of the fibers (e.g. mean diameter and orientation), their mechanical properties, and the characteristics of the inter-facial adhesion have significant influence on the deformation behavior, failure dynamics, and hence on the overall behavior of the composite (cf. KOWALEWSKI et al. [200]). Since the matrix around the fibers cannot move freely and its motion is constrained by the adjacent fibers, the process of plasticity, crack propagation and fracture are highly influenced by the given micro-structural circumstances. The macroscopic permanent deformation can be explained by processes happening on the micro level, namely plastic flow and debonding damage processes of the matrix between inclusions. In literature, several authors claim that debonding processes between fibers and the thermoplastic matrix are the main initial damage mechanisms, finally leading to failure. Especially composites consisting of matrix material with a low polarity, as it is the case for polypropylene, inter-facial adhesion is usually weak and separation of matrix and fiber takes place more easily. Hence, several approaches to address the issue of fiber matrix debonding are proposed in literature (cf. PUKANSZKY et al. [322], ZHUK et al. [434], VAN HARTINGSVELDT & VAN AARTSEN [399]). In order to estimate the influence of the interface debonding for the present composite, the fracture surfaces of unidirectional specimens were investigated, using a scanning electron microscope (SEM). Figure 3.28 shows a series of representative micro scans of specimens subjected to cryo-fracture using a high magnification. In contrast to the expected results, no clear interface debonding between matrix and fiber is observed, but the fibers are impregnated with a thin layer of matrix. It is assumed here, that additives are responsible for the improved adherence. As expected, tension loading in fiber direction results in a linear stress-strain relation. In this material direction,

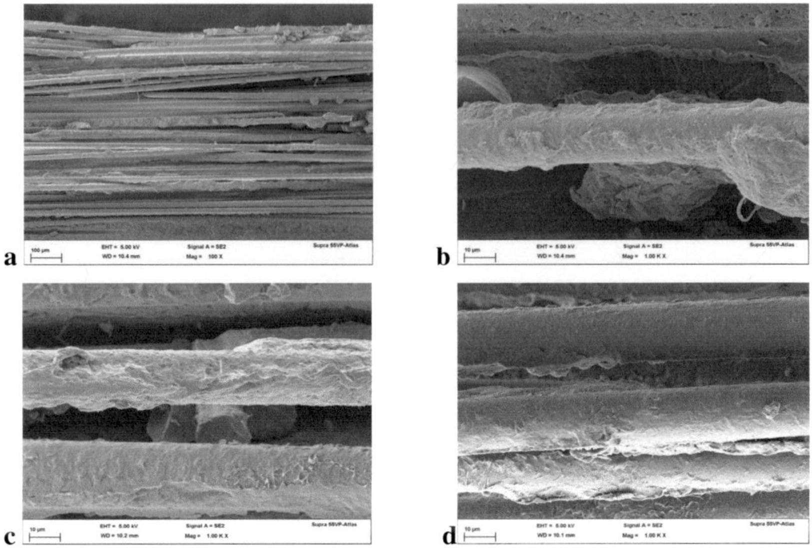

Figure 3.28: Series of SEM scans of unidirectional specimens subjected to cryo-fracture. (a) Fracture plane at a magnification of 100. (b)-(d) Detailed views of single fibers (1000x magnification). All examined fibers show pronounced matrix adhesion.

the fibers are the constituents which mainly carry the load. Failure occurs in an abrupt manner and the integrity of the specimen is completely lost and no distinct fracture plane is recognizable. Just before the final failure of the specimen, a series of sound emissions is perceptible, indicating a successive failure of single fibers up to the point that the active load exceeds the remaining load-bearing capacity and the specimen fails catastrophically. The average failure strain is slightly below $\varepsilon^{t}_{\parallel} = 3\%$. Even though unexpected, a slightly rate-dependent material behavior was observed. Due to the processing, fibers were not aligned perfectly parallel and a certain effect of the viscoelastic behavior of the matrix could be observed. Compressive loading in fiber direction leads to a matrix dominated failure behavior, even though no significant dependency on the loading rate was detected. Slight initial misalignments in the fibers lead to buckling which consequently yields to shear load exposure and finally to failure in the adjacent matrix regions followed by fiber fracture due to excessive bending. Transverse to the fiber direction, the material shows very brittle behavior under tension forming a macroscopic crack perpendicular to the loading direction. The composite failed under tension at about $\varepsilon_{\perp}^{t} = 0.5\%$, even though a thermoplastic matrix is present between the single fibers, that however failed in a brittle manner but reached relatively high failure strains as shown in Section 3.2.2. It is assumed that due to the high stiffness of the glass fibers, the macroscopic deformation accumulates in the softer matrix. Locally, the material is strained to a larger extent compared to uniaxial test in unreinforced matrix. It is noticeable that the values for the material strengths of the UD materials are lower than in the neat matrix. Due to the fibrous inclusions, the stress state in the material as a result of external uniaxial loading gets disturbed and hence becomes multi-axial on a microscopic level with high volumetric shares. Especially under tension, these favor damage initiation and inhibit plastic deformation (cf. Asp et al. [15], Kim et al. [191]). Therefore, the fibers introduce a local notch effect, disturbing a smooth strain field. Compressive loading induces the typical failure patterns known from thermoset composites, being reproducible also within the thermoplastic matrix. A macroscopically planar but inclined fracture plane characterizes a shear failure in the matrix around the fibers. In contrast to the very brittle deformation behavior in longitudinal and transverse direction, loading in a ±45°-direction allows the material to develop high strains. The pronounced non-linearity of the stress-strain relation indicates a high mobility of the matrix. This allows the fibers to re-orientate in the load direction during the course of loading, leading to a successive stiffening in the material response, visible towards the end of the experiment. The reached high strains of more than $\varepsilon_{s} \geq 25\%$ allow for the assumption that the deformation in the matrix is accompanied by high plastic shares. Regarding the application of such materials in structural parts, this large deformation is not expected due to mixed stackings of UD layers. In a similar manner, in woven structures large deformations will be inhibited by the interweaving of the rovings. Even though the deformation behavior is assumed to be matrix-dominated, no strain rate dependency was observed. In general, the viscous behavior in the UD composite becomes less visible than in the neat matrix due to multi-axial strain and stress states in the matrix phase.

3.4 Woven fabric reinforced polypropylene

This section is concerned with the characterization of the full composite, consisting of a polypropylene matrix and a glass fiber twill weave as reinforcement structure (PP/GF twill weave). The organo sheet with the typecast TEPEX®dynalite 104-RG600(n)/47%, where n specifies the number of layers, was purchased from Bond Laminates (cf. B.2 for the data sheet). It is suitable for further processing supported by the application of temperature and pressure. The overall fiber content in the material is $v_{\mathrm{fw}} = 47\%$ and the interwoven

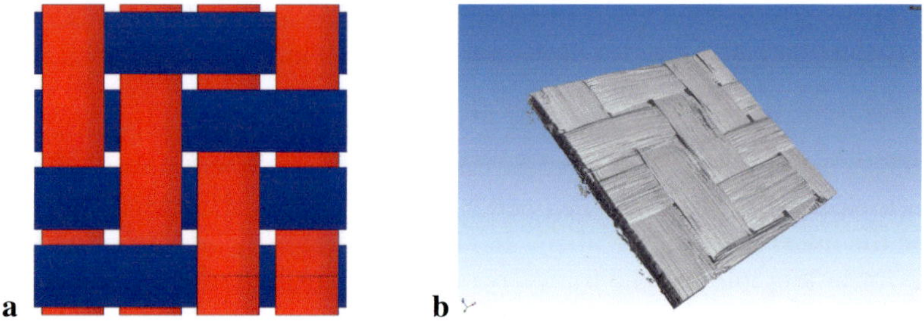

Figure 3.29: Twill weave pattern. (a) Schematic display of twill weave from *TexGen* (Sherburn [350]) and (b) Depiction of twill weave pattern in real specimen taken from μCT scan.

rovings are assembled in a balanced twill weave pattern. Figure 3.29 shows a schematic and a real view taken from μCT data of the weave. In both cases the matrix phase is masked out and only the roving structures are visible for a better presentation of the weave pattern. The procedure to measure the fiber volume fracture inside the UD material was also applied here, using microsections as depicted in Fig. 3.30. Image segmentation yields an average volume fraction of $\kappa = 63\%$ within the rovings. According to $v_{\mathrm{tow}} = v_{\mathrm{fw}}/\kappa$ this yields a tow volume fraction of $v_{\mathrm{tow}} = 74.6\%$. Analyses of the fiber diameters were performed subsequently and an average of $\bar{d}_F = 13\,\mu\mathrm{m}$ could be confirmed here. The relevant dimensions introduced and summarized in Fig. 2.8 and Table 2.1 were determined and are given in Table 3.7. Semi-finished sheet material with a dimension of $620 \times 1300 \times d\,\mathrm{mm}^3$ was delivered. With a thickness per layer of $d_0 = 0.5\,\mathrm{mm}$, the variable amount of n layers defines the complete thickness of the compound d. In the case of stacked layers, single lamina were arranged with aligning preferred directions. Even if the preferred directions of the single

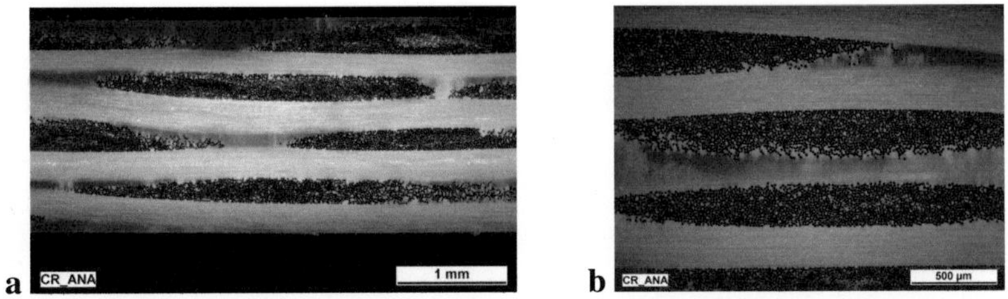

Figure 3.30: Microsections of multi-layer weave. (a) Exemplary microsection that was used for the determination of out-of-plane weave geometries and (b) detailed view on single rovings.

lamina coincide, the mesoscopical geometrical conditions within a laminate can severely differ. Describing the undulation of a roving according to a sinusoidal curve, top and bottom turning points (peak P and valley V) can be identified. The relative positioning of the local extrema of the undulated fiber rovings is known as nesting behavior and has significant influence on the mechanical behavior of the laminate. Figure 3.31 depicts some possible nesting configurations in a four-layer laminate. The middle sections illustrated in Fig. 3.31a and b show the two extreme examples of nesting, namely the combinations of peaks (P) and valleys (V) in the setting [P, V, P, V] for section (a) and [V, V, V, V] followed by [P, P, P, P] in section (b). These configurations are identified as *symmetric* and *simple* stacking. In general, more or less expressed intermediate states of nesting can be found (cf. Fig. 3.31 c and d). The analysis of micro sections of the composite showed no microvoids in matrix rich areas

Table 3.7: Measured average dimensions of one-layer twill weaves (cf. Fig. 2.8 and Table 2.1).

Dimension	Value		
Geometrical (mm)	λ	15.667	± 0.723
	h	0.543	± 0.014
	p	3.264	± 0.141
	t_w	0.256	± 0.019
	t_f	0.255	± 0.024
	g_1	0.596	± 0.221
	g_2	0.573	± 0.237
Volume fraction (-)	κ	0.631	
	v_{fw}	0.471	
	$v_{tow} = \frac{v_{fw}}{\kappa}$	0.746	

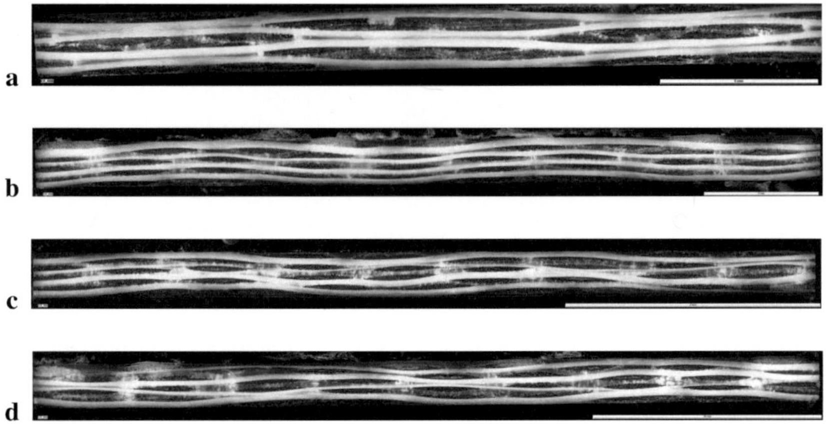

Figure 3.31: Microsections of woven specimens in warp direction. Various characteristics of layer phase shift are recognizable. (a) *Symmetric* stacking configuration: peaks *P* and valleys *V* are opposing, (b) *Simple* stacking configuration: peaks and valleys are in-phase, and (c) and (d) intermediate topologies of the two extreme states.

between the impregnated rovings and a complete coverage of the fibers with plastic matrix. In the following, all specimens were cut by waterjetting from the corresponding plates taking into consideration the direction of the reinforcements. Ruler specimens with the dimensions of 250 x 25 x 2 mm^3 were used for the tensile quasi-static experiments according to DIN EN ISO 527-4 [95]. The measurement length was set to $L_0 = 50$ mm. For the experiments with a loading under an angle of 45°, additional specimens with one layer with thickness $d = 0.5$mm were manufactured. Compression tests were conducted using specimens with a dimension of 100 x 15 x 4 mm^3 and a measurement length of $L_0 = 20$ mm in a similar fashion to DIN EN ISO 14126 [92]. In order to understand the deformation behavior and the influence of the loading rate, it is necessary to investigate the glass fiber woven fabric reinforced thermoplastic experimentally. The experimental results are summarized in the following subsections.

3.4.1 Longitudinal tensile tests

Tensile tests in fiber direction were performed at three different position-controlled cross beam velocities. Table 3.8 shows the corresponding strain

Table 3.8: Cross beam velocities and corresponding strain rates for tensile tests of woven specimens (measurement length $L_0 = 50\,\mathrm{mm}$).

Cross beam velocity v_T (mm/min)	Strain rate $\dot{\varepsilon}$ (1/s)
5	$1.67 \cdot 10^{-3}$
50	$1.67 \cdot 10^{-2}$
500	$1.67 \cdot 10^{-1}$

rates. Figure 3.32 shows the respective material responses. The presented results were obtained from a ruler specimen with one layer $n = 1$. After a linear pathway, the material fails instantaneously for all cases. Figures A.26

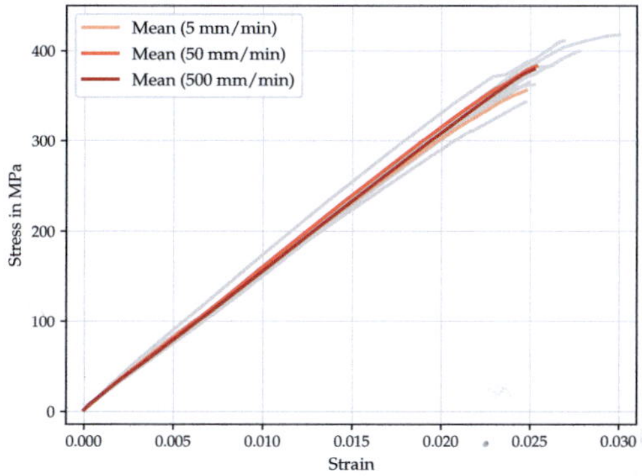

Figure 3.32: Stress-strain relations of quasi-static tensile test in fiber direction of UD specimen at different loading rates.

to A.28 give more detailed results and summarize the corresponding measured stiffness $E^t_{\parallel}$, material strength $R^t_{\parallel}$ and strain at failure $\varepsilon^t_{\parallel}$. No significant strain rate effect could be observed regarding stiffness, strength, or failure strain when comparing the single curves for the two higher velocities. Merely, the tensile test at $v_T = 5\,\mathrm{mm/min}$ indicates, that the time-dependent properties of the matrix and the woven structure and geometry of the composite have an influence. Under quasi-static loading, the stress-strain response is approximately linear elastic up to the maximum bearable stress followed by abrupt failure. The fracture strain lies at about $\varepsilon^t_{\parallel} = 2.5\%$ and the corresponding strength is in the range $R^{t}_{\parallel} = 350\,\mathrm{MPa}$ to $R^{t}_{\parallel} = 390\,\mathrm{MPa}$, which agrees approximately with the indicated values in the data sheet. The expected progressive stiffening of the material due to straightening of the undulated rovings was not observed. Figure 3.33 shows a top view of a one-layer woven specimen that failed under

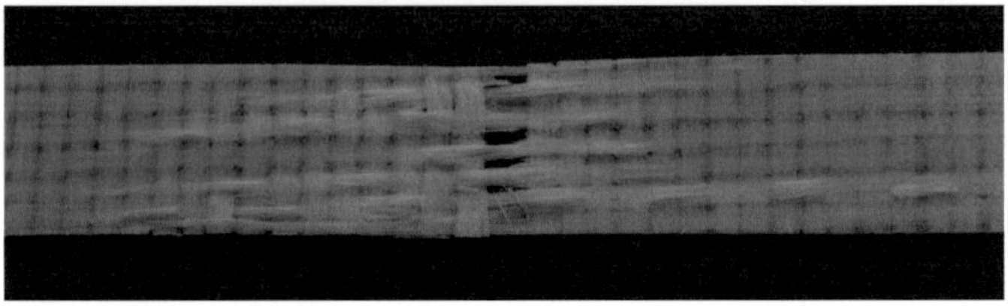

Figure 3.33: Top view of a one-layer woven specimen failed under longitudinal tension.

longitudinal tension loading. The damage mechanisms generally include matrix cracking, localized warp fiber fracture, weft fiber pull-out, and in case of a multilayer specimen, delamination along the middle plies. Matrix cracking was observed dominantly in the weft fiber rovings, oriented perpendicular to the loading direction. These cracks propagate in an opening mode due to the normal stresses acting in the load direction. When reaching the roving interfaces, they can cause interfacial debonding within the lamina or interply delamination. The tensile test in fiber direction has been conducted in both warp and weft direction. However, for both directions comparable results could be achieved. No significant differences were observed when testing layered specimens.

3.4.2 Longitudinal compression tests

Compression tests in longitudinal fiber direction were performed at three different position-controlled cross beam velocities. The corresponding strain rates are given in Table 3.9. Figure 3.34 shows the respective material responses. In Figs. A.29 to A.31 more detailed results together with the corresponding measured stiffness $E^{c}_{\parallel}$, material strength $R^{c}_{\parallel}$, and strain at failure $\varepsilon^{c}_{\parallel}$ are given. Under compression, the material response is approximately linear elastic up to the point of instantaneous failure. Comparing the stiffness values in longitudinal direction obtained under tension, a less stiff response is observed under compression. This behavior can be explained by the geometrical fact that

Table 3.9: Cross beam velocities and corresponding strain rates for compression tests of woven specimens (measurement length L_0 = 20 mm).

Cross beam velocity v_T (mm/min)	Strain rate $\dot{\varepsilon}$ (1/s)
1	$8.33 \cdot 10^{-4}$
10	$8.33 \cdot 10^{-3}$
100	$8.33 \cdot 10^{-2}$

the fiber rovings are impregnated in the thermoplastic matrix in an undulated way. Loaded under compression, the main deformation is not carried by the fibers in longitudinal direction directly, but leads to a buckling motion of the already deflected bundles. Consequently, not the entire stiffness of the fibers becomes effective but rather a combination of flexural (bending) modulus of the undulated rovings and matrix, yielding a more compliant material response. Due to the high shares of matrix being loaded, time-dependent behavior can be observed with regard to the stiffness values. With increasing loading rate, the material responds in a stiffer manner. The averaged strengths $R_{\parallel}^{c}$ = 125 MPa and failure strains $\varepsilon_{\parallel}^{c}$ = 1% are similar for the different compressive loading rates. High scattering of the material strength could be an indication of the influence of nesting effects. Under compression, the material fails at strains that are significantly lower compared to tensile tests. This behavior has already

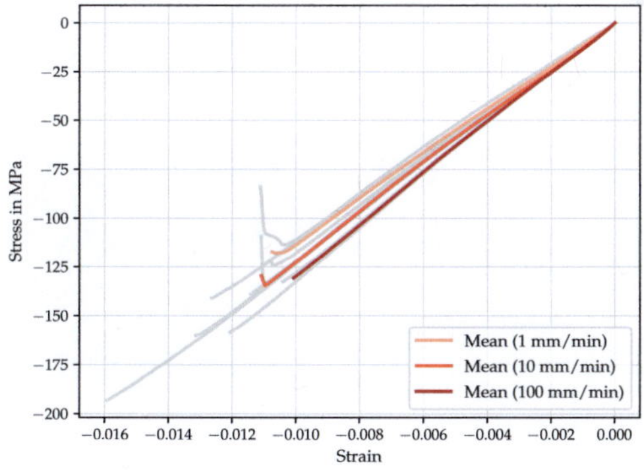

Figure 3.34: Stress-strain relations of quasi-static compression test in fiber direction of woven specimens at different loading rates.

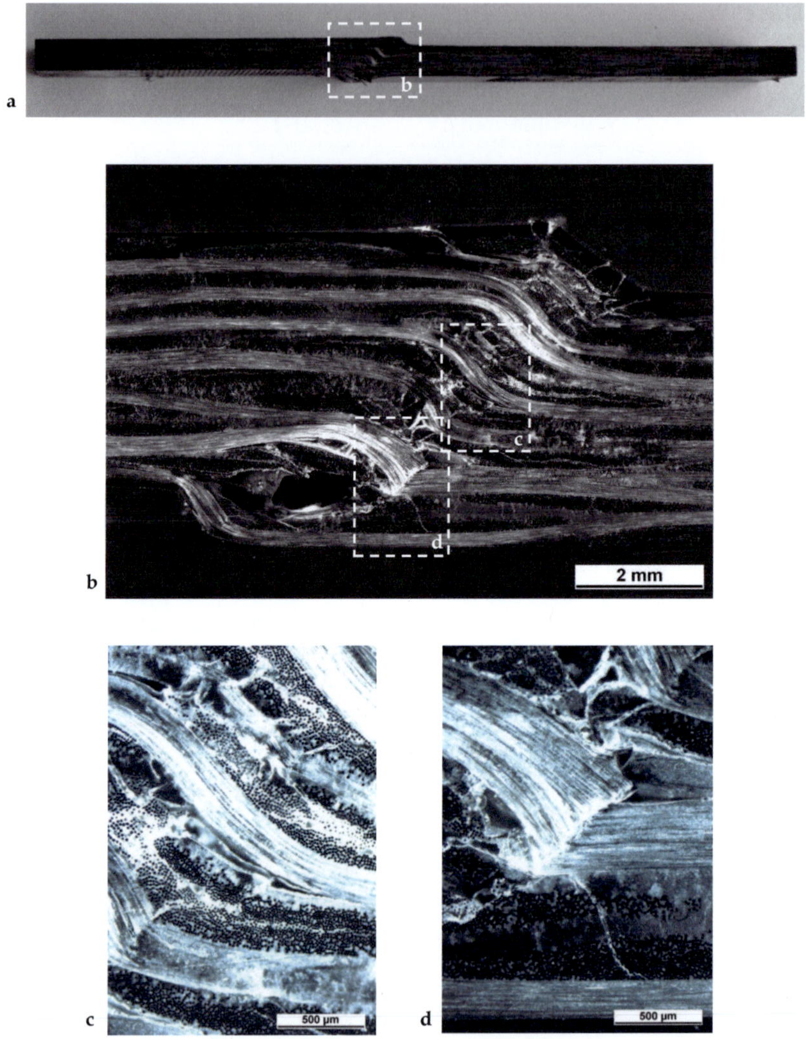

Figure 3.35: Typical failure manifestation under compressive load in a woven fabric reinforced composite. (a) Side view photograph of a woven specimen failed under longitudinal compression, (b) Detailed section of a woven specimen failed under longitudinal compression, (c) detailed microscopy of section of a woven specimen failed under longitudinal compression showing fiber kinking and (d) fiber shear fracture.

been observed in the UD matrix and is accredited again to the brittle matrix failure, which is responsible for the failure behavior. Figure 3.35a and a more detailed view in Fig. 3.35b show a typical failure of woven composite specimens that failed under longitudinal compressive loading. The whole specimen failed under the formation of a localized through thickness shear band that becomes macroscopically visible. The shear failure bands range from 30° to 57° for all specimens, with an average of about 45°. Failure in the form of a shear band can be explained by the woven nature of the material, especially the preexisting deflections in the undulated rovings, as suggested by HARDING [149]. Taking a closer look on the micrograph of the damaged area in Fig. 3.35b, several failure modes become visible. In general, fiber micro-buckling leads to fiber kinking followed by failure of the fibers and matrix through the thickness of the specimen along the distinct shear band. As a result of the buckling motion of the tows within the material, matrix areas positioned in between two rovings are sheared locally to a high extent. This results in high volumetric strain and hence failing prematurely in a brittle manner (cf. GROVES et al. [135]). Reducing locally the stabilizing effect of the matrix surrounding the fiber bundles, micro-buckling is caused. Subsequently fiber kinking (cf. Fig. 3.35c) and fiber breakage (cf. Fig. 3.35d) favor the formation of the macroscopic shear band.

3.4.3 Tensile tests of off-axis specimens

Off-axis ±45° tensile tests were performed at three different position-controlled cross beam velocities. The corresponding strain rates are given in Table 3.10. Figure 3.36 shows the respective material responses. The single material

Table 3.10: Cross beam velocities and corresponding strain rates for tension tests of woven ±45° off-axis specimens (measurement length L_0 = 20 mm).

Cross beam velocity v_T (mm/min)	Strain rate $\dot{\varepsilon}$ (1/s)
5	$1.67 \cdot 10^{-3}$
50	$1.67 \cdot 10^{-2}$
500	$1.67 \cdot 10^{-1}$

responses as well as the corresponding measured stiffness E_s, material strength R_s, and strain at failure ε_s as extracted from the experimental results are summarized in more detail in Figs. A.32 to A.34. The shear stress-strain behavior is dominated by the matrix material and exhibits significant non-linearity over a large strain range. Analyzing the stress-strain response in terms of strain rate dependency, several conclusions can be drawn. The initial stiffness is independent of the loading rate, however, the non-linear pathway changes with the strain rate. The results are inconclusive regarding a valid statement for the failure strain, showing high scatter within the single velocity sets. Merely the shear strength shows an increasing trend for increasing shear rates. Such a behavior is also reported in Bonnet [39]. Photographs of

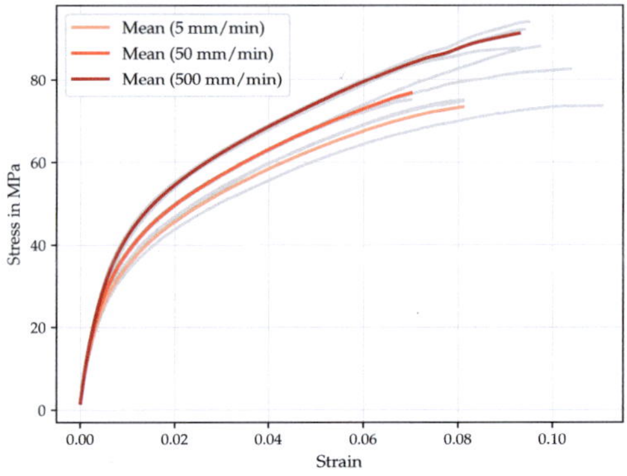

Figure 3.36: Stress-strain relations of quasi-static ±45° off-axis tension tests of woven specimens at different loading rates.

the post-test in-plane shear specimens are depicted in Fig. 3.37. The major

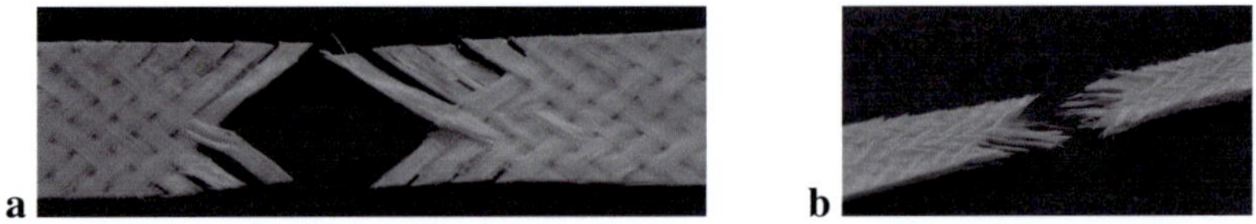

Figure 3.37: Detailed views of (a) one-layer and (b) multi-layer ±45° off-axis woven specimen failed under tension. Roving/Fiber pullout in conjunction with matrix cracking is identified as the dominant failure mode in both cases.

macroscopic failure modes were matrix cracking, fiber pull-out, fiber bundle pull-out and delamination. Similar to the ±45° off-axis UD specimen before, also in the weave, fiber reorientation (scissoring) in the direction of the external load direction was observed. According to KELLAS et al. [190] this process induces interlaminar stresses at the ply interfaces resulting in delamination. The final failure is accompanied by the formation of a V-shaped in-plane fracture surface (±45° to fiber axis). This plane indicates at the same time the plane where the critical shear load is active.

3.4.4 Discussion

The previous sections presented the experimental results obtained from woven fabric reinforced polypropylene subjected to tension and compression load. Additionally, ±45° off-axis tension tests have been performed. Special attention was paid to the influence of time-dependent viscous behavior of the composite, taking into consideration the anisotropic structure. It could be shown that tensile tests in fiber direction of the woven composite yield a slightly time-dependent behavior. Hence, the strain rate effects in the material are influenced primarily by the matrix viscoelasticity, fiber-matrix interfacial properties, the composite woven reinforcement architecture, and the time-dependent nature of damage accumulation. Mainly the matrix component, which takes a high volume fraction in the composite, is responsible for this behavior. The share of the matrix with regard to the overall stiffness can therefore not be neglected. Due to the undulation of the reinforcement structures, matrix-related effects are measured in all directions of the composite. GUAGLIANO & RIVA [136] and NICOLETTO & RIVA [297] showed that nesting effects within the stacking of a woven composite have significant influence on the mechanical behavior in terms of stiffness and strength. Comparing the compression and tension tests in fiber direction, a tension/compression asymmetry could be observed with respect to stiffness and strength. Both entities yielded considerably lower values under compression, which was entirely related to effects happening in the thermoplastic matrix, stabilizing the impregnated tows. Undulation triggers first local matrix failure and therefore a disturbance on the stabilizing effect, leading to matrix damage. The continuous fibers tend to be prone for kinking due to their high aspect ratio and are hence highly dependent on the supporting effect of the surrounding matrix. Increasing undulation results in decreasing strength and stiffness, as reported in GUAGLIANO & RIVA [136], NICOLETTO & RIVA [297] as well as GARNICH & KARAMI [120] and KARAMI & GARNICH [189]. Losing this support, the fibers fail collectively within the roving, accompanied by concurrent formation of matrix failure and delamination. The final failure becomes visible in the form of a macroscopical shear band. Off-axis ±45° specimen loaded under tension showed a significant non-linearity, leading to large deformations within the material. Scissoring effects within the material enable a reorientation of the reinforcement structures with the external loading direction. The specimen finally failed due to delamination caused by foregone matrix cracking and fiber pullout. Large rotations of the fibers yield high inter-ply shear loads which finally lead to delamination and separation of the crack fronts. However, taking into account the results obtained by ROESNER [339], the measured strength values are dependent on the specimen dimensions and are therefore not to be taken as material parameters. Regarding the appearance of failure within the woven specimen as well as the determination of material parameters, no difference concerning the loading in warp or weft direction could be observed. This is due to the fact that a balanced weave under waiver of the use of binders was investigated.

In this chapter, experimental results have been presented for the constituents and subcomposites appearing in organo sheets. The assessment of the deformation behavior, especially of anisotropic materials, is very challenging and an urgent issue in current research. Due to their complex inner structure and hence direction-dependent behavior, many tests have to be conducted in order to understand the processes taking place internally. However, superimposed loading states are often hard or impossible to achieve. Therefore, computer aided techniques deliver significant benefits. Microstructures can be generated virtually and loaded in arbitrary loading combinations. In the upcoming chapters a procedure is presented that enables a virtual characterization of organo sheets on different levels of geometrical resolution, introducing both generation of microstructures and necessary material models.

4 Constitutive modeling of constituents and composites

This chapter demonstrates the derivations of the three required constitutive models in order to run the multiscale approach presented in Fig. 1.2. Section 4.1 addresses the rate-dependent definition of the elastoplastic and damage behavior of the neat thermoplastic matrix, characterized in Section 3.2. With regard to later developments, the initially defined small strain framework is extended towards a finite setting formulated with the help of logarithmic strains. The transition is done by purely geometrical operations applied on the strain tensor. In order to perform the first scale-bridging step, a fully three-dimensional, transversely isotropic continuum damage model is introduced in Section 4.2. Representing the impregnated rovings, this constitutive model is used in the virtual unit cell, i.e. the mesoscopic model. Incorporating the anisotropic nature of unidirectionally reinforced materials, a transversely isotropic ground-state elasticity law is presented in a finite strain framework in order to consider large rotations that can occur in weave reinforced composites. The preferred direction is therefore characterized by one vector. Depending on the orientation, experimentally observed predominant emerging failure mechanisms are integrated by the definition of suitable criteria for damage onset. Further evolution of directional damage parameters is driven by a thermodynamically consistent formulation using the theory of maximum dissipation. The last constitutive model presented here enables the transition from a mesoscopic to a macroscopic formulation, hence representing the weave reinforced material on part level. Section 4.3 demonstrates the composition of the two aforementioned material models, creating a new straightforward formulation for the macroscopic material behavior, incorporating plastic and anisotropic damage effects by combining analytical and numerical homogenization steps. In the present case of a twill weave reinforced composite, the preferred directions of the woven composite coincide conveniently with the principal directions of the reinforcing systems. However, due to external loading, the initially known preferred directions change as the material deforms. Therefore, an intermediate configuration based on the plastic share of the deformation gradient is introduced in order to take these kinematic particularity into consideration. By extending the given definitions by separate damage formulations for matrix and rovings, the transfer of dominant material effects onto the macroscopic scale is enabled. While Chapter 4 is limited entirely to the theoretical derivation of the constitutive models including some remarks on the algorithmic treatment and implementation of the governing equations of all scales, Chapter 5 presents the numerical application and results.

4.1 Constitutive modeling of polypropylene

This section focuses on the modeling of the rate-dependent elastoplastic and damage behavior of the matrix material for isothermal conditions. The formulation of elastoplasticity in a finite framework has been worked on extensively during the last decades. For a thorough overview, the interested reader is referred to the works of NAGHDI [286], or XIAO et al. [425]. Basically, two main approaches exist for modeling large-strain inelastic material models. The first approach bases on a multiplicative decomposition of the deformation gradient into elastic and plastic contributions $\boldsymbol{F} = \boldsymbol{F}^{\mathrm{e}}\boldsymbol{F}^{\mathrm{p}}$, also referred to as *Kröner-Lee decomposition* (cf. KRÖNER [207], LEE [216], MANDEL [250]). The origin of this definition can be found in the micromechanical description of plasticity in crystalline materials, but is frequently used for the phenomenological description of plasticity. The second approach bases on the *Green-Naghdi theory* and uses the Lagrangean plastic deformation measure $\boldsymbol{E}^{\mathrm{p}}$ (cf. GREEN & NAGHDI [131]). For the present work, the formulation introduced by MIEHE [273, 274] together with the additive framework for the plastic strains (cf. MIEHE et al. [268], PAPADOPOULOS & LU [306, 307], MIEHE & APEL [275]) is used to set up the constitutive model in the logarithmic strain space. This framework allows for the formulation of a geometrically linear plasticity model in the core that is embedded into a purely geometric pre- and post-processing algorithm, rendering the model geometrically nonlinear. In the present work, first a material model for the thermoplastic matrix is derived in the context of a geometrically linear frame using the additive decomposition of the strains which is then embedded into the aforementioned logarithmic framework. The experimental behavior under quasi-static tension and compression at different strain rates of the present thermoplastic polypropylene matrix is demonstrated in Section 3.2. For the model here it is assumed that the time-dependency arises purely due to time-dependent plastic effects in the matrix. In this section, we first introduce the overall constitutive equations of the elasto-viscoplastic material model, including an isotropic formulation for damage. A second part describes in more detail the modular structure of the kinematic approach required for the implementation in Abaqus.

4.1.1 Basic kinematics and state variables

Aiming at a continuum mechanical description of a non-associative elasto-plasticity coupled to continuum damage mechanics at small strain, the basic kinematics need to be defined. The macroscopic strain $\boldsymbol{\varepsilon} = \nabla^{\mathrm{s}}\boldsymbol{u}$ can be decomposed into elastic and inelastic parts, which allows for the introduction of the stress producing elastic strains.

$$\boldsymbol{\varepsilon}^{\mathrm{e}} = \boldsymbol{\varepsilon} - \boldsymbol{\varepsilon}^{\mathrm{p}} \tag{4.1}$$

The elastic strain is used to described energy storage mechanisms that are related to elastic macroscopic distortions of the material. In addition, hardening effects are captured by the hardening variables α_{c} and α_{t} taking into account hardening effects under compression and tension, respectively. For modeling damage effects, i.e. a softening of the material, a scalar damage variable d is introduced, modeling a gradual degradation of the material. Thus, the total constitutive state can be summarized by

$$\{\boldsymbol{\varepsilon}; \boldsymbol{\varepsilon}^{\mathrm{p}}, \alpha_{\mathrm{c}}, \alpha_{\mathrm{t}}, d\} \tag{4.2}$$

which builds the kinematic framework for the formulation of the constitutive equations. Regarding the later implementation, the coupled plasticity and damage partition will not be treated monolithically. In a typical loading phase, the material will be loaded plastically up to a critical state with the damage partition not being active. After the turning point, the plastic state of the material is frozen and damage evolves.

4.1.2 Non-associative elasto-plasticity

4.1.2.1 Energy storage mechanisms and stresses

The energy storage mechanism is determined by the free energy function of the specific form

$$\begin{aligned}\Psi &= \Psi(\boldsymbol{\varepsilon};\boldsymbol{\varepsilon}^{\mathrm{p}},\alpha_{\mathrm{c}},\alpha_{\mathrm{t}},d)\\ &= (1-d)\Psi_0(\boldsymbol{\varepsilon},\boldsymbol{\varepsilon}^{\mathrm{p}},\alpha_{\mathrm{c}},\alpha_{\mathrm{t}})\\ &= (1-d)\left[\Psi_0^{\mathrm{e}}(\boldsymbol{\varepsilon}^{\mathrm{e}}(\boldsymbol{\varepsilon};\boldsymbol{\varepsilon}^{\mathrm{p}})) + \Psi_0^{\mathrm{p,c}}(\alpha_{\mathrm{c}}) + \Psi_0^{\mathrm{p,t}}(\alpha_{\mathrm{t}})\right].\end{aligned} \tag{4.3}$$

In this representation, damage effects enter the free energy by a classical $(1-d)$-*ansatz* which reduces the stiffness in the material. Focusing on a decoupled plasticity and damage formulation, the latter is not active during plastic loading of the material. Thus, evaluation of the *Clausius-Plank inequality* for the internal dissipation with $\dot{d}=0$ takes the form

$$\mathcal{D} = \boldsymbol{\sigma}_0 : \dot{\boldsymbol{\varepsilon}} - \dot{\Psi}_0 \geq 0 \tag{4.4}$$

in the isothermal case. From Eq. 4.3 we obtain by application of the chain rule

$$\dot{\Psi}_0 = \frac{\partial \Psi^{\mathrm{e}}}{\partial \boldsymbol{\varepsilon}^{\mathrm{e}}} : \dot{\boldsymbol{\varepsilon}}^{\mathrm{e}} + \frac{\partial \Psi_0^{\mathrm{p,c}}}{\partial \alpha_{\mathrm{c}}}\dot{\alpha}_{\mathrm{c}} + \frac{\partial \Psi_0^{\mathrm{p,t}}}{\partial \alpha_{\mathrm{t}}}\dot{\alpha}_{\mathrm{t}}. \tag{4.5}$$

Insertion into the dissipation inequality gives

$$\mathcal{D} = \left[\boldsymbol{\sigma}_0 - \frac{\partial \Psi_0^{\mathrm{e}}}{\partial \boldsymbol{\varepsilon}^{\mathrm{e}}}\right] : \dot{\boldsymbol{\varepsilon}} + \frac{\partial \Psi_0^{\mathrm{e}}}{\partial \boldsymbol{\varepsilon}^{\mathrm{e}}} : \dot{\boldsymbol{\varepsilon}}^{\mathrm{p}} - \frac{\partial \Psi_0^{\mathrm{p,c}}}{\partial \alpha_{\mathrm{c}}}\dot{\alpha}_{\mathrm{c}} - \frac{\partial \Psi_0^{\mathrm{p,t}}}{\partial \alpha_{\mathrm{t}}}\dot{\alpha}_{\mathrm{t}} \geq 0. \tag{4.6}$$

Standard arguments, see e.g. Coleman & Gurtin [75] or Lubliner [241], give the definition of stresses

$$\boldsymbol{\sigma}_0 = \frac{\partial \Psi_0^{\mathrm{e}}}{\partial \boldsymbol{\varepsilon}^{\mathrm{e}}} \tag{4.7}$$

and the reduced dissipation inequality

$$\mathcal{D}^{\mathrm{red}} = \boldsymbol{\sigma}_0 : \dot{\boldsymbol{\varepsilon}}^{\mathrm{p}} + Y_{\mathrm{c}}\dot{\alpha}_{\mathrm{c}} + Y_{\mathrm{t}}\dot{\alpha}_{\mathrm{t}} \geq 0 \tag{4.8}$$

in terms of the driving forces

$$Y_c = -\frac{\partial \Psi_0^{p,c}}{\partial \alpha_c} \quad \text{and} \quad Y_t = -\frac{\partial \Psi_0^{p,t}}{\partial \alpha_t} \tag{4.9}$$

for hardening mechanisms under compression and tension, respectively. In what follows, the quadratic form of the free energy function is chosen

$$\Psi_0^e(\boldsymbol{\varepsilon}^e) = \frac{1}{2}\kappa \mathrm{tr}^2[\boldsymbol{\varepsilon}^e] + \mu \|\mathrm{dev}[\boldsymbol{\varepsilon}^e]\|^2 \tag{4.10}$$

in terms of the compression modulus κ and the shear modulus μ. Equation 4.7 yields the stresses

$$\boldsymbol{\sigma}_0 = \kappa \mathrm{tr}[\boldsymbol{\varepsilon}^e]\,\mathbb{1} + 2\mu \mathrm{dev}[\boldsymbol{\varepsilon}^e] = \mathbb{C}_0 : \boldsymbol{\varepsilon}^e \tag{4.11}$$

expressed by the isotropic elastic stiffness tensor $\mathbb{C}_0 = \kappa \mathbb{1} \otimes \mathbb{1} + 2\mu \mathbb{P}^{\mathrm{sym}}$ in terms of the fourth-order deviatoric projection tensor $\mathbb{P}^{\mathrm{sym}} = \mathbb{I}^{\mathrm{sym}} - \frac{1}{3}\mathbb{1} \otimes \mathbb{1}$. The hardening mechanisms are taken into account by the energetic contributions

$$\begin{aligned} \Psi_0^{p,c}(\alpha_c) &= \left(\sigma_c^\infty - \sigma_c^0\right)\left(\alpha_c + \frac{1}{\omega_c}\exp[-\omega_c \alpha_c]\right) \\ \Psi_0^{p,t}(\alpha_t) &= \left(\sigma_t^\infty - \sigma_t^0\right)\left(\alpha_t + \frac{1}{\omega_t}\exp[-\omega_t \alpha_t]\right) \end{aligned} \tag{4.12}$$

for compressive and tensile response. Equation 4.9 gives the driving forces

$$\begin{aligned} Y_c &= -\left(\sigma_c^\infty - \sigma_c^0\right)(1 - \exp[-\omega_c \alpha_c]) \\ Y_t &= -\left(\sigma_t^\infty - \sigma_t^0\right)(1 - \exp[-\omega_t \alpha_t]) . \end{aligned} \tag{4.13}$$

4.1.2.2 Non-associative plastic flow

For the present material, the elastic domain $\mathbb{E}$ in stress-space is a smooth domain

$$\mathbb{E} := \{(\boldsymbol{\sigma}_0, Y_c, Y_t) \mid \Phi(\boldsymbol{\sigma}_0, Y_c, Y_t) < 0\} \tag{4.14}$$

bounded by the yield function Φ. An associative form of the evolution equations for the plastic strain $\boldsymbol{\varepsilon}^{\mathrm{p}}$ and the hardening variables $\alpha_{\mathrm{c/t}}$ follow from the principal of maximum dissipation $\left(\boldsymbol{\sigma}_0 - \boldsymbol{\sigma}_0^*\right) : \dot{\boldsymbol{\varepsilon}}^{\mathrm{p}} - \left(Y_{\mathrm{c}} - Y_{\mathrm{c}}^*\right)\dot{\alpha}_{\mathrm{c}} - \left(Y_{\mathrm{t}} - Y_{\mathrm{t}}^*\right)\dot{\alpha}_{\mathrm{t}} \geq 0$ for all $\left(\boldsymbol{\sigma}_0^*, Y_{\mathrm{c}}^*, Y_{\mathrm{t}}^*\right) \in \mathbb{E}$. See also HILL [162], DRUCKER [102], and others. It directly induces the associative evolution equations

$$\dot{\boldsymbol{\varepsilon}}^{\mathrm{p}} = \dot{\gamma}\partial_{\boldsymbol{\sigma}_0}\Phi \quad \text{and} \quad \dot{\alpha}_{\mathrm{c}} = \dot{\gamma}\partial_{Y_{\mathrm{c}}}\Phi \quad \text{and} \quad \dot{\alpha}_{\mathrm{t}} = \dot{\gamma}\partial_{Y_{\mathrm{t}}}\Phi \tag{4.15}$$

along with the loading/unloading conditions (*Karush-Kuhn-Tucker* conditions)

$$\dot{\gamma} \geq 0 \quad \text{and} \quad \Phi \leq 0 \quad \text{and} \quad \dot{\gamma}\Phi = 0 \tag{4.16}$$

where $\dot{\gamma}$ denotes the plastic multiplier. Carefully note that within the associative theory, the yield function Φ plays the role of a so-called plastic potential which governs the flow rule. The generalization towards a non-associative flow response bases on the introduction of independent flow directions $\boldsymbol{N}$

$$\dot{\boldsymbol{\varepsilon}}^{\mathrm{p}} = \dot{\gamma}\boldsymbol{N} \quad \text{and} \quad \dot{\alpha}_{\mathrm{c}} = \dot{\gamma}N_{\mathrm{c}} \quad \text{and} \quad \dot{\alpha}_{\mathrm{t}} = \dot{\gamma}N_{\mathrm{t}}. \tag{4.17}$$

The flow directions are assumed to be derived from a plastic potential function $\Xi\left(\boldsymbol{\sigma}_0, Y_{\mathrm{c}}, Y_{\mathrm{t}}\right)$ in the sense

$$\boldsymbol{N} := \frac{\partial \Xi}{\partial \boldsymbol{\sigma}_0} \quad \text{and} \quad N_{\mathrm{c}} = N_{\mathrm{t}} = \sqrt{k}\,||\boldsymbol{N}||. \tag{4.18}$$

The variable k, controlling the hardening mechanism, depends on the used yield criterion but can generally be defined by

$$k = \frac{1}{1 + 2\nu_{\mathrm{p}}^2} \tag{4.19}$$

using the plastic pendant of the Poisson's ratio ν_{p}. Clearly, definition Eq. 4.18_2 identifies the internal variables α_{c} and α_{t} with the accumulated plastic strain with the evolution equation

$$\dot{\alpha}_{\mathrm{c}} = \dot{\alpha}_{\mathrm{t}} = \sqrt{k}\,||\dot{\boldsymbol{\varepsilon}}^{\mathrm{p}}|| =: \dot{\varepsilon}^{\mathrm{p}}_{\mathrm{acc}}. \tag{4.20}$$

To be specific, for the present material model, the pressure dependent yield function according to Tschoegl [393] is chosen and reads

$$\begin{aligned}\Phi\left(\boldsymbol{\sigma}_0, Y_c, Y_t\right) = {} & 6J_2(\boldsymbol{\sigma}_0) \\ & + 2\left[\sigma_c\left(Y_c\right) - \sigma_t\left(Y_t\right)\right] I_1(\boldsymbol{\sigma}_0) \\ & - 2\sigma_c\left(Y_c\right)\sigma_t\left(Y_t\right)\end{aligned} \tag{4.21}$$

in terms of the yielding parameters σ_c and σ_t and the invariants

$$J_2 = \frac{1}{2}\|\mathrm{dev}\left[\boldsymbol{\sigma}_0\right]\|^2 \quad \text{and} \quad I_1 = \mathrm{tr}\left[\boldsymbol{\sigma}_0\right]. \tag{4.22}$$

A graphical representation of the yield function in the stress space is depicted in Fig. 4.1. The hardening mechanisms under compression and tension take

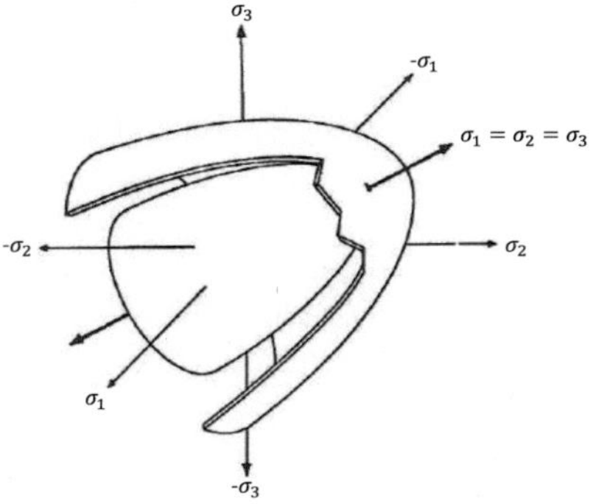

Figure 4.1: Paraboloidal yield surface Φ. Image taken and modified from Tschoegl [393].

the form

$$\sigma_c\left(Y_c\right) := \sigma_c^0 - Y_c \quad \text{and} \quad \sigma_t\left(Y_t\right) := \sigma_t^0 - Y_t \tag{4.23}$$

in terms of the driving forces Eq. 4.13. The flow direction from Eq. 4.18 is characterized by the plastic potential which, according to Melro [260], reads

$$\Xi\left(\boldsymbol{\sigma}_0\right) = 3J_2\left(\boldsymbol{\sigma}_0\right) + \frac{1}{9}\alpha I_1^2\left(\boldsymbol{\sigma}_0\right) \tag{4.24}$$

again in terms of the invariants from Eq. 4.22. Making use of equations Eq. 4.18, the flow direction can be identified as

$$\boldsymbol{N} = \boldsymbol{N}_{\text{iso}} + \boldsymbol{N}_{\text{vol}} = 3\text{dev}\left[\boldsymbol{\sigma}_0\right] + \frac{2}{9}\alpha\text{tr}\left[\boldsymbol{\sigma}_0\right]\mathbb{1} \tag{4.25}$$

in terms of a variable controlling the volumetric plastic flow

$$\alpha = \frac{9}{2}\frac{1-2\nu_{\text{p}}}{1+\nu_{\text{p}}}. \tag{4.26}$$

4.1.3 Modeling ultimate damage

This section serves to extend the introduced elasto-plastic response by a stress-based damage formulation. Physically speaking, damage is interpreted as the result, growth and coalescence of microcracks (cf. SIMO & JU [353, 354]). Introducing a material model within the framework of continuum mechanics, damage is often treated as an evolving internal variable which can be a scalar or tensorial entity. In the following, a scalar damage variable is assumed which does not take into account tension/compression asymmetry and is independent of the loading direction. Hence, damage is defined as completely isotropic. In this work, the hypothesis of strain equivalence is applied (cf. LEMAITRE [218, 220]). This method favors the implication that the evolution of damage is directly linked to the history of total strain, as stated by SIMO & JU [353,354]. In order to derive the model in a thermodynamically consistent manner, the complementary free energy potential, the Gibb's potential written as

$$\Lambda(\boldsymbol{\sigma}, \boldsymbol{\varepsilon}^{\text{p}}, Y_{\text{c}}, Y_{\text{t}}, d_{\sigma}) = d_{\sigma}\Lambda_0(\boldsymbol{\sigma}_0) + \boldsymbol{\sigma} : \boldsymbol{\varepsilon}^{\text{p}} - \Lambda^{\text{p}}(Y_{\text{c}}, Y_{\text{t}}, \boldsymbol{\varepsilon}^{\text{p}}) \tag{4.27}$$

is introduced. Hereby, the stress based damage variable $d_{\sigma} = (1-d)^{-1}$, the thermodynamic driving forces for hardening Y_{c} and Y_{t}, together with the elastic contribution

$$\Lambda_0(\boldsymbol{\sigma}) = \frac{1}{2}\boldsymbol{\sigma}_0^{\text{T}} : \mathbb{H}_0 : \boldsymbol{\sigma}_0 \tag{4.28}$$

and complementary plastic potential Λ^{p}, analogously to the general formulation of SIMO & JU [353, 354] are used. $\mathbb{H}_0 = (\mathbb{C}_0)^{-1}$ denotes the undamaged compliance tensor, that can be expressed in an analogous way to the pristine stiffness tensor $\mathbb{C}_0$ in terms of the isotropic stiffness parameters E and ν. Consequently, the subsequent focus lies on the damage mechanisms solely. For an isothermal case the Clausius-Duhem inequality (cf. COLEMAN & NOLL [76], COLEMAN & GURTIN [75]) takes the form

$$\mathcal{D} = \dot{\Lambda} - \dot{\boldsymbol{\sigma}} : \boldsymbol{\varepsilon} \geq 0. \tag{4.29}$$

with

$$\dot{\Lambda} = \dot{\Lambda}(\boldsymbol{\sigma}, \boldsymbol{\varepsilon}^{\mathrm{p}}, \boldsymbol{q}, d_\sigma) \tag{4.30}$$

for any admissible process. For a decoupled plastic and damage formulation with frozen plastic state variables during damage loading, the application of the chain rule of Eq. 4.29 takes a simplified form

$$\frac{\partial \Lambda}{\partial \boldsymbol{\sigma}} : \dot{\boldsymbol{\sigma}} + \frac{\partial \Lambda}{\partial d_\sigma} \dot{d}_\sigma - \dot{\boldsymbol{\sigma}} : \boldsymbol{\varepsilon} \geq 0. \tag{4.31}$$

With standard arguments together with the additional assumption that unloading processes are always elastic we obtain

$$\boldsymbol{\varepsilon} = \frac{\partial \Lambda}{\partial \boldsymbol{\sigma}} = d_\sigma \frac{\partial \Lambda_0}{\partial \boldsymbol{\sigma}} + \boldsymbol{\varepsilon}^{\mathrm{p}} \tag{4.32}$$

along with the dissipative inequality

$$\mathcal{D}^{\mathrm{red}} = \Lambda_0(\boldsymbol{\sigma}_0) \dot{d}_\sigma \geq 0 \tag{4.33}$$

for dissipative contributions during the evolution of damage. MALVERN [249] states, that the positiveness of the dissipated energy is required by any constitutive model, since it represents the second law of thermodynamics. Carefully note, that the additive decomposition for the strains is recovered from Eq. 4.32. It follows from Eq. 4.27 and Eq. 4.33 that

$$Y = \frac{\partial \Lambda(\boldsymbol{\sigma}, d_\sigma)}{\partial d_\sigma} = -\Lambda_0(\boldsymbol{\sigma}_0). \tag{4.34}$$

Therefore, the initial elastic complimentary energy $\Lambda_0(\boldsymbol{\sigma}_0)$ is the thermodynamic force $-Y$ conjugate to the damage variable d_σ, which intrinsically fulfills the condition of irreversibility of damage $\dot{d} \geq 0$. Note that for the linear case

$\Lambda_0(\boldsymbol{\sigma}_0) = \frac{1}{2}\boldsymbol{\sigma}_0^{\mathrm{T}} : \mathbb{H}_0 : \boldsymbol{\sigma}_0$, where $\mathbb{H}_0$ is the undamaged compliance matrix. It follows from Eq. 4.32 and

$$\Lambda_0(\mathbf{0}) = 0 \quad \text{and} \quad \left.\frac{\partial \Lambda_0(\boldsymbol{\sigma}_0)}{\partial \boldsymbol{\sigma}}\right|_{\boldsymbol{\sigma}_0=\mathbf{0}} = 0 \tag{4.35}$$

that the plastic strain $\boldsymbol{\varepsilon}^{\mathrm{p}}$ is precisely the residual strain obtained upon (local) unloading. Thus, identifying the elastic strain with recoverable strain after unloading, i.e. $\boldsymbol{\varepsilon}^{\mathrm{e}} = \boldsymbol{\varepsilon} - \boldsymbol{\varepsilon}^{\mathrm{p}}$, from 4.32

$$\boldsymbol{\varepsilon}^{\mathrm{e}} \equiv d_{\sigma}\frac{\partial \Lambda_0}{\partial \boldsymbol{\sigma}} \Rightarrow \tilde{\boldsymbol{\varepsilon}}^{\mathrm{e}} = (1-d)\boldsymbol{\varepsilon}^{\mathrm{e}} = \frac{\partial \Lambda_0}{\partial \boldsymbol{\sigma}} \tag{4.36}$$

is obtained with $\tilde{\boldsymbol{\varepsilon}}^{\mathrm{e}}$ being the effective elastic strain. Therefore, the updated damaged stiffness tensor

$$\mathbb{C} = (1-d)\mathbb{C}_0 = \left[d_{\sigma}\frac{\partial^2 \Lambda_0}{\partial \boldsymbol{\sigma}^2}\right]^{-1} \tag{4.37}$$

follows, since from the concept of stress equivalence ("*the stress associated with a damaged state under the applied strain is equivalent to the stress associated with its undamaged state under the effective strain*" cf. Simo & Ju [353, 354]) follows

$$\boldsymbol{\sigma} = \mathbb{C}_0 : \tilde{\boldsymbol{\varepsilon}}^{\mathrm{e}} \overset{!}{=} \mathbb{C} : \boldsymbol{\varepsilon}^{\mathrm{e}} = (1-d)\mathbb{C}_0 : \boldsymbol{\varepsilon}^{\mathrm{e}}. \tag{4.38}$$

An equivalent derivation based on the equivalence of the Helmholtz free energy yields the same results due to the special position of damage in the respective potentials as well as the assumption of independence between the evolution of the damage variable d from plastic processes. In the following, the hypothesis of strain equivalence is used, introducing the relation

$$\tilde{\boldsymbol{\sigma}} = \mathbb{M}^{-1} : \boldsymbol{\sigma} \tag{4.39}$$

with $\boldsymbol{\sigma}$ as the homogenized stress tensor and $\tilde{\boldsymbol{\sigma}}$ denoting the effective stress tensor (cf. Carol et al. [52, 53]). The fourth-order tensor $\mathbb{M}$ characterizing the damage state, simplifies for the present isotropic case to $\mathbb{M} = (1-d)\mathbb{I}^{\mathrm{sym}}$, with $\mathbb{I}^{\mathrm{sym}}$ being the rank four identity tensor and a formulation for the stresses equivalent to the one in Eq. 4.38 is valid,

$$\boldsymbol{\sigma} = \mathbb{C} : \boldsymbol{\varepsilon}^{\mathrm{e}} = (1 - d)\mathbb{C}_0 : \boldsymbol{\varepsilon}^{\mathrm{e}}. \tag{4.40}$$

The evolution of damage in the material is characterized by means of a stress-based damage criterion and a damage rule. Here, a similar criterion as for the presented plasticity formulation from earlier is used (cf. TSCHOEGL [393], MELRO [260]), whereby the yield strengths are replaced by failure strengths. The damage activation function F^{d} defines the elastic domain under a general stress state, using the tensile and compressive strengths of the material

$$F^{\mathrm{d}} = \phi^{\mathrm{d}} - r = \begin{cases} < 0, & \text{elastic domain} \\ = 0, & \text{damage criterion activated} \end{cases} \tag{4.41}$$

where r is the internal variable controlled by the damage evolution law (threshold variable) and ϕ^{d} is the loading function defined by

$$\phi^{\mathrm{d}} = \frac{3}{X_{\mathrm{c}} X_{\mathrm{t}}} \tilde{J}_2 + \frac{(X_{\mathrm{c}} - X_{\mathrm{t}})}{X_{\mathrm{c}} X_{\mathrm{t}}} \tilde{I}_1 \tag{4.42}$$

with X_{c} as the materials compressive and X_{t} as tensile strength, respectively. The definitions for the invariants here are given analogously to Eq. 4.22 but as functions of the effective stress $\tilde{\boldsymbol{\sigma}}$. The Karush-Kuhn-Tucker conditions for the evolution of damage are given as

$$\dot{r} \geq 0 \quad \text{and} \quad F^{\mathrm{d}} \leq 0 \quad \text{and} \quad \dot{r} F^{\mathrm{d}} = 0. \tag{4.43}$$

In order to distinguish loading from unloading situations and determine if damage evolution takes place, the rate of the loading function $\dot{\phi}^{\mathrm{d}}$ must be evaluated. If $\dot{\phi}^{\mathrm{d}} \leq 0$ the state is one of unloading, otherwise damage evolution is taking place and the following consistency condition is valid

$$\dot{F}^{\mathrm{d}} = \dot{\phi}^{\mathrm{d}} - \dot{r} = 0 \tag{4.44}$$

under the condition that (a) the internal variable r depends only on the damage variable and (b) the loading function is defined in terms of the strain tensor. If these conditions hold true, the constitutive model can be integrated explicitly (cf. SIMO & JU [353, 354]). From the consistency condition ($F^{\mathrm{d}} = \dot{F}^{\mathrm{d}} = 0$) we get

$$r = \max \left\{ r^0, \max_{t \to \infty} \left\{ \phi_{\mathrm{t}}^{\mathrm{d}} \right\} \right\}. \tag{4.45}$$

For the described model, the initial internal threshold parameter is set to $r^0 = 1$. The missing constitutive relation describes the mapping of $r \in [1, \infty]$ on a discrete damage variable $d \in [0, 1]$. In a general approach it is easily found that damage states can basically be divided in three stages that formulate boundary conditions for a valid choice of the damage evolution law:

1. Pristine state: Material is undamaged $r = 1 \to d = 0$.
2. Damage progress: The change rate of damage is greater than zero ($\dot{r} \geq 0 \to \dot{d} \geq 0$). This is fulfilled if the damage evolution law satisfies the condition

   $$\frac{\partial d}{\partial r} \geq 0$$

   since

   $$\dot{d} = \frac{\partial d}{\partial r} \dot{r} \geq 0.$$

3. Fully damaged material: The threshold variable strives towards $r \to \infty$ and the damage variable takes the value $d = 1$.

The *ansatz* for damage evolution law is therefore chosen as

$$d = 1 - \frac{1}{r} \exp \left(A_{\mathrm{m}}(1 - r) \right) \tag{4.46}$$

fulfilling all required conditions. Figure 4.2 shows the evolution of the damage variable with increasing threshold variable $r > 1$. To integrate a development of a suitable strain localization limiter that minimizes mesh-sensibility associated with strain-softening, Bažant's crack band model (cf. BAŽANT & OH [19]) has been applied and integrated in the definitions of the damage evolution law. Hereby, A_{m} is the adjustment parameter, dependent on the characteristic length of the particular observed element l^{e}, that will be determined under the assumption of a uniaxial tensile load. Without that, damage would localize in a narrow band with the same thickness as the element where damage was activated. This would cause a dependency of the structural response on the mesh size. The smaller the element in the band of localized damage, the less

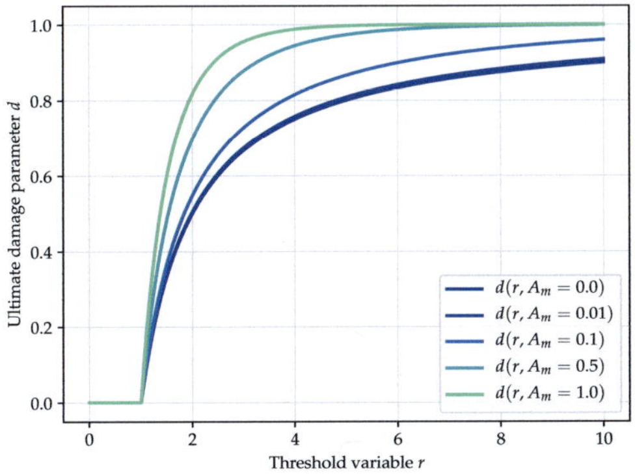

Figure 4.2: Mapping function of the damage variable $d \in [0, 1]$ as a function of the threshold variable $r \in [1, \infty]$ and exemplary adjustment parameters A_m accounting for the element size.

the computed dissipated energy. The solution for this localization problem lies in a regularization of the computed dissipated energy by the characteristic length l^e of each element obtaining the dissipative energy density

$$\psi^\mathrm{D} = \int_0^\infty -Y \dot{d}_\sigma \mathrm{d}t = \int_1^\infty \Lambda_0 (\boldsymbol{\sigma}) \frac{\partial d_\sigma}{\partial r} \mathrm{d}r = \frac{G_\mathrm{f}}{l^\mathrm{e}} \tag{4.47}$$

where G_f is the critical energy release rate. Solving Eq. 4.47 for A_m enables the subsequent damage update. Remarks on the numerical implementation are given in Appendix C. From here the damaged stresses can be computed. The approach of strain equivalence states that

$$\boldsymbol{\sigma} = \mathbb{C} : \boldsymbol{\varepsilon}^\mathrm{e} \quad \text{and} \quad \boldsymbol{\sigma}_0 = \mathbb{C}_0 : \boldsymbol{\varepsilon}^\mathrm{e}. \tag{4.48}$$

Simply plugging the above equations together, yields

$$\boldsymbol{\sigma} = \mathbb{C} : (\mathbb{C}_0)^{-1} : \tilde{\boldsymbol{\sigma}} = (1 - d)\boldsymbol{\sigma}_0 \tag{4.49}$$

in terms of the undamaged elastic stresses defined in Eq. 4.11 and the damage variable d.

4.1.4 Extension towards rate-dependency

Recollecting the experimental results for the thermoplastic matrix from Section 3.2 suggests that the material exhibits a pronounced rate sensitivity in terms of both plastic and damage response. A simple phenomenological approach to cover this behavior in terms of plasticity is obtained by means of a viscous regularization of the evolution parameters of the presented rate-independent strain-based plasticity as presented by PERZYNA [313]. This method has the property of introducing a viscosity coefficient η_{vp} as one additional parameter for a rate-dependent plasticity. Depending on its value, the non-linearity during plastic evolution becomes rate-dependent in the sense that plastic effects are decelerated at higher strain rates. The governing rate equations for viscous plasticity are different from their inviscid counterparts and, following the arguments of PERZYNA, can be explicitly rewritten

$$\dot{\gamma} = \frac{1}{\eta_{\mathrm{vp}}} \langle \Phi(\sigma) \rangle_{+}. \tag{4.50}$$

However, an alternative approach is chosen, introducing rate-dependent hardening terms, yield strengths, and failure strengths. For all cases, a similar exponential formulation as for the hardening is chosen (cf. Eq. 4.23), introducing the time-dependent definition for the corresponding parameters $\mathcal{P} \in \left\{\sigma^{0}_{\mathrm{c,t}}, \sigma^{\infty}_{\mathrm{c,t}}, \omega_{\mathrm{c,t}}, X_{\mathrm{c,t}}\right\}$ of the form

$$\mathcal{P} = \mathcal{P}^{0} + (\mathcal{P}^{\infty} - \mathcal{P}^{0}) \left(1 - \exp\left(-\eta^{\mathcal{P}}_{\mathrm{v}} \|\dot{\varepsilon}\|^{\alpha}\right)\right), \tag{4.51}$$

where $\mathcal{P}^{0}$ corresponds to the parameter at low and $\mathcal{P}^{\infty}$ at high strain rates respectively. The viscous parameter $\eta^{\mathcal{P}}_{\mathrm{v}}$ controls the transition from $\mathcal{P}^{0}$ to $\mathcal{P}^{\infty}$.

4.1.5 Algorithmic treatment of constitutive equations

The upcoming section will describe the algorithmic setting that is necessary for a numerical implementation of the visco-plasticity and damage formulation defined above. For the present case, the model is implemented in a user-defined material subroutine (*UMAT*) for the implicit solver Abaqus (cf. SIMULIA [355]). In Fig. 4.3, the proposed model is visualized in a one-dimensional rheological system, consisting of a series of one spring, responsible for ground-state elasticity, and a parallel setting of a friction element and a damper, representing the viscoplastic part. Affecting the material stiff-

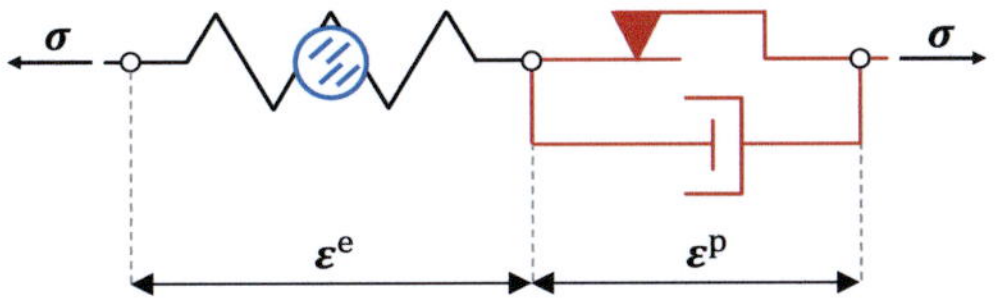

Figure 4.3: One-dimensional representation of the proposed matrix material model, depicting the strain split into elastic (*black*) and plastic (*red*) contributions. The stiffness is degraded by the progressive damage variable (*blue*).

ness, the damage variable is responsible for the ultimate damage representation. Plasticity and damage related terms are rate-dependent. This model induces the calculation of stresses according to

$$\boldsymbol{\sigma} = (1 - d\,(\boldsymbol{\varepsilon}, \dot{\boldsymbol{\varepsilon}}))\,\mathbb{C}_0 : \boldsymbol{\varepsilon}^{\mathrm{e}}\,(\boldsymbol{\varepsilon}, \boldsymbol{\varepsilon}^{\mathrm{p}}\,(\boldsymbol{\varepsilon}, \dot{\boldsymbol{\varepsilon}}))\,. \tag{4.52}$$

For the following derivations it is assumed however, that (a) one constant strain rate $\dot{\boldsymbol{\varepsilon}}$ is active and (b) plasticity and damage evolution are not inter-dependent and will therefore be calculated in a *staggered* manner, i.e. consecutively. In other words, plasticity will evolve ($\dot{\gamma} > 0$) only as long as macro damage has not started to evolve ($d = 0$). As soon as the material strength in terms of failure is reached, plastic strains will be frozen. That is the reason why the algorithmic setting of plastic evolution and damage can be looked at separately. In this setting, the plastic strain rate defined in Eq. 4.17_1 is integrated yielding

$$\boldsymbol{\varepsilon}^{\mathrm{p}} = \boldsymbol{\varepsilon}^{\mathrm{p}}_{\mathrm{n}} + \int_{t_{\mathrm{n}}}^{\mathrm{t}} \dot{\boldsymbol{\varepsilon}}^{\mathrm{p}} \mathrm{d}t = \boldsymbol{\varepsilon}^{\mathrm{p}}_{\mathrm{n}} + \int_{t_{\mathrm{n}}}^{\mathrm{t}} \dot{\gamma} \boldsymbol{N} \mathrm{d}t. \tag{4.53}$$

Application of implicit backward Euler gives

$$\boldsymbol{\varepsilon}^{\mathrm{p}} = \boldsymbol{\varepsilon}^{\mathrm{p}}_{\mathrm{n}} + \dot{\gamma} \Delta t \boldsymbol{N} = \boldsymbol{\varepsilon}^{\mathrm{p}}_{\mathrm{n}} + \Delta\gamma \boldsymbol{N}. \tag{4.54}$$

By definition the elastic trial strains are introduced by

$$\boldsymbol{\varepsilon}^{\mathrm{tr}} = \boldsymbol{\varepsilon} - \boldsymbol{\varepsilon}^{\mathrm{p}}_{\mathrm{n}}. \tag{4.55}$$

The assumption that the complete strain increment $\Delta\boldsymbol{\varepsilon}$ made in the current time increment is purely elastic for the moment holds. Based on this trial state we obtain an alternative representation of the current elastic strains, namely

$$\begin{aligned}\operatorname{dev}\left[\boldsymbol{\varepsilon}^{\mathrm{e}}\right] &= \operatorname{dev}\left[\boldsymbol{\varepsilon}^{\mathrm{tr}}\right] - \Delta\gamma\boldsymbol{N}_{\mathrm{iso}}\\ \frac{1}{3}\operatorname{tr}\left[\boldsymbol{\varepsilon}^{\mathrm{e}}\right] &= \frac{1}{3}\operatorname{tr}\left[\boldsymbol{\varepsilon}^{\mathrm{tr}}\right] - \Delta\gamma\boldsymbol{N}_{\mathrm{vol}}\end{aligned} \tag{4.56}$$

With these current strains, the current (undamaged) stresses can be determined according to Eq. 4.11 yielding

$$\begin{aligned}\operatorname{dev}\left[\boldsymbol{\sigma}_0\right] &= 2\mu\operatorname{dev}\left[\boldsymbol{\varepsilon}^{\mathrm{tr}}\right] - 2\mu\Delta\gamma\boldsymbol{N}_{\mathrm{iso}}\\ \frac{1}{3}\operatorname{tr}\left[\boldsymbol{\sigma}_0\right] &= \kappa\operatorname{tr}\left[\boldsymbol{\varepsilon}^{\mathrm{tr}}\right] - 3\kappa\Delta\gamma\boldsymbol{N}_{\mathrm{vol}}.\end{aligned} \tag{4.57}$$

Combining equations 4.57_1 and 4.57_2 the classical formulation for an elastic predictor/plastic corrector algorithm can be recovered, which reads

$$\boldsymbol{\sigma}_0 = \mathbb{C}_0 : \left(\boldsymbol{\varepsilon}^{\mathrm{e}}_{\mathrm{n}} + \Delta\boldsymbol{\varepsilon}\right) - \Delta\gamma\mathbb{C}_0 : \boldsymbol{N}. \tag{4.58}$$

The first part of the equation is called elastic predictor $\boldsymbol{\sigma}^{\mathrm{tr}}=\mathbb{C}_0 : \left(\boldsymbol{\varepsilon}^{\mathrm{e}}_{\mathrm{n}} + \Delta\boldsymbol{\varepsilon}\right)$ which estimates the stresses at the end of each loading increment $t_{\mathrm{n+1}}$ assuming the step to be purely elastic, and the second part plastic corrector. Latter is zero if the step is elastic and greater than zero in the case of plastic flow in the time increment, thereby correcting the elastic predictor. To determine the plastic corrector, the set of equations in Box 4.1 has to be solved simultaneously in the return mapping scheme. Taking into consideration the definitions from Eq. 4.25 and performing a stress split into deviatoric and volumetric parts, it is possible to rewrite the stresses as

$$\begin{aligned}\operatorname{dev}\left[\boldsymbol{\sigma}_0\right] &= 2\mu\operatorname{dev}\left[\boldsymbol{\varepsilon}^{\mathrm{tr}}\right] - 6\mu\Delta\gamma\operatorname{dev}\left[\boldsymbol{\sigma}_0\right]\\ \frac{1}{3}\operatorname{tr}(\boldsymbol{\sigma}_0) &= \kappa\operatorname{tr}(\boldsymbol{\varepsilon}^{\mathrm{tr}}) - \frac{2}{3}\kappa\alpha\Delta\gamma\operatorname{tr}(\boldsymbol{\sigma}_0)\end{aligned} \tag{4.59}$$

and hence, solving for the current stress

$$\operatorname{dev}\left[\boldsymbol{\sigma}_0\right] = \frac{2\mu}{\zeta_{\mathrm{s}}}\operatorname{dev}\left[\boldsymbol{\varepsilon}^{\mathrm{tr}}\right] \quad \text{and} \quad \frac{1}{3}\operatorname{tr}(\boldsymbol{\sigma}_0) = \frac{\kappa}{\zeta_{\mathrm{p}}}\operatorname{tr}\left[\boldsymbol{\varepsilon}^{\mathrm{tr}}\right] \tag{4.60}$$

with the two coefficients $\zeta_s = 1 + 6\mu\Delta\gamma$ and $\zeta_p = 1 + 2\kappa\alpha\Delta\gamma$ as the plastic corrector terms which reduce the elastic predictor stresses (cf. MELRO [260]). Note that these two terms depend on the plastic multiplier solely. Having formulated all the above information, it is now possible to formulate the radial return mapping scheme in order to identify the plastic multiplier $\Delta\gamma$.

Box 4.1: Evolution equations necessary for the definition of the plastic return mapping algorithm.

$$\Delta\boldsymbol{\varepsilon}^{\mathrm{p}} = \Delta\gamma\boldsymbol{N}(\boldsymbol{\sigma}_0, \boldsymbol{q})$$

$$\Delta\alpha_{\mathrm{c}} = \Delta\alpha_{\mathrm{t}} = \Delta\varepsilon^{P}_{\mathrm{acc}} = \sqrt{k}||\Delta\boldsymbol{\varepsilon}^{\mathrm{p}}||$$

$$\boldsymbol{\sigma}_0 = \boldsymbol{\sigma}^{\mathrm{tr}} - \mathbb{C}_0 : \Delta\boldsymbol{\varepsilon}^{\mathrm{p}}$$

$$\Phi(\boldsymbol{\sigma}_0, Y_{\mathrm{c}}, Y_{\mathrm{t}}) = 0$$

4.1.5.1 Return mapping algorithm for update of plastic strains

The starting point of the return mapping algorithm is the stress split described in Eq. 4.58 where the current stresses can be written as a combination of an elastic predictor and a plastic corrector that becomes active depending on the current loading situation. In the case of plastic loading, the consistency condition needs to be evaluated in order to find the correct plastic multiplier $\Delta\gamma$ which yields the correct plastic material response. From the definition for the yield function 4.21 and the substituting formulations for the stresses from Eq. 4.60, we obtain the nonlinear equation

$$\begin{aligned} \Phi\left(\Delta\gamma\right) = 3\left(\frac{2\mu}{\zeta_{\mathrm{s}}\left(\Delta\gamma\right)}\right)^2 ||\mathrm{dev}\left[\boldsymbol{\varepsilon}^{\mathrm{tr}}\right]||^2 \\ + \frac{6\kappa(\sigma_{\mathrm{c}}\left(\Delta\gamma\right) - \sigma_{\mathrm{t}}\left(\Delta\gamma\right))}{\zeta_{\mathrm{p}}\left(\Delta\gamma\right)}\mathrm{tr}\left[\boldsymbol{\varepsilon}^{\mathrm{tr}}\right] \\ - 2\sigma_{\mathrm{c}}\left(\Delta\gamma\right)\sigma_{\mathrm{t}}\left(\Delta\gamma\right) \stackrel{!}{=} 0 \end{aligned} \tag{4.61}$$

that has to be solved for the plastic multiplier $\Delta\gamma$. This is done, using a local Newton iteration scheme taking into consideration the other equations from

Box 4.1. Coming back to the elastic-plastic stress split (cf. Eq. 4.58), the most apparent relation is the definition of the increment of plastic strain that can be read as

$$\Delta\boldsymbol{\varepsilon}^{\mathrm{p}} = \Delta\gamma \left(3\frac{\mathrm{dev}\left[\boldsymbol{\sigma}^{\mathrm{tr}}\right]}{\zeta_{\mathrm{s}}} + \frac{2}{9}\alpha\frac{\mathrm{tr}(\boldsymbol{\sigma}^{\mathrm{tr}})}{\zeta_{\mathrm{p}}}\mathbb{1} \right) \tag{4.62}$$

using the definition for the plastic flow direction $\boldsymbol{N}$ (cf. Eq. 4.25). With this, the increment of the accumulated plastic strains that have been initially introduced in rate form (cf. Eq. 4.20), can now be written as

$$\Delta\varepsilon^{P}_{\mathrm{acc}} = \sqrt{k}\|\Delta\boldsymbol{\varepsilon}^{\mathrm{p}}\| \tag{4.63}$$

The above formulation is important later to describe the materials hardening behavior. Note that so far all quantities are only dependent on the trial strain $\boldsymbol{\varepsilon}^{\mathrm{tr}}$ and the plastic multiplier $\Delta\gamma$. Thus, since the trial strains are constant over one increment and there exists no closed form solution for $\Delta\gamma$, a local Newton-Raphson iterative update scheme has to be devised. Following the reasoning of the return mapping algorithm, where the equilibrium state $\Phi(\Delta\gamma) = 0$ is in demand, we define for the present case the residual r_Φ for a frozen deformation state $\boldsymbol{\varepsilon}$ at time $t_{\mathrm{n+1}}$ as

$$r_\Phi = \Phi(\Delta\gamma) \stackrel{!}{=} 0 \tag{4.64}$$

The basis for a Newton iteration is the linearization of the residual function

$$\mathrm{Lin}\left\{r_\Phi\right\}(\Delta\gamma, \varDelta\Delta\gamma) = r_\Phi(\Delta\gamma) + \mathbb{C}\varDelta\Delta\gamma \stackrel{!}{=} 0. \tag{4.65}$$

where the local tangent of the Newton iteration is defined as

$$\mathbb{C} = \frac{\partial r_\Phi\left(\Delta\gamma, \sigma_{\mathrm{c}}\left(\Delta\gamma\right), \sigma_{\mathrm{t}}\left(\Delta\gamma\right)\right)}{\partial\Delta\gamma}. \tag{4.66}$$

The plastic multiplier is then updated according to

$$\Delta\gamma \Leftarrow \Delta\gamma + \Delta\Delta\gamma = \Delta\gamma - r_\Phi(\Delta\gamma)\mathbb{C}^{-1} \tag{4.67}$$

until convergence is obtained, i.e. $\|r_\Phi(\Delta\gamma)\| \lesssim \varepsilon_{\mathrm{tol}}$.

4.1.5.2 Update of macroscopic damage variable

Parallely to the update of the plastic evolution, the damage activation criterion from Eq. 4.41 is checked in each increment. Since plasticity and damage are assumed not to be coupled in terms of dissipative processes, the material point is evaluated in an elastic-plastic-damage order. As soon as damage has been activated, the calculation of plastic evolution is skipped and the plastic strain remains constant. Up to that point, the material behaves purely elasto-viscoplastic. Analogously to the return mapping algorithm for plastic strain, an update of elastic strain

$$\boldsymbol{\varepsilon}^{\mathrm{e}} = \boldsymbol{\varepsilon}^{\mathrm{n}} + \Delta\boldsymbol{\varepsilon} - \boldsymbol{\varepsilon}^{\mathrm{p}}\left(\Delta\gamma\right) \tag{4.68}$$

and resulting new trial stress

$$\boldsymbol{\sigma}_0 = \mathbb{C}_0 : \boldsymbol{\varepsilon}^{\mathrm{e}} \tag{4.69}$$

is performed, that is used to evaluate damage evolution. In the case of active damage propagation, the internal variables r and consequently d are updated according to Eqs. 4.45 and 4.46. For a geometrically non-linear analyses the mesh adjustment parameter A_{m} has to be calculated in every time increment. A more detailed description of the integration method is not given at this point, but it is referred to the implementation presented in the work of Maimí et al. [245] (cf. also Appendix C). With the damage variable being known, the damaged stiffness tensor can be updated according to Eq. 4.37 and finally the damaged stresses are computed by Eq. 4.49.

4.1.5.3 Determination of the consistent algorithmic tangent operator

In order to obtain a robust convergence behavior in the commercial software package Abaqus, the global Newton-type iterative scheme requires a proper definition of the global algorithmic tangent. With the definition of the tangent being the sensitivity of the stresses with respect to the strains the operator can be defined by

$$\mathbb{C}^{\mathrm{t}} = \Delta_{\boldsymbol{\varepsilon}}\boldsymbol{\sigma} = \frac{\mathrm{d}\boldsymbol{\sigma}}{\mathrm{d}\boldsymbol{\varepsilon}}. \tag{4.70}$$

In terms of the stresses defined in Eq. 4.52 the tangent can be rewritten as

$$\mathbb{C}^{\mathrm{t}} = (1-d)\,\mathbb{C}_0 : (\mathbb{I}^{\mathrm{sym}} - \frac{\partial \boldsymbol{\varepsilon}^{\mathrm{p}}}{\partial \boldsymbol{\varepsilon}}) - (\mathbb{C}_0 : \boldsymbol{\varepsilon}^{\mathrm{e}}) \otimes \frac{\partial d}{\partial \boldsymbol{\varepsilon}} \tag{4.71}$$

being able to separate again plastic and damage effects. Knowing that the plastic strains defined in Eq. 4.54 are a function of the plastic multiplier and the flow direction, we can extend the derivation

$$\frac{\partial \boldsymbol{\varepsilon}^{\mathrm{p}}}{\partial \boldsymbol{\varepsilon}} = \frac{\partial \left[\boldsymbol{\varepsilon}^{\mathrm{p}}(\Delta\gamma, \boldsymbol{N})\right]}{\partial \boldsymbol{\varepsilon}} = \underbrace{\frac{\partial \boldsymbol{\varepsilon}^{\mathrm{p}}}{\partial \Delta\gamma}}_{\text{(i)}} \otimes \underbrace{\frac{\partial \Delta\gamma}{\partial \boldsymbol{\varepsilon}}}_{\text{(iv)}} + \underbrace{\frac{\partial \boldsymbol{\varepsilon}^{\mathrm{p}}}{\partial \boldsymbol{N}}}_{\text{(ii)}} : \underbrace{\frac{\partial \boldsymbol{N}}{\partial \boldsymbol{\varepsilon}}}_{\text{(iii)}} . \tag{4.72}$$

From here it becomes clear that Eq. $4.72_{\text{(i-iii)}}$ are rather easy to find

$$\begin{aligned} \frac{\partial \boldsymbol{\varepsilon}^{\mathrm{p}}}{\partial \Delta\gamma} &= \boldsymbol{N} + \Delta\gamma \frac{\partial \boldsymbol{N}}{\partial \Delta\gamma} \\ \frac{\partial \boldsymbol{\varepsilon}^{\mathrm{p}}}{\partial \boldsymbol{N}} &= \Delta\gamma \mathbb{I}^{\mathrm{sym}} \\ \frac{\partial \boldsymbol{N}}{\partial \boldsymbol{\varepsilon}} &= \mathbb{C}_0 : \left(3\mathbb{P}^{\mathrm{sym}} + \frac{2}{9}\alpha\,(\mathbb{1} \otimes \mathbb{1})\right) \end{aligned} \tag{4.73}$$

with $\mathbb{P}^{\mathrm{sym}}$ being the symmetric deviatoric projection tensor, whereas the determination of $4.72_{\text{(iv)}}$ is problematic since $\Delta\gamma$ is solved iteratively in the Newton-Raphson scheme described in Section 4.1.5.1 and thus no analytical formulations for the plastic multiplier and its derivations exists. Exploiting however the persistence condition implying the steady fulfillment of vanishing residuum defined in Eq. 4.64, this situation can be alleviated by applying the implicit functional theorem (in analogy to Goektepe [126]). Therefore, the total derivative of the residuum must also be zero at any given instant of the deformation,

$$\mathrm{d}_{\varepsilon} r_{\Phi} \stackrel{!}{=} 0. \tag{4.74}$$

Extending the formulation of Eq. 4.74

$$\mathrm{d}_{\varepsilon}\Phi = \frac{\partial \Phi}{\partial \boldsymbol{\varepsilon}} + \frac{\partial \Phi}{\partial \Delta\gamma}\frac{\partial \Delta\gamma}{\partial \boldsymbol{\varepsilon}} \stackrel{!}{=} 0 \tag{4.75}$$

a substitute for the wanted expression

$$\frac{\partial \Delta\gamma}{\partial \boldsymbol{\varepsilon}} = -\frac{\partial \Phi}{\partial \boldsymbol{\varepsilon}} \left[\frac{\partial \Phi}{\partial \Delta\gamma} \right]^{-1} \tag{4.76}$$

is obtained. This can be inserted into Eq. 4.72. The derivation of the yield surface Φ being defined in Eq. 4.61 with respect to the total strain $\boldsymbol{\varepsilon}$ reads

$$\begin{aligned}
\frac{\partial \Phi}{\partial \boldsymbol{\varepsilon}} = & \frac{6}{\zeta_s^2(\Delta\gamma)} \frac{\partial J_2^{\text{tr}}(\boldsymbol{\varepsilon}^{\text{tr}})}{\partial \boldsymbol{\varepsilon}} \\
& + 2 \frac{I_1^{\text{tr}}(\boldsymbol{\varepsilon}^{\text{tr}})}{\zeta_p(\Delta\gamma)} \left(\frac{\partial \sigma_c(\boldsymbol{\varepsilon}^{\text{tr}}, \Delta\gamma)}{\partial \boldsymbol{\varepsilon}} - \frac{\partial \sigma_t(\boldsymbol{\varepsilon}^{\text{tr}}, \Delta\gamma)}{\partial \boldsymbol{\varepsilon}} \right) \\
& + 2 \frac{\left(\sigma_c\left(\boldsymbol{\varepsilon}^{\text{tr}}, \Delta\gamma\right) - \sigma_t\left(\boldsymbol{\varepsilon}^{\text{tr}}, \Delta\gamma\right)\right)}{\zeta_p\left(\Delta\gamma\right)} \frac{\partial I_1^{\text{tr}}\left(\boldsymbol{\varepsilon}^{\text{tr}}\right)}{\partial \boldsymbol{\varepsilon}} \\
& - 2 \left(\sigma_c\left(\boldsymbol{\varepsilon}^{\text{tr}}, \Delta\gamma\right) \frac{\partial \sigma_t\left(\boldsymbol{\varepsilon}^{\text{tr}}, \Delta\gamma\right)}{\partial \boldsymbol{\varepsilon}} + \frac{\partial \sigma_c\left(\boldsymbol{\varepsilon}^{\text{tr}}, \Delta\gamma\right)}{\partial \boldsymbol{\varepsilon}} \sigma_t\left(\boldsymbol{\varepsilon}^{\text{tr}}, \Delta\gamma\right) \right)
\end{aligned} \tag{4.77}$$

where

$$\frac{\partial J_2^{\text{tr}}}{\partial \boldsymbol{\varepsilon}} = 4\mu^2 \operatorname{dev}\left[\varepsilon^{\text{tr}}\right] \quad \text{and} \quad \frac{\partial I_1^{\text{tr}}}{\partial \boldsymbol{\varepsilon}} = 3\kappa \mathbb{1}. \tag{4.78}$$

The derivation of the yield strengths ($\sigma_{c,t}$) with respect to total strain $\boldsymbol{\varepsilon}$ are based on the definition for hardening, which is a function of the increment of accumulated plastic strain (cf. Eq. 4.63). The part of the tangent concerning macro damage has to be further investigated. The total derivations of d with respect to total strains can be rewritten as

$$\frac{\partial d}{\partial \boldsymbol{\varepsilon}} = \frac{\partial d}{\partial r} \frac{\partial r}{\partial \boldsymbol{\varepsilon}}. \tag{4.79}$$

In the case of active damage evolution ($F^{\text{d}} = 0$) it becomes obvious from equation 4.41 and the definition of the threshold parameter r (cf. Eq. 4.45) that

$$r = \Phi^{\text{d}}(\boldsymbol{\varepsilon}^{\text{e}}). \tag{4.80}$$

Therefore, the missing derivation $\partial_{\varepsilon} r$ can be determined as

$$\frac{\partial r}{\partial \boldsymbol{\varepsilon}} = \frac{3}{X_c X_t} \frac{\partial J_2^{\text{tr}}}{\partial \boldsymbol{\varepsilon}} + \frac{(X_c - X_t)}{X_c X_t} \frac{\partial I_1^{\text{tr}}}{\partial \boldsymbol{\varepsilon}} \tag{4.81}$$

4.1.6 Transformation to a large deformation setting

Decomposing total strains additively into elastic and plastic parts is a typical feature of the geometrically linear theory of plasticity. The classical approach using the Lee-Kröner decomposition of the deformation gradient yields a stress-free relaxed intermediate configuration described by its metric tensor $\bar{\boldsymbol{G}}$. Consistent with this multiplicative decomposition, MIEHE et al. [268] propose an additive definition of total strains in the logarithmic strain space, hence giving a possibility to transfer material definitions given in Hencky strains to a logarithmic setting. A comparison shows the closeness of solutions obtained by the additive finite plasticity in the logarithmic strain space and the results given by classic multiplicative approaches. Hereby, both isotropic and anisotropic cases have been investigated. With the assumption that total and plastic deformations are coaxial, i.e. $\boldsymbol{C}$ and $\boldsymbol{G}^{\text{p}}$ commute, this approximation yields close results to a fully multiplicative form $\boldsymbol{\varepsilon}^{\text{e}} \approx \bar{\boldsymbol{\varepsilon}}^{\text{e}}=\frac{1}{2}\ln[\boldsymbol{F}^{\text{p-T}}\boldsymbol{C}\boldsymbol{F}^{\text{p}-1}]$ with the right Cauchy-Green strain measure $\boldsymbol{C}=\boldsymbol{F}^{\text{T}}\boldsymbol{g}\boldsymbol{F}$ and the plastic part of the deformation gradient $\boldsymbol{F}^{\text{p}}$. The framework defined in the following was used throughout this work, extending the original form by its application in the context of viscoplastic material as shown in MIEHE [277]. For the present case, the logarithmic Lagrangian elastic strain split can be defined by

$$\boldsymbol{\varepsilon}^{\text{e}} = \frac{1}{2}\ln[\boldsymbol{C}] - \boldsymbol{\varepsilon}^{\text{p}} \quad \text{with} \quad \boldsymbol{\varepsilon} = \frac{1}{2}\ln[\boldsymbol{C}] \,. \tag{4.82}$$

The internal variable of plastic strains is defined as

$$\boldsymbol{\varepsilon}^{\text{p}} = \frac{1}{2}\ln[\boldsymbol{G}^{\text{p}}] \tag{4.83}$$

where $\boldsymbol{G}^{\text{p}}=\boldsymbol{F}^{\text{p}\,\text{T}}\bar{\boldsymbol{G}}\boldsymbol{F}^{\text{p}}$ is the plastic metric equivalent to the Cauchy-Green strain measure and within this framework the primary variable for plastic deformation. $\bar{\boldsymbol{G}}$ is the metric tensor on the intermediate plastic configuration. The major characteristics of the proposed logarithmic framework is its modular structure:

1. *Geometric pre-processor*: Strain measures for total and plastic deformations are defined in the logarithmic space according to Eqs. 4.82 and 4.83.

2. *Constitutive model*: Here, constitutive equations in accordance to the geometric linear theory as described above are defined. The strain measure $\boldsymbol{\varepsilon}$ and the set of internal variables $\boldsymbol{q}$ are used as input for the proposed model, yielding finally stress and tangent moduli entities in the logarithmic space.

3. *Geometric post-processor*: After the computation of the relevant objects in the logarithmic domain, a mapping to their desired stress measures and tangent moduli is required. This is made possible by purely geometrical considerations and taking into account the consistency of local stress powers

$$w_{\text{int}} = \boldsymbol{g} : \boldsymbol{P} : \dot{\boldsymbol{F}} = \boldsymbol{\sigma} : \dot{\boldsymbol{\varepsilon}}. \tag{4.84}$$

MIEHE & LAMBRECHT [276] show that by applying the chain rule for the time derivation

$$\dot{\boldsymbol{\varepsilon}} = \frac{\partial \boldsymbol{\varepsilon}}{\partial \boldsymbol{F}} : \dot{\boldsymbol{F}} = \mathbb{P} : \dot{\boldsymbol{F}} \tag{4.85}$$

a fourth order stress projection tensor $\mathbb{P}{=}\partial_{\boldsymbol{F}}\boldsymbol{\varepsilon}$ can be identified. A de novo time derivation of the projection tensor

$$\dot{\mathbb{P}} = \mathbb{L} : \dot{\boldsymbol{F}} \tag{4.86}$$

yields the sixth order projection tensor $\mathbb{L}{=}\partial_{\boldsymbol{F}\boldsymbol{F}}\boldsymbol{\varepsilon}$ such that the tangent moduli from the logarithmic domain $\mathbb{C}^{\text{t}}$ are projected by

$$\mathbb{C} = \mathbb{P}^{\text{T}} : \mathbb{C}^{\text{t}} : \mathbb{P} + \boldsymbol{\sigma} : \mathbb{L}. \tag{4.87}$$

MIEHE & LAMBRECHT [276] provide a thorough presentation of the necessary algorithms to determine the projectors $\mathbb{P}$ and $\mathbb{L}$.

4.2 Material model for unidirectional fiber reinforced plastics

Woven materials consist of two basic constituents, fibers and matrix. A main issue in modeling these kind of materials is the incomplete set of test data due to deformation modes not accessible through experiments. To overcome this problem, a virtual material characterization framework is set up, taking into account the microstructure of the materials (cf. Fig. 1.1). For the upcoming steps in the multi-scale analysis, a homogenized material model considering anisotropic damage initiation and progression is necessary to predict the behavior of the impregnated rovings within the woven composite. By definition, the embedded fiber yarn is considered to be initially transversely isotropic with the preferred direction along the fiber axes. However, the symmetry class changes due to introduction of anisotropic damage. The relevant failure modes found in literature and during the experimental analysis are considered via direction-dependent failure criteria. This formulation allows furthermore the distinction of tensile/compressive asymmetric failure behavior. This section introduces therefore a fully three-dimensional continuum damage model for impregnated rovings, representing parallel fiber arrangement embedded in a thermoplastic matrix. Since on the macroscopic level the woven structure experiences large deformations, especially in the form of rotations, these motions have to be accounted for and reasonably described on the mesoscopic scale. To this end, a thermodynamically consistent constitutive model in the finite strain setting with the capability of predicting damage onset and evolution is derived here. A general framework to formulate elastic degradation and damage at small strain was presented by CAROL et al. [51] as an example. Extensions towards a formulation of large strains in a *Lagrangian* setting is presented in the following section. Assuming an initially transversal isotropic material behavior while being undamaged, an elastic constitutive formulation for large strain deformations is derived.

4.2.1 Transverse isotropic ground-state elasticity

Transversal isotropy is defined by the invariance of constitutive equations describing a material as a result of rotations around a principal symmetry axis $\boldsymbol{A}$ acting as the normal of the transversely isotropic plane. For the present case, this direction can conveniently be identified by the fiber axis. The principal symmetry axis in the reference configuration $\mathcal{B}_{\mathrm{m}}$ is described by a normed contra-variant vector $\boldsymbol{A} = \{A^I\}_{I=1,2,3}$; $\|\boldsymbol{A}\| = 1$. Its deformed counterpart in the current configuration $\mathcal{B}_{\mathrm{s}}(\mathcal{B}_0)$ can be written as a contra-variant vector $\boldsymbol{a} = \{a^i\}_{i=1,2,3}$. The two representations are connected via the deformation gradient $\boldsymbol{F}$ by the tangent map

$$\boldsymbol{a} = \boldsymbol{F}\boldsymbol{A}. \tag{4.88}$$

Figure 4.4 depicts the valid kinematics of the principal symmetry axis of a transversal isotropic material in the framework of large deformations. The

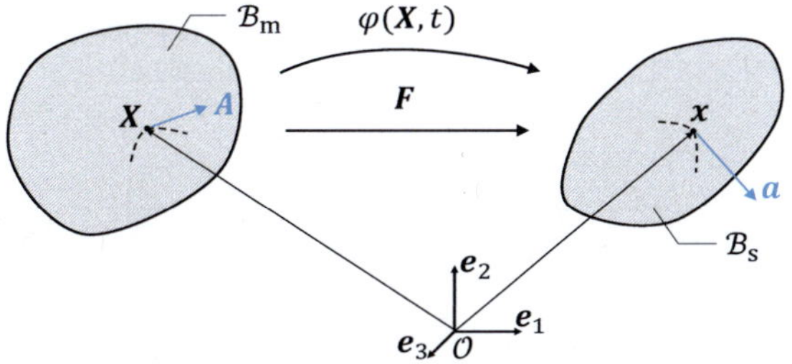

Figure 4.4: Kinematics of transversal isotropy at large deformations. Due to a deformation, the initial fiber direction $\boldsymbol{A}$ is transformed to a direction $\boldsymbol{a}$ in the current configuration. The deformation is described by the tangent map with the deformation gradient $\boldsymbol{F}$.

knowledge of the direction of local reinforcements allows the formulation of the (transversal isotropic) *structural tensors*

$$\boldsymbol{M} = \boldsymbol{A} \otimes \boldsymbol{A} \quad \text{or} \quad \boldsymbol{m} = \boldsymbol{a} \otimes \boldsymbol{a} \tag{4.89}$$

which are in general a function of the chosen coordinate system and the symmetry properties of the material under investigation. In a general formulation, the purely elastic free energy function Ψ^0 is introduced as a function of the right Cauchy-Green deformation tensor $\boldsymbol{C}$ and the structural tensor $\boldsymbol{M}$, thus directly containing the information about the materials orientation (cf. Schröder [346]). The requirement of objectivity states, that the free energy function is invariant with regard to orthonormal transformations $\boldsymbol{R}$ and therefore has to fulfill the condition (cf. TRUESDELL & NOLL [389], LURIE [242], SUHUBI [365], RIVLIN & ERICKSEN [336])

$$\Psi^0(\boldsymbol{C}, \boldsymbol{M}) = \Psi^0(\boldsymbol{R} * \boldsymbol{C}, \boldsymbol{R} * \boldsymbol{M}) \quad \forall \quad \boldsymbol{R} \in \mathbb{R}^3 \tag{4.90}$$

where the operator $*$ denotes the Rayleigh product. The first step in order to obtain a geometrically non-linear transversely isotropic material formulation is to substitute the deformation tensor of a linear formulation by the Green deformation tensor $\boldsymbol{E} = \frac{1}{2}(\boldsymbol{C} - \boldsymbol{G})$. $\boldsymbol{G}$ denotes the metric tensor of the underlying base system. For an isotropic case, this corresponds to the *St. Venant material*, according to SCHROEDER [346]. Furthermore, the author provides the extended formulation of the free energy function for a transversal isotropic material in a *Lagrangian setting* as a function of the Cauchy-Green strain

$$\Psi^0(\boldsymbol{C}, \boldsymbol{M}) = \Psi^0_{\text{iso}}(\boldsymbol{C}) + \Psi^0_{\text{aniso}}(\boldsymbol{C}, \boldsymbol{M}) \tag{4.91}$$

where the isotropic and anisotropic contributions can be written as

$$\begin{aligned} \Psi^0_{\text{iso}}(\boldsymbol{C}) &= \frac{1}{8}\lambda\left(\text{tr}[\boldsymbol{C}] - 3\right)^2 + \frac{1}{4}\mu_T\left(\text{tr}\left[\boldsymbol{C}^2\right] - 2\text{tr}[\boldsymbol{C}] + 3\right) \\ \Psi^0_{\text{aniso}}(\boldsymbol{C}, \boldsymbol{M}) &= \frac{1}{4}\alpha\left(\boldsymbol{ACA} - 1\right)\left(\text{tr}[\boldsymbol{C}] - 3\right) \\ &+ \frac{1}{8}\beta\left(\boldsymbol{ACA} - 1\right)^2 \\ &+ \frac{1}{2}\left(\mu_L - \mu_T\right)\left(\boldsymbol{AC}^2\boldsymbol{A} - 2\boldsymbol{ACA} + 1\right). \end{aligned} \tag{4.92}$$

The five independent material parameters λ, μ_T, μ_L, α and β hereby completely describe transversal isotropy. Applying *Coleman's method*, the second Piola-Kirchhoff stresses $\boldsymbol{S}^0$ can be identified following the standard arguments by

$$\begin{aligned} \boldsymbol{S}^0 &= 2\frac{\partial \Psi(\boldsymbol{C}, \boldsymbol{M})}{\partial \boldsymbol{C}} \\ &= 2\frac{\partial \Psi^0_{\text{iso}}(\boldsymbol{C})}{\partial \boldsymbol{C}} + 2\frac{\partial \Psi^0_{\text{aniso}}(\boldsymbol{C}, \boldsymbol{M})}{\partial \boldsymbol{C}} \\ &= \boldsymbol{S}^0_{\text{iso}} + \boldsymbol{S}^0_{\text{aniso}}. \end{aligned} \tag{4.93}$$

The contributions in the reference configuration read then

$$\begin{aligned} \boldsymbol{S}^0_{\text{iso}} &= \frac{1}{2}\lambda\left(\text{tr}\left[\boldsymbol{C}\right] - 3\right)\left(\boldsymbol{G}^{-1}\right) + \mu_T\left(\left(\boldsymbol{G}^{-1}\right)\boldsymbol{C}\left(\boldsymbol{G}^{-1}\right) - \left(\boldsymbol{G}^{-1}\right)\right) \\ \boldsymbol{S}^0_{\text{aniso}} &= \left[\frac{1}{2}\alpha\left(\text{tr}\left[\boldsymbol{C}\right] - 3\right) - \left(\mu_L - \mu_T\right) + \frac{1}{2}\beta\left(\boldsymbol{ACA} - 1\right)\right]\boldsymbol{A}\otimes\boldsymbol{A} \\ &\quad + \frac{1}{2}\alpha\left(\boldsymbol{ACA} - 1\right)\left(\boldsymbol{G}^{-1}\right) + \frac{1}{2}(\mu_L - \mu_T)\Xi_{\text{S}} \end{aligned} \tag{4.94}$$

with

$$\begin{aligned} \Xi^{EF}_{\text{S}} &= \left(A^E\left(G^{-1}\right)^{FC} + A^F\left(G^{-1}\right)^{EC}\right)C_{CD}A^D \\ &\quad + A^{\text{A}}C_{AB}\left(\left(G^{-1}\right)^{BE}A^F + \left(G^{-1}\right)^{BF}A^E\right). \end{aligned} \tag{4.95}$$

The Kirchhoff stresses are obtained by a *push-forward* operation ($\triangleright$) of $\boldsymbol{S}$ onto the current configuration

$$\begin{aligned} \boldsymbol{\tau}^0(\boldsymbol{b}, \boldsymbol{m}) &= \boldsymbol{F} \triangleright \boldsymbol{S}^0(\boldsymbol{C}, \boldsymbol{M}) \\ &= \boldsymbol{F}\boldsymbol{S}^0(\boldsymbol{C}, \boldsymbol{M})\boldsymbol{F}^{\text{T}} \\ &= \boldsymbol{\tau}^0_{\text{iso}}(\boldsymbol{b}) + \boldsymbol{\tau}^0_{\text{aniso}}(\boldsymbol{b}, \boldsymbol{m}). \end{aligned} \tag{4.96}$$

A detailed notation of $\boldsymbol{\tau}^0_{\text{iso}}$ and $\boldsymbol{\tau}^0_{\text{aniso}}$ as a function of the *finger tensor* $\boldsymbol{b}$ and the deformed principal symmetry axis $\boldsymbol{a}$ is given in SCHRÖDER [346] and is not repeated here. Further differentiation of Eq. 4.93 with respect to the deformation measure gives the transversal isotropic elasticity tensor

$$\mathbb{C}^0 = 4\frac{\partial^2 \Psi^0(\boldsymbol{C},\boldsymbol{M})}{\partial \boldsymbol{C}^2} = \mathbb{C}^0_{\text{iso}} + \mathbb{C}^0_{\text{aniso}}. \tag{4.97}$$

The contributions can be written as

$$\begin{aligned}
\mathbb{C}^0_{\text{iso}} &= \lambda\left(\boldsymbol{G}^{-1}\right)\otimes\left(\boldsymbol{G}^{-1}\right) + 2\mu_T\,\mathbb{I}^{\text{sym}} \\
\mathbb{C}^0_{\text{aniso}}(\boldsymbol{M}) &= \alpha\left(\boldsymbol{A}\otimes\boldsymbol{A}\otimes\left(\boldsymbol{G}^{-1}\right)+\left(\boldsymbol{G}^{-1}\right)\otimes\boldsymbol{A}\otimes\boldsymbol{A}\right) \\
&\quad + \beta\boldsymbol{A}\otimes\boldsymbol{A}\otimes\boldsymbol{A}\otimes\boldsymbol{A} + \frac{1}{2}\left(\mu_L-\mu_T\right)\mathbb{X}
\end{aligned} \tag{4.98}$$

with

$$\left(\mathbb{I}^{\text{sym}}\right)^{EFGH} = \frac{1}{2}\left(\left(G^{-1}\right)^{EH}\left(G^{-1}\right)^{FG} + \left(G^{-1}\right)^{EG}\left(G^{-1}\right)^{FH}\right) \tag{4.99}$$

and

$$\begin{aligned}
X^{EFGH} &= A^E\left(G^{-1}\right)^{FG}A^H + A^E\left(G^{-1}\right)^{FH}A^G \\
&\quad + A^F\left(G^{-1}\right)^{EG}A^H + A^F\left(G^{-1}\right)^{EH}A^G \\
&\quad + A^G\left(G^{-1}\right)^{HE}A^F + A^G\left(G^{-1}\right)^{HF}A^E \\
&\quad + A^H\left(G^{-1}\right)^{GE}A^F + A^H\left(G^{-1}\right)^{GF}A^E.
\end{aligned} \tag{4.100}$$

The spatial counterparts are again obtained by performing a *push-forward* operation according to SCHRÖDER [346] as

$$\mathbb{c}(\boldsymbol{b},\boldsymbol{m}) = \boldsymbol{F} \rhd \mathbb{C}^0(\boldsymbol{C},\boldsymbol{M}) = \mathbb{c}_{\text{iso}} + \mathbb{c}_{\text{aniso}}. \tag{4.101}$$

The formulation of the elastic moduli according to Eq. 4.97 yields a structure known from *Hooke's law* at small deformations

$$\boldsymbol{S}^0 = \mathbb{C}^0(\boldsymbol{M}) : \boldsymbol{E}. \tag{4.102}$$

Equation 4.102 allows the computation of the linear elastic material behavior in a geometrically non-linear framework. $\mathbb{C}^0(\boldsymbol{M})$ depicts the standard representation of a pristine transversely isotropic stiffness tensor, specified by the five independent material parameters and the structural tensor $\boldsymbol{M}$. Written in Nye notation, and assuming the preferred direction to be $A^I = [1, 0, 0]^{\mathrm{T}}$, the symmetry properties become apparent

$$\underline{\underline{C}}^0 = \begin{bmatrix} \lambda + \beta - 2(\mu_T + \alpha + 2\mu_L) & \lambda + \alpha & \lambda + \alpha & 0 & 0 & 0 \\ \lambda + \alpha & \lambda + 2\mu_T & \lambda & 0 & 0 & 0 \\ \lambda + \alpha & \lambda & \lambda + 2\mu_T & 0 & 0 & 0 \\ 0 & 0 & 0 & \mu_L & 0 & 0 \\ 0 & 0 & 0 & 0 & \mu_L & 0 \\ 0 & 0 & 0 & 0 & 0 & \mu_T \end{bmatrix}. \tag{4.103}$$

A comparison of the coefficients with the stiffness tensor consisting of engineering constants ($E_{\parallel} = E_1$, $E_{\perp} = E_2 = E_3$, $G_{\parallel} = G_{12} = G_{13}$, $G_{\perp} = G_{23}$, and the in-plane Poisson's ratio $\nu = \nu_{12} = \nu_{13}$ and the isotropy assumption $\nu_{\perp} = \nu_{23} = \nu_{32}$) yields the following relations

$$\begin{aligned} \alpha &= -\frac{2\left(\left(1 + \dfrac{(\nu - 1)}{2G_{\perp}}\right) E_{\parallel} - E_{\perp}\nu^2\right) G_{\perp}}{\left(\frac{1}{2}\dfrac{E_{\perp}}{G_{\perp}} - 2\right) E_{\parallel} + 2E_{\perp}\nu^2} \\ \beta &= \frac{2G_{\perp}\beta^*}{\left(\left(\frac{1}{2}\dfrac{E_{\perp}}{G_{\perp}} - 2\right) E_{\parallel} + 2E_{\perp}\nu^2\right) E_{\perp}} \\ \lambda &= \frac{2\left(E_{\perp}\nu^2 + E_{\parallel}\left(\dfrac{1}{2}\dfrac{E_{\perp}}{G_{\perp}} - 1\right)\right) G_{\perp}}{\left(\dfrac{1}{2}\dfrac{E_{\perp}}{G_{\perp}} - 2\right) E_{\parallel} + 2E_{\perp}\nu^2} \\ \mu_L &= G_{\parallel}, \quad \mu_T = G_{\perp}. \end{aligned} \tag{4.104}$$

with

$$\beta^* = \left(\frac{1}{4}\frac{E_\perp^2}{G_\perp^2} - \frac{E_\perp}{G_\perp}\right) E_\parallel^2 + \nu^2 \left(1 - 4\frac{G_\parallel}{G_\perp}\right) E_\perp^2 + \left(\left(\frac{\nu}{G_\perp} - \frac{G_\parallel}{G_\perp^2}\right) E_\perp^2 + \left(4\frac{G_\parallel}{G_\perp} - 1\right) E_\perp\right) E_\parallel . \tag{4.105}$$

With the elastic moduli, the corresponding stresses can be computed.

4.2.2 Failure mechanisms and 3D failure criteria

The evolution of damage depends on the strain states during loading and is therefore a continuous process. The onset of this progression is determined by the evaluation of failure criteria, which indicate a single event of failure in the material. Such formulations allow the extension towards a progressive damage formulation, thus introducing corresponding damage modes. The mechanical failure behavior of unidirectional composites and major failure modes have been described in Section 2.3.2. As a result taken from these observations, it can be concluded that failure occurs in different forms and sizes, strongly depending on the active loading condition and the loading direction. One of the major issues for the homogenized modeling of the onset and propagation of damage in fiber reinforced plastics is the setup of constitutive equations that capture the materials microstructural behavior and associated anisotropic failure behavior. The main task of a realistic computational model is therefore to establish if, when and how damage occurs and to give a sound prediction of how the material behavior changes with further damage evolution. For the present case the anisotropic manner of failure has to be respected and reflected in the choice of relevant failure criteria. Hereby, four different failure modes are considered which are summarized in Table 4.1. The manifestation of each mode has to be predictable for any given three-dimensional loading case and is treated as a distinct state in the load history, where a certain combination of stress components reaches a maximum. Therefore, in analogy to the procedure already introduced in Section 4.1, scalar functions (criteria) are formulated for each failure mode. Due to their predominant importance, a combination of the three-dimensional failure criteria by Puck [319] and the LaRC04 criterion (cf. Pinho [315]), modified according to Maimí et al. [247], has

Table 4.1: Considered failure modes in the material model for unidirectional fiber-reinforced composite. Display of nomenclature used throughout this work as well as a model representation of the failure manifestation. Images taken and modified from KNOPS [195].

Notation	Failure mode	Manifestation
FF^+	Tensile fiber failure	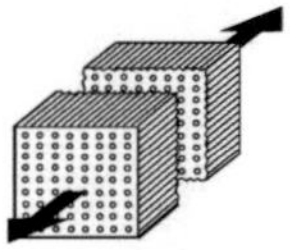
FF^-	Compressive fiber failure	
IFF^+	Tensile inter-fiber failure	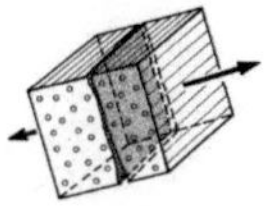
IFF^-	Compressive inter-fiber failure	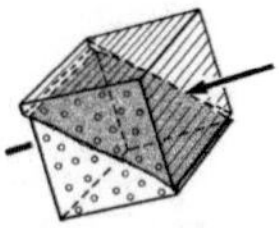

been chosen. In order to distinguish whether a certain loading scenario leads to failure, the stress-dependent scalar *effort value* ϕ_i is introduced, where $i \in \{FF^+, FF^-, IFF^+, IFF^-\}$ represents the different failure modes. The failure planes for modes I and II are related to failure in fiber direction whereas modes III and IV are associated with the matrix dominated failure occurring on one distinct *action plane*. The latter has to be chosen from the set of the transversal isotropic planes and is identified as the plane with the highest probability of failure, following the theory of Puck's action plane concept. The stresses acting on the action plane are displayed in Fig. 2.9. The orientation of the potential fracture plane is characterized by the action plane angle Θ, giving the inclination of the normal $\boldsymbol{x}_n$. Table 4.2 summarizes the implemented effort functions that characterize fiber and inter-fiber failure. Please note, that for the given case of rotations around the fiber axis, the normal and shear stress

Table 4.2: Implemented effort functions to characterize fiber and inter-fiber failure. In the present model four different failure mechanisms are considered: Fiber failure due to tensile and compressive load, ϕ_{FF^+} and ϕ_{FF^-} (cf. MAIMÍ et al. [247], DEUSCHLE [90]), and inter-fiber failure due to tensile and compressive load, ϕ_{IFF^+} and ϕ_{IFF^-} (cf. PUCK [319]).

Mode	Effort value	Condition
FF^+	$\phi_{FF^+} = \left(\frac{S_{11}}{R_{\parallel}^{+}}\right)^2$	$S_{11} \geq 0$
FF^-	$\phi_{FF^-} = \left(\frac{S_{11}}{R_{\parallel}^{-}}\right)^2$	$S_{11} < 0$
IFF^+	$\phi_{IFF^+} = \sqrt{\left(\frac{1}{R_{\perp}^{+}} - \frac{p_{\perp\Psi}^{+}}{R_{\perp\Psi}^{A}}\right)^2 S_{nn}^2 + \left(\frac{S_{nt}}{R_{\perp\perp}^{A}}\right)^2 + \left(\frac{S_{n1}^{-}}{R_{\perp\parallel}^{A}}\right)^2} + \frac{p_{\perp\Psi}^{+}}{R_{\perp\Psi}^{A}} S_{nn}$	$S_{nn} \geq 0$
IFF^-	$\phi_{IFF^-} = \sqrt{\left(\frac{p_{\perp\Psi}^{-}}{R_{\perp\Psi}^{A}}\right)^2 S_{nn}^2 + \left(\frac{S_{nt}}{R_{\perp\perp}^{A}}\right)^2 + \left(\frac{S_{n1}^{-}}{R_{\perp\parallel}^{A}}\right)^2} + \frac{p_{\perp\Psi}^{-}}{R_{\perp\Psi}^{A}} S_{nn}$	$S_{nn} < 0$

components S_{nn}, S_{n1}, and S_{nt} on the tangential action planes are functions of S_{22}, S_{33}, S_{12}, S_{13}, and S_{23} but not S_{11}. Hence failure modes III and IV are independent of stresses in fiber direction. On the other hand, failure in fiber direction (I and II) is solely caused by S_{11}. In order to take into account differing tensile and compressive failure behavior, the given criteria are formulated unilaterally. These equations put into context the active stress state $\boldsymbol{S}$ with a set of strength parameters, valid in different directions depending on the orientation of the action plane. In fiber direction the damage onset is characterized by the tensile and compressive strength, $R_{\parallel}^{+}$ and $R_{\parallel}^{-}$ respectively. The choice of such effort functions is a strong simplification, neglecting kinking and buckling effects of fibers, especially under compression. Perpendicular to the fibers, Puck's proposition of three-dimensional effort functions are used, incorporating several strength parameters valid on the transversal isotropic planes. According to PUCK [319] $R_{\perp}^{+}$ is the resistance of the action plane against failure due to normal tensile loading S_{nn}^{+}, $R_{\perp\perp}^{\mathrm{A}}$ the resistance of the action plane against failure due to shear on the transversal plane S_{nt}, and $R_{\perp\parallel}^{\mathrm{A}}$ the resistance of the action plane against failure due to shear in fiber direction S_{n1}. A graphical display of the failure surfaces regarding inter-fiber failure in the domain of second Piola-Kirchhoff stresses as well as the indication of the corresponding strength parameters is given in Fig. 4.5. Additionally, an

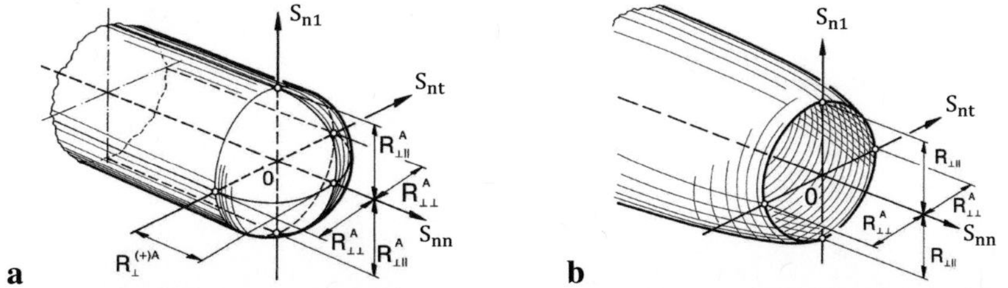

Figure 4.5: Failure surfaces for inter-fiber fracture formulated in terms of second Piola-Kirchhoff stresses and depiction of material strength parameters. (a) Capped failure surface under tension. (b) Paraboloidal failure surface for compressive failure. Images taken and modified from PUCK [319].

interpolation term including $R_{\perp\Psi}^{\mathrm{A}}$ is defined by

$$\frac{p_{\perp\Psi}^{\pm}}{R_{\perp\Psi}^{\mathrm{A}}} = \frac{p_{\perp\perp}^{\pm}}{R_{\perp\perp}^{\mathrm{A}}} \cos^2 \Psi + \frac{p_{\perp\parallel}^{\pm}}{R_{\perp\perp}^{\mathrm{A}}} \sin^2 \Psi \tag{4.106}$$

with

$$\Psi = \arctan \frac{S_{n1}}{S_{nt}} \quad \text{and} \quad p_{\perp\perp}^{-} = -\frac{\cos\left[2\hat{\Theta}_{\mathrm{fp}}\right]}{1+\cos\left[2\hat{\Theta}_{\mathrm{fp}}\right]}. \tag{4.107}$$

Hereby $\hat{\Theta}_{\mathrm{fp}}$ corresponds to the measured action plane angle appearing under transverse compressive failure. Furthermore, it is assumed that $p_{\perp\perp}^{+} = p_{\perp\perp}^{-}$. $p_{\perp\parallel}^{\pm}$ are the inclination parameters introduced by Puck [319].

4.2.3 Modelling ultimate damage

In order to assess damage initiation, the *damage activation function* F_i is introduced, corresponding to the formulation of the failure criteria

$$F_i = \phi_i - r_i \leq 0 \quad \forall \quad i \in \left\{FF^+, FF^-, IFF^+, IFF^-\right\}. \tag{4.108}$$

If F_i is less than 0, the material sees no further damage due to the corresponding failure mode (i). When F_i reaches 0, damage evolution is active and a set of damage variables is computed. The damage threshold value r_i, initially 1, must satisfy the Karush-Kuhn-Tucker conditions

$$\dot{r}_i \geq 0, \quad \text{and} \quad F_i \leq 0, \quad \text{and} \quad \dot{r}_i F_i = 0 \tag{4.109}$$

in order to guarantee thermodynamic consistency. Furthermore, the consistency condition

$$\dot{F}_i = 0 \quad \text{if} \quad F_i = 0 \tag{4.110}$$

must be fulfilled at all times. For the evolution of the threshold values, the above equations can be summarized by

$$\dot{r}_i = \begin{cases} 0 & \text{if} \quad F_i < 0 \\ \dot{\phi}_i & \text{if} \quad F_i = 0 \end{cases}, \tag{4.111}$$

which according to SIMO & JU [353,354] can be integrated explicitly yielding the representation

$$r_i = \max\left\{1, \max_{\tau\in[0,t]} \{\phi_i^\tau\}\right\} \tag{4.112}$$

with t being the total time. In analogy to a plasticity formulation, the failure criteria may be interpreted as loading criteria, with the threshold variables r_i substituting the yield stress (cf. MATZENMILLER [257]). The maximum value of ϕ_i in time will define the value of the corresponding damage variable as outlined later on.

4.2.3.1 Damage effect and the concept of effective stresses

Degradation of material stiffness due to damage effects in a continuum framework is understood as the average effect of microcracks. In accordance with the work of KACHANOV [183, 184], only the undamaged material between the voids is subjected to load and therefore transmit stresses. *Effective stresses* $\tilde{\boldsymbol{S}}$ and *effective strains* $\tilde{\boldsymbol{E}}$ are introduced to quantify those entities. To this end, the behavior of the remaining (undamaged) material skeleton can be written as

$$\tilde{\boldsymbol{S}} = \mathbb{C}^0 : \tilde{\boldsymbol{E}} \quad \text{and} \quad \tilde{\boldsymbol{E}} = \mathbb{H}^0 : \tilde{\boldsymbol{S}}, \tag{4.113}$$

where $\mathbb{H}^0 = \left(\mathbb{C}^0\right)^{-1}$ corresponds to the undamaged compliance tensor and $\mathbb{C}^0$ is the undamaged stiffness tensor already defined in Eq. 4.102. According to MATZENMILLER [257] the formulation of the failure criteria should consequently depend on the effective entities rather than their nominal counterparts. The task of the damage variables is to relate effective quantities to their externally measured (*nominal*) counterparts. Basically three possibilities are available to establish this relationship, strain, stress and energy equivalence (cf. LEMAÎTRE & CHABOCHE [220,223], CORDEBOIS & SIDOROFF [77], CHOW & WANG [72] among others). Most common is the usage of the theory of strain equivalence, where it is assumed that effective and nominal strains are equal and stresses differ. Therefore, Eq. 4.113 simplifies to

$$\tilde{\boldsymbol{S}} = \mathbb{C}^0 : \boldsymbol{E} \quad \text{and} \quad \boldsymbol{E} = \mathbb{H}^0 : \tilde{\boldsymbol{S}}. \tag{4.114}$$

The effect of damage is introduced by the fourth-order *damage effect tensors* $\mathbb{M}\,(\mathcal{D})$ and $\mathbb{B}\,(\mathcal{D})$ as functions of the damage state $\mathcal{D}$, yielding

$$\boldsymbol{S} = \left[(\mathbb{M})^{-1} : \mathbb{C}^0\right] : \boldsymbol{E} \quad \text{and} \quad \boldsymbol{E} = \left[\mathbb{H}^0 : \mathbb{M}\right] : \boldsymbol{S}, \tag{4.115}$$

since

$$\tilde{\boldsymbol{S}} = \mathbb{M} : \boldsymbol{S} \quad \text{and} \quad \boldsymbol{S} = (\mathbb{M})^{-1} : \tilde{\boldsymbol{S}} = \mathbb{B} : \tilde{\boldsymbol{S}} \tag{4.116}$$

define the simple relationship between effective and nominal stresses. Despite the known shortcoming in terms of symmetry of the secant stiffness tensor (cf. Carol et al. [52,53]), this approach is pursued throughout this work due to a possible straightforward derivation and implementation. The notion of unilateral failure criteria motivates the introduction of two different crack systems with associated action planes, active under tensile ($\mathcal{D}^+$) and compressive loading ($\mathcal{D}^-$) respectively. In this way the asymmetric behavior under tension and compression is considered. Thus, Eq. 4.115_1 can be rewritten as

$$\begin{aligned} \boldsymbol{S} &= \left[\left(\mathbb{P}^+_{\boldsymbol{E}} : \mathbb{B}^+\left(\mathcal{D}^+\right) + \mathbb{P}^-_{\boldsymbol{E}} : \mathbb{B}^-\left(\mathcal{D}^-\right)\right) : \mathbb{C}^0\right] : \boldsymbol{E} \\ &= \left[\mathbb{B}^+\left(\mathcal{D}^+\right) : \mathbb{C}^0\right] : \boldsymbol{E}^+ + \left[\mathbb{B}^-\left(\mathcal{D}^-\right) : \mathbb{C}^0\right] : \boldsymbol{E}^-, \end{aligned} \tag{4.117}$$

with the projection tensor

$$\mathbb{P}^\pm_{\boldsymbol{E}} = \frac{\partial}{\partial \boldsymbol{E}}\left[\sum_{I=1}^{3} \langle E_I \rangle_\pm \boldsymbol{n}_I \otimes \boldsymbol{n}_I\right] \tag{4.118}$$

yielding positive or negative contributions of the strain tensor $\boldsymbol{E}$

$$\boldsymbol{E}^\pm = \mathbb{P}^\pm_{\boldsymbol{E}} : \boldsymbol{E}. \tag{4.119}$$

Hereby, E_i and $\boldsymbol{n}_i$ are the eigenvalues and eigenvectors of the strain tensor $\boldsymbol{E}$. Hence, damage effects active under tensile load manifest only if the strain tensor contains tensile entries and vice versa. Following the arguments of Carol et al. [52], Murakami et al. [285], Cordebois & Sidoroff [77] and Murakami [283], a symmetric second-order damage tensor is chosen to characterize the internal damage state $\mathcal{D} \equiv \boldsymbol{D}^\pm$. The index $\pm$ allows again the distinction between damage states active under tensile or compressive conditions. Modes I and III are respected in $\boldsymbol{D}^+$, whereas $\boldsymbol{D}^-$ contains damage variables that evolve due to damage progression as a result of damage modes II and IV. However, the notation is neglected for the next few derivations. $\boldsymbol{D}$ varies between zero at the beginning, corresponding to an undamaged state and unity at the fully damaged state. An equivalent notation is introduced by the *integrity tensor*

$$\boldsymbol{\Upsilon} = \mathbb{1} - \boldsymbol{D} \mathrel{\hat{=}} \delta_{ij} - D_{ij} \tag{4.120}$$

behaving in the inverted manner (cf. Betten [34] and Valanis [398]). With the nominal-effective relations given in Eq. 4.116 the damage effect tensors can be identified. In order to ensure symmetry of these terms, a product-type symmetrization is applied, originally proposed by Cordebois & Sidoroff [77]. This induces the introduction of the square root terms $\boldsymbol{\omega}(\boldsymbol{D})$ of the integrity tensor

$$\boldsymbol{\Upsilon} = \boldsymbol{\omega}^2 \mathrel{\hat{=}} \omega_{ik}\omega_{kj}. \tag{4.121}$$

Rewriting Eq. 4.116_2 yields therefore

$$S_{ij} = \omega_{ik}\tilde{S}_{kl}\omega_{lj}. \tag{4.122}$$

Taking into consideration the symmetry of $\tilde{\boldsymbol{S}}$, the damage effect tensor $\mathbb{B}$ with all minor and major symmetries is obtained by

$$\mathbb{B} \mathrel{\hat{=}} \frac{1}{2}\left(\omega_{ik}\omega_{jl} + \omega_{il}\omega_{jk}\right). \tag{4.123}$$

Carefully note the similarity with the symmetric fourth-order identity tensor. The damage effect tensor in the principal axes of damage in Nye's notation then reads

$$\underline{\underline{\mathbb{B}}} = \begin{bmatrix} \Upsilon_1 & & & & & \\ & \Upsilon_2 & & & & \\ & & \Upsilon_3 & & & \\ & & & \sqrt{\Upsilon_1\Upsilon_2} & & \\ & & & & \sqrt{\Upsilon_1\Upsilon_3} & \\ & & & & & \sqrt{\Upsilon_2\Upsilon_3} \end{bmatrix} \tag{4.124}$$

The damaged stiffness tensor can then be formulated as

$$\mathbb{C} = \mathbb{B} : \mathbb{C}^0 = \left[\mathbb{P}_{\boldsymbol{E}}^{+} : \mathbb{B}^{+}\left(\boldsymbol{D}^{+}\right) + \mathbb{P}_{\boldsymbol{E}}^{-} : \mathbb{B}^{-}\left(\boldsymbol{D}^{-}\right)\right] : \mathbb{C}^0. \tag{4.125}$$

4.2.3.2 Damage evolution law and damage update

Having selected the form of the damage effect tensors, the upcoming derivations deal with the definition of damage evolution. Following the example of the damage evolution law already defined for the matrix model (cf. Eq. 4.46), the damage variables $\bar{d}_i^*$ valid on the action plane Θ (indicated by $\bar{\cdot}$) are calculated according to

$$\bar{d}_i^* = \bar{d}_i^*\left(\phi_i\left(\Theta\right)\right) = 1 - \frac{1}{\phi_i}\exp\left(A_i(\Theta)\left(1-\phi_i\right)\right). \tag{4.126}$$

A reduction of mesh sensitivity during damage localization is obtained by application of the crack band model according to Bažant & Oh [19]. Therefore, the adjustment parameters A_i are calculated in such a manner that the equation for dissipated energy density g_i for a monotonous uniaxial test

$$g_i = \int_0^\infty \frac{\partial\Psi}{\partial\bar{d}_i^*}\dot{\bar{d}}_i^*\mathrm{d}t = \int_1^\infty \frac{\partial\Psi}{\partial\bar{d}_i^*}\frac{\partial\bar{d}_i^*}{\partial\phi_i}\mathrm{d}\phi_i = \frac{G_i}{l^*} \tag{4.127}$$

is fulfilled at all times. Hereby, l^* is the characteristic length of the respective finite element and G_i is the fracture toughness for the associated loading case. The numerical implementation is described in more detail in Maimí et al. [245] and outlined in Appendix C. Using the damage variables $\bar{d}_i^*$ computed on the active action plane, the trial damage tensors $\bar{\boldsymbol{D}}^*$ is obtained. It has three orthogonal principal directions $\bar{\boldsymbol{n}}_I$ with $I \in [1,2,3]$ and the corresponding principal values $\bar{D}_I^*$. This formulation allows the interpretation of the damage variables as a measure for the net area reduction due to damage. It reads

$$\bar{\boldsymbol{D}}^* = \sum_{I=1}^{3} D_I^*\boldsymbol{n}_I^* \otimes \boldsymbol{n}_I^* = \bar{D}_{ij}^*\bar{\boldsymbol{e}}_i \otimes \bar{\boldsymbol{e}}_j = \begin{pmatrix} \bar{d}_1^* & 0 & 0 \\ 0 & \bar{d}_n^* & 0 \\ 0 & 0 & 0 \end{pmatrix} \bar{\boldsymbol{e}}_i \otimes \bar{\boldsymbol{e}}_j. \tag{4.128}$$

The base vectors $\bar{\boldsymbol{n}}_I = \{\bar{\boldsymbol{e}}_i\}_{i=1,2,3}$ define the coordinates of the action plane system inclined by the angle Θ and correspond to the principal axes of the trial damage. The damage variable $\bar{d}_1^*$ corresponds to the failure modes I and II, $\bar{d}_n^*$ describes the damage state due to inter-fiber damage evolution (modes III and IV). The damage tensors from Eq. 4.128 are then compared to the damage tensors $\boldsymbol{D}^{\mathrm{n}}$ induced during the load history. In order to obtain thermodynamic consistency ($\dot{\boldsymbol{D}} > \boldsymbol{0}$), the damage update has the following form in a general case

$$\begin{aligned}\boldsymbol{D} \Leftarrow \boldsymbol{D}^{\mathrm{n}} + \Delta \boldsymbol{D}\left(\bar{\boldsymbol{D}}^*\right) &= \boldsymbol{D}^{\mathrm{n}} + \begin{pmatrix} \langle \bar{d}_1^* - \bar{d}_1^{\mathrm{n}} \rangle_+ & 0 & 0 \\ 0 & \langle \bar{d}_n^* - \bar{d}_n^{\mathrm{n}} \rangle_+ & 0 \\ 0 & 0 & 0 \end{pmatrix} \bar{\boldsymbol{e}}_i \otimes \bar{\boldsymbol{e}}_j \\ &= \begin{pmatrix} \bar{d}_1 & 0 & 0 \\ 0 & \bar{d}_n & \bar{d}_{nt}^{\mathrm{n}} \\ 0 & \bar{d}_{nt}^{\mathrm{n}} & \bar{d}_t^{\mathrm{n}} \end{pmatrix} \bar{\boldsymbol{e}}_i \otimes \bar{\boldsymbol{e}}_j .\end{aligned} \tag{4.129}$$

Together with the knowledge of Θ, $\Delta\bar{\boldsymbol{D}}$ does not only contain information about the magnitude, but also the direction of damage evolution.Therefore, $\bar{\boldsymbol{D}}$ describes the change of symmetry class and increasing anisotropy. Rewriting the damage tensor from Eq. 4.129 in the material base coordinates and in the principal system yields

$$\boldsymbol{D} = \begin{pmatrix} d_1 & 0 & 0 \\ 0 & d_n & d_{nt} \\ 0 & d_{nt} & d_t \end{pmatrix} \boldsymbol{e}_i \otimes \boldsymbol{e}_j = \sum_{I=1}^{3} D_I \boldsymbol{n}_I \otimes \boldsymbol{n}_I . \tag{4.130}$$

At this point it is also considered that the damage effect caused under compressive load is activated under tension as well, since the introduced cracks open under tension, whereas tensile damage is passive under compressive load. Depending on the previous state of damage $\boldsymbol{D}^{\mathrm{n}}$ the resulting principal system of damage does not have to be conform with the principal system of the trial damage, i.e. in general $\boldsymbol{n}_I \neq \boldsymbol{n}_I^*$. Carefully note, that the index $\pm$ has not been written here for reasons of better readability, but applies to all damage entities. In analogy to Eq. 4.125 the damaged stiffness tensor on the action plane can thus be formulated according to

$$\bar{\mathbb{C}} = \left[\bar{\mathbb{P}}_{\bar{\boldsymbol{E}}}^+ : \bar{\mathbb{B}}^+\left(\bar{\boldsymbol{D}}^+\right) + \bar{\mathbb{P}}_{\bar{\boldsymbol{E}}}^- : \bar{\mathbb{B}}^-\left(\bar{\boldsymbol{D}}^-\right)\right] : \mathbb{C}^0 \tag{4.131}$$

with $\bar{\mathbb{C}} = \mathbb{C}^0$ for $\boldsymbol{D} = 0$. Back-transformation yields the damaged stiffness tensor in the material system

$$\mathbb{C} = \boldsymbol{R}^{\mathrm{T}}(\Theta) * \bar{\mathbb{C}}. \tag{4.132}$$

The damaged stresses can then finally be written as

$$\boldsymbol{S} = \mathbb{C} : \boldsymbol{E}. \tag{4.133}$$

4.2.4 Algorithmic treatment

The proposed continuum damage model is implemented in the implicit finite element software Abaqus/Standard using the user-defined material subroutine *UMAT* (cf. Simulia [355]). An algorithmically simplified counterpart of the stress response from Eq. 4.133 is employed. Hereby, the structural tensor $\boldsymbol{M}$ with the corresponding direction of the reinforcement $\boldsymbol{A}$ is handled as an input variable (cf. Eq. 4.89). The material stiffness takes the specific form

$$\mathbb{C}\left(\boldsymbol{E}, \boldsymbol{E}^{\mathrm{n}}\right) = \boldsymbol{R}^{\mathrm{T}}\left(\Theta\left(\boldsymbol{E}^{\mathrm{n}}\right)\right) * \bar{\mathbb{C}}\left(\boldsymbol{E}, \boldsymbol{E}^{\mathrm{n}}\right). \tag{4.134}$$

The projection tensors from Eq. 4.131 are redefined as $\mathbb{P}^{\pm}_{\boldsymbol{E}^{\mathrm{n}}}$, being a function of the deformation state $\bar{\boldsymbol{E}}^{\mathrm{n}}$ on the action plane from the previous time increment n. The analogous assumption applies for the computation of the action plane angle $\Theta = \Theta\left(\bar{\boldsymbol{E}}^{\mathrm{n}}\right) = \Theta_{\mathrm{n}}$, and thus also the adjustment parameters $A_i\left(\Theta_{\mathrm{n}}\right)$, resulting in a significantly simplified formulation for the global tangent operator as demonstrated later on. Schirmaier et al. [344] recently proposed an enhanced method to accelerate the search for the action plane angle Θ. However, in this implementation an arbitrary number of $m = 360$ effort values $\phi_j\left(\bar{\boldsymbol{E}}^{\mathrm{n}}, \Theta_j\right)$ with $j \in [IFF^+, IFF^-]$ at different given $\Theta_j \in [-90°, 90°]$ are calculated for every time step. Finally, the configuration resulting in the maximum ϕ_j is chosen to be the valid action plane angle Θ. A straight forward convergence of the calculation in the implicit framework of Abaqus necessitates the definition of the global tangent operator, defined as

$$\begin{aligned}\mathbb{C}^{\mathrm{t}} = \Delta_{\boldsymbol{E}}\boldsymbol{S} &= \frac{\mathrm{d}\boldsymbol{S}\left(\boldsymbol{E},\boldsymbol{E}^{\mathrm{n}}\right)}{\mathrm{d}\boldsymbol{E}} \\ &= \frac{\mathrm{d}}{\mathrm{d}\boldsymbol{E}}\left[\mathbb{C}\left(\boldsymbol{E},\boldsymbol{E}^{\mathrm{n}}\right):\boldsymbol{E}\right] \\ &= \mathbb{C}\left(\boldsymbol{E},\boldsymbol{E}^{\mathrm{n}}\right) + \frac{\partial\mathbb{C}\left(\boldsymbol{E},\boldsymbol{E}^{\mathrm{n}}\right)}{\partial\boldsymbol{E}}:\boldsymbol{E}.\end{aligned} \tag{4.135}$$

The partial derivation of the damaged stiffness tensor $\mathbb{C}$ with respect to the deformation yields a sixth-order tensor

$$\begin{aligned}\frac{\partial\mathbb{C}}{\partial\boldsymbol{E}} &= \frac{\partial}{\partial\boldsymbol{E}}\left[\boldsymbol{R}^{\mathrm{T}}\left(\Theta_{\mathrm{n}}\right) * \bar{\mathbb{C}}\left(\boldsymbol{E},\boldsymbol{E}_{\mathrm{n}}\right)\right] \\ &\mathrel{\hat{=}} \frac{\partial}{\partial E^{AB}}\left[\left(R^{C}{}_{D}\right)^{\mathrm{T}}\left(R^{E}{}_{F}\right)^{\mathrm{T}}\left(R^{G}{}_{H}\right)^{\mathrm{T}}\left(R^{I}{}_{J}\right)^{\mathrm{T}}\bar{\mathbb{C}}_{CEGI}\right] \\ &= \left(R^{C}{}_{D}\right)^{\mathrm{T}}\left(R^{E}{}_{F}\right)^{\mathrm{T}}\left(R^{G}{}_{H}\right)^{\mathrm{T}}\left(R^{I}{}_{J}\right)^{\mathrm{T}}\frac{\partial\bar{\mathbb{C}}_{CEGI}}{\partial E^{AB}} \\ &= \left(R^{C}{}_{D}\right)^{\mathrm{T}}\left(R^{E}{}_{F}\right)^{\mathrm{T}}\left(R^{G}{}_{H}\right)^{\mathrm{T}}\left(R^{I}{}_{J}\right)^{\mathrm{T}}\mathbb{A}_{CEGIAB}.\end{aligned} \tag{4.136}$$

Keep in mind that the transformation tensors $\boldsymbol{R}\left(\Theta_{\mathrm{n}}\right)$ are independent of the current deformation state. The partial derivation of the active stiffness tensor on the action plane in Eq. 4.136 can be further simplified to

$$\frac{\partial\bar{\mathbb{C}}}{\partial\boldsymbol{E}} = \frac{\partial\bar{\mathbb{C}}}{\partial\bar{\boldsymbol{E}}}:\frac{\partial\bar{\boldsymbol{E}}}{\partial\boldsymbol{E}} \tag{4.137}$$

with

$$\frac{\partial\bar{\boldsymbol{E}}}{\partial\boldsymbol{E}} = \frac{\partial\left(\boldsymbol{R}\left(\Theta_{\mathrm{n}}\right) * \boldsymbol{E}\right)}{\partial\boldsymbol{E}} \mathrel{\hat{=}} R^{O}{}_{I}R^{P}{}_{J}\mathbb{I}_{OPMN}. \tag{4.138}$$

With $\bar{\mathbb{C}}\left(\bar{\boldsymbol{E}}, \bar{\boldsymbol{E}}_{\mathrm{n}}\right) = \left[\bar{\mathbb{P}}^{+}_{\bar{\boldsymbol{E}}_{\mathrm{n}}} : \bar{\mathbb{B}}^{+}\left(\bar{\boldsymbol{E}}\right) + \bar{\mathbb{P}}^{-}_{\bar{\boldsymbol{E}}_{\mathrm{n}}} : \bar{\mathbb{B}}^{-}\left(\bar{\boldsymbol{E}}\right)\right] : \mathbb{C}^{0}$, only the derivations of the damage effect tensor $\bar{\mathbb{B}}^{\pm}$ with respect to the deformation (on the action plane) stay undetermined

$$\frac{\partial \bar{\mathbb{C}}\left(\bar{\boldsymbol{E}}, \bar{\boldsymbol{E}}_{\mathrm{n}}\right)}{\partial \bar{\boldsymbol{E}}} = \left[\bar{\mathbb{P}}^{+}_{\bar{\boldsymbol{E}}_{\mathrm{n}}} : \frac{\partial \bar{\mathbb{B}}^{+}\left(\bar{\boldsymbol{E}}\right)}{\partial \bar{\boldsymbol{E}}} + \bar{\mathbb{P}}^{-}_{\bar{\boldsymbol{E}}_{\mathrm{n}}} : \frac{\partial \bar{\mathbb{B}}^{-}\left(\bar{\boldsymbol{E}}\right)}{\partial \bar{\boldsymbol{E}}}\right] : \mathbb{C}^{0}. \tag{4.139}$$

Analogously for damage effects active under compression and tension, the exploitation of the chain rule then yields

$$\frac{\partial \bar{\mathbb{B}}\left(\bar{\boldsymbol{E}}\right)}{\partial \bar{\boldsymbol{E}}} = \frac{\partial \bar{\mathbb{B}}\left(\bar{\boldsymbol{D}}\right)}{\partial \bar{\boldsymbol{D}}} : \frac{\partial \bar{\boldsymbol{D}}\left(\bar{\boldsymbol{E}}\right)}{\partial \bar{\boldsymbol{E}}} \tag{4.140}$$

and further

$$\frac{\partial \bar{\mathbb{B}}\left(\bar{\boldsymbol{E}}\right)}{\partial \bar{\boldsymbol{D}}} = \frac{\partial \bar{\mathbb{B}}\left(\bar{\boldsymbol{E}}\right)}{\partial \bar{\omega}} : \frac{\partial \bar{\omega}}{\partial \bar{\boldsymbol{\Upsilon}}} : \frac{\partial \bar{\boldsymbol{\Upsilon}}}{\partial \bar{\boldsymbol{D}}}. \tag{4.141}$$

Taking into consideration Eqs. 4.120 and 4.125, the derivations

$$\begin{aligned} \frac{\partial \bar{\mathbb{B}}\left(\bar{\boldsymbol{E}}\right)}{\partial \bar{\omega}} \mathrel{\hat{=}} \frac{1}{2}\Bigg(& \frac{\partial \bar{\omega}_{IK}}{\partial \bar{\omega}_{MN}} \bar{\omega}_{JL} + \bar{\omega}_{IK} \frac{\partial \bar{\omega}_{JL}}{\partial \bar{\omega}_{MN}} \\ & + \frac{\partial \bar{\omega}_{IL}}{\partial \bar{\omega}_{MN}} \bar{\omega}_{JK} + \bar{\omega}_{IL} \frac{\partial \bar{\omega}_{JK}}{\partial \bar{\omega}_{MN}} \Bigg) \end{aligned} \tag{4.142}$$

and

$$\frac{\partial \bar{\boldsymbol{\Upsilon}}}{\partial \bar{\boldsymbol{D}}} = -\mathbb{I} \tag{4.143}$$

are rather easy to determine. The remaining contribution in Eq. 4.141 can be rewritten using the eigenvalue decomposition of $\bar{\boldsymbol{\Upsilon}}$ as

$$\frac{\partial \bar{\omega}}{\partial \bar{\boldsymbol{\Upsilon}}} = \frac{\partial}{\partial \bar{\boldsymbol{\Upsilon}}} \left[\sum_{I=1}^{3} \left(\bar{\Upsilon}_{I}\right)^{1/2} \boldsymbol{n}_{I} \otimes \boldsymbol{n}_{I} \right]. \tag{4.144}$$

An algorithmic treatment for the derivation of isotropic tensor functions can be found in MIEHE [274]. The missing contribution in Eq. 4.140 can be identified by

$$\frac{\partial \bar{\boldsymbol{D}}\left(\phi_i\left(\bar{\boldsymbol{E}}\right)\right)}{\partial \bar{\boldsymbol{E}}} = \frac{\partial \bar{\boldsymbol{D}}}{\partial \phi_i} : \frac{\partial \phi_i}{\partial \bar{\boldsymbol{E}}} = \frac{\partial \bar{\boldsymbol{D}}}{\partial \phi_i} : \frac{\partial \phi_i}{\partial \tilde{\boldsymbol{S}}} : \mathbb{C}^0 \tag{4.145}$$

using the corresponding case dependent definitions for the effort functions ϕ_i which were formulated in the effective stress $\tilde{\boldsymbol{S}}\left(\bar{\boldsymbol{E}}\right)$ on the action plane. After the computation of the damaged stresses and moduli a push-forward operation is performed, yielding the corresponding entities ($\boldsymbol{\tau}$ and $\mathbb{c}$) on the current configuration. Regarding an implementation of the material model using an Abaqus user material subroutine (*UMAT*), the true stresses (Cauchy stresses)

$$\boldsymbol{\sigma} = \frac{\boldsymbol{\tau}}{J} \tag{4.146}$$

and the corresponding moduli associated to the Jaumann derivation are needed (cf. MIEHE [266] and BAASER et al. [20]). The latter can be written as

$$\tilde{\mathbb{c}}^{abcd} = \left[\mathbb{c}^{abcd} + \delta^{ac}\tau^{bd} + \tau^{ac}\delta^{bd}\right] / J \tag{4.147}$$

where $J = \det \boldsymbol{F}$ is the determinant of the deformation gradient $\boldsymbol{F}$.

4.3 Constitutive modeling of weave reinforced plastics

Organo sheets consist of a fabric structure embedded in a thermoplastic matrix material. These are layered composites, where each layer consist of two sets of interwoven rovings (weft and warp), set up in a twill weave pattern. The undulated glass-fiber reinforcement structures define preferred directions in the material as depicted in Fig. 4.6. Materials with preferred directions perpendicular to each other are associated with orthotropic symmetry. Deviations from this lead to different groups of symmetry, e.g. monoclinic symmetry. The principal directions of the reinforcements are assumed to be initially known, e.g.

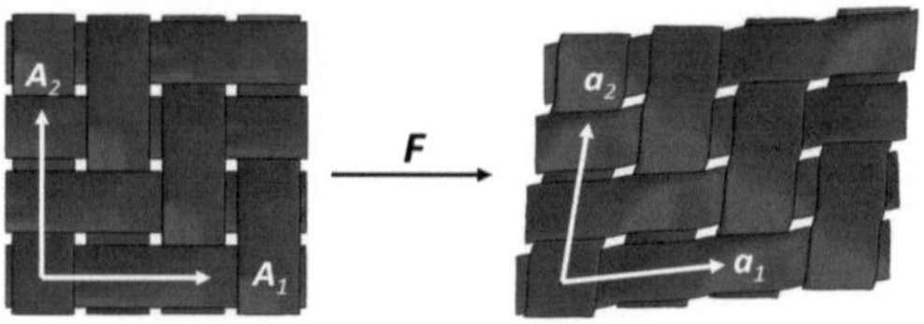

Figure 4.6: Change of material orientation due to deformation of a weave reinforced composite. The display of the matrix domain is omitted.

by drape simulation or assumptions. However, the application of external mechanical loads results in a finite change of reinforcement orientation, containing both reversible and irreversible contributions and hence establishing a differing stress-free state upon unloading. Hereby, the matrix behavior is related to plastic deformation processes in the matrix-rich regions of the composite material, i.e. in the interstices of the rovings. The work proposed here concentrates on the composite formulation of a constitutive model taking into account the aforementioned kinematic observations. In crystal plasticity it is often assumed that the initially known preferred directions (i.e. the crystal latices) do not change with the material directly (cf. Rice [334], Kröner & Teodosiu [206] and Mandel [250]). In terms of finite plasticity, this means that the structural tensors in the intermediate configuration stay constant and the corresponding yield condition is formulated in terms of the *Schmidt stresses* $\boldsymbol{\Sigma}$ (cf. Hill [162], Asaro [14], Cuitino & Ortiz [83] or Miehe [265], among others). For the present material system, the preferred directions coincide with the directions of the rovings which change during loading continuously due to material deformation. In his work, Miehe [268] states that for such cases a finite plasticity formulation based on the *plastic metric*

$$\boldsymbol{G}^{\mathrm{p}} = \boldsymbol{F}^{\mathrm{p}^{\mathrm{T}}} \bar{\boldsymbol{G}} \boldsymbol{F}^{\mathrm{p}} \tag{4.148}$$

can be formulated. Due to the occurring anisotropy within the chosen material class, it is hardly possible to find a sound overall formulation for plasticity and damage onset of the material. To overcome this problem, the approach of the model presented here is to superimpose two angled preferred directions, coupled through an isotropic matrix by means of analytical homogenization technique. Moreover, using the damage models presented before (cf. Section 4.1 and Section 4.2), the incorporation of mechanism-based damage formulations in both the reinforcements and the thermoplastic matrix is enabled. The overall consistency in terms of thermodynamics (dissipation inequality) is assumed to be preserved since the formulations for the phases were developed along these guidelines. To the knowledge of the author such an approach, combining analytical homogenization techniques with damage formulations within the single phases, is the first of its kind, addressing the damage modeling of textile woven composites.

4.3.1 Material orientation and intermediate configuration

In general, the initial structural tensors of a weave can be written as $\boldsymbol{M}_i = \boldsymbol{A}_i \otimes \boldsymbol{A}_i$ for each preferred direction $i \in [1, 2]$ as displayed in the undeformed weave in Fig. 4.6. Given an arbitrary deformation expressed in terms of the deformation gradient $\boldsymbol{F} = \boldsymbol{F}^{\mathrm{e}} \boldsymbol{F}^{\mathrm{p}}$, the material orientation may change and will now be aligned with the deformed structural tensors $\boldsymbol{m}_i = \boldsymbol{a}_i \otimes \boldsymbol{a}_i$. The vectors $\boldsymbol{a}_i$ correspond to a deformed preferred direction (cf. Fig. 4.6, right) and are defined by

$$\boldsymbol{a}_i = \boldsymbol{F} \boldsymbol{A}_i. \tag{4.149}$$

Figure 4.7 depicts that behavior schematically. For thermoplastic basic con-

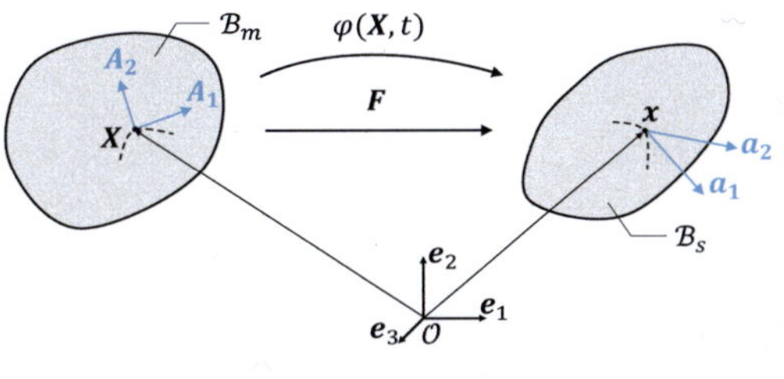

Figure 4.7: Weave kinematics at large deformations. The initial material orientation $\boldsymbol{A}_i$ in the reference configuration is mapped on its deformed representation $\boldsymbol{a}_i$ on the current configuration by means of the deformation gradient $\boldsymbol{F}$.

stituents, inelastic behavior during loading prior damage effects is observed. In the following it is assumed that this behavior is attributed solely to plastic effects in the matrix (phase index m), resulting in a plastic share of the deformation gradient $\boldsymbol{F}^{\mathrm{p}}_{\mathrm{m}}$. In the present case, plasticity in the matrix model is introduced additively according to Eq. 4.1. Consulting again Section 4.1.6, a relation of the logarithmic plastic strains $\boldsymbol{\varepsilon}^{\mathrm{p}} = \sum_{I=1}^{3} \varepsilon_I^{\mathrm{p}} \boldsymbol{n}_I \otimes \boldsymbol{n}_I$ and the plastic metric tensor $\boldsymbol{G}^{\mathrm{p}}$ (cf. Eqs. 4.83 and 4.148) was defined by MIEHE [277]. Combining the above mentioned equations, a formulation for the plastic contribution of the deformation gradient can be found. With the assumption that only the thermoplastic matrix (α =m) outside of the tows is objected to plastic effects, it reads

$$\boldsymbol{F}^{\mathrm{p}} = \boldsymbol{F}_{\mathrm{m}}^{\mathrm{p}} = \sum_{I=1}^{3} F_I^{\mathrm{p}} \boldsymbol{n}_I \otimes \boldsymbol{n}_I = \sqrt{\boldsymbol{G}^{\mathrm{p}}} \tag{4.150}$$

with the eigenvalues

$$F_I^{\mathrm{p}} = \left[\exp\left[2\varepsilon_I^{\mathrm{p}}\right]\right]^{1/2}. \tag{4.151}$$

This kind of formulation yields a symmetric form for $\boldsymbol{F}^{\mathrm{p}}$, an observation that was also reported by REESE [333]. As a consequence, the plastic spin remains undetermined, similar to isotropic elastoplasticity formulations (cf. REESE [332, 333]). The possibility to split the deformation gradient in an elastic and plastic contribution gives rise to the introduction of the plastic intermediate configuration in the *co-/contravariant* domains $\bar{\mathcal{B}}$ and $\bar{\mathcal{B}}^*$ for both metrics and stresses (cf. Fig. 4.8). Please recall that $\mathcal{B}_{\mathrm{m}}$ is the reference or

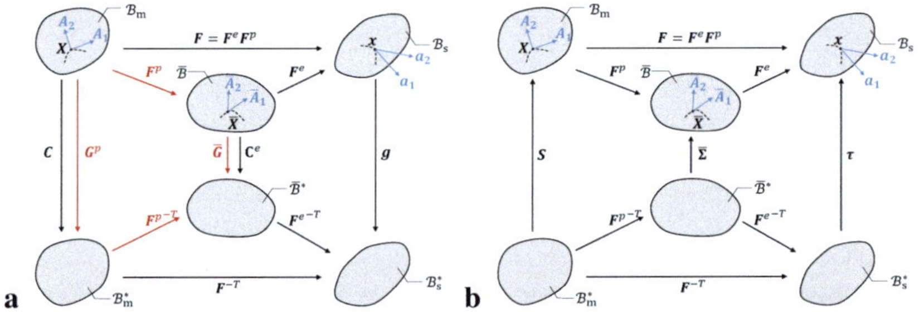

Figure 4.8: Schematic display of *co-/contravariant* domains for (a) metric and (b) stresses in analogy to Fig. 2.2. The plastic intermediate configuration is a function of the plastic share of the deformation gradient $\boldsymbol{F}^{\mathrm{p}} = \boldsymbol{F}_{\mathrm{m}}^{\mathrm{p}}$ due to inelastic effects in the matrix material. Failure of tows is evaluated on this configuration in terms of the *Schmidt stresses* $\bar{\boldsymbol{\Sigma}}$.

material configuration, whereas $\mathcal{B}_{\mathrm{s}}$ denotes the current or spatial configuration. Both entities are defined in the *co-variant* domain. Their counterparts in the *contra-variant* domain can be written as $\mathcal{B}_{\mathrm{m}}^*$ and $\mathcal{B}_{\mathrm{s}}^*$. The key concept of the presented model is the idea of embedding reinforcement structures in a thermoplastic matrix by separating the plastic deformation and the definition of the intermediate configuration. The elastic response and damage evaluation of the matrix is computed in the reference configuration. By contrast the elastic response of the reinforcement structures (index r_i) are calculated on the intermediate configuration. In Eq. 4.97 it has been shown that the elastic response of a transversal isotropic roving is a function of its structural tensor. For the present case, the normalized preferred directions $\bar{\boldsymbol{A}}_i^{\mathrm{n}}$ are used therefore for the weft and warp yarn ($i \in [1, 2]$), respectively. They can be denoted as

$$\bar{\boldsymbol{A}}_i^{\mathrm{n}} = \frac{\boldsymbol{F}^{\mathrm{p}} \boldsymbol{A}_i}{\|\boldsymbol{F}^{\mathrm{p}} \boldsymbol{A}_i\|}. \tag{4.152}$$

The right Cauchy-Green tensor $\boldsymbol{C} = \boldsymbol{F}^{\mathrm{T}} \boldsymbol{g} \boldsymbol{F}$ is chosen to describe the deformation for the overall definition of the continuous weave model in the reference configuration, where $\boldsymbol{g}$ is the metric tensor in the Eulerian configuration (cf. Fig. 4.8).

4.3.2 Combination of models from lower scales

Having defined the assumed geometrical framework, the computation of stresses and moduli of the composite follows. The scalar strain energy potential

$$\begin{aligned} \Psi = f\Big(& \nu^{\mathrm{m}}, \Psi^{\mathrm{m}} \left(\boldsymbol{C}, \boldsymbol{G}^{\mathrm{p}}(\boldsymbol{C}), d(\boldsymbol{C})\right), \\ & \nu^{\mathrm{r}_i}, \Psi^{\mathrm{r}_i} \left(\boldsymbol{C}^{\mathrm{e}}(\boldsymbol{G}^{\mathrm{p}}), \bar{\boldsymbol{A}}_i^{\mathrm{n}}(\boldsymbol{G}^{\mathrm{p}}), \boldsymbol{D}^{\pm}(\boldsymbol{C}^{\mathrm{e}})\right)\Big) \end{aligned} \tag{4.153}$$

is assumed, describing the overall composite as a function of matrix and reinforcement behavior. Hereby, Ψ^{m} corresponds to the strain energy function of the matrix defined within the finite strain framework extension defined in Section 4.1.6 and Ψ^{r_i} as the strain energy functions of the reinforcements r_i formulated on the plastic intermediate configuration dependent on $\boldsymbol{C}^{\mathrm{e}}$ (cf. Fig. 4.8). Carefully note that, while Ψ^{m} is independent of the preferred directions, Ψ^{r_i} takes $\bar{\boldsymbol{A}}_i^{\mathrm{n}}$ as input, hence, introducing preferred directions, since the ground state elasticity of the reinforcement structures are assumed to be initially transversely isotropic. v^{m} and v^{r_i} in Eq. 4.153 denote the volume fractions of the matrix and the two reinforcement phases respectively, with

$$v^{\mathrm{m}} + \sum_{i=1}^{n} v^{\mathrm{r}_i} = 1. \tag{4.154}$$

For the present work a balanced weave is assumed, yielding $v^{\mathrm{r}_i} = v_{\mathrm{tow}}/2$ (cf. Table 3.7). Standard arguments yield the second Piola-Kirchhoff formulation defined on the reference configuration for the stresses in the matrix according to

$$\boldsymbol{S}^{\mathrm{m}} = 2\frac{\partial \Psi^{\mathrm{m}}}{\partial \boldsymbol{C}} = \boldsymbol{S}^{\mathrm{m}}\left(\boldsymbol{C}, \boldsymbol{G}^{\mathrm{p}}\left(\boldsymbol{C}\right), d_{\mathrm{m}}\left(\boldsymbol{C}\right)\right). \tag{4.155}$$

Analogously, the global tangent operator can be expressed as

$$\mathbb{C}^{\mathrm{m}} = 4\frac{\partial^2 \Psi^{\mathrm{m}}}{\partial \boldsymbol{C}^2} = \mathbb{C}^{\mathrm{m}}\left(\boldsymbol{C}, \boldsymbol{G}^{\mathrm{p}}\left(\boldsymbol{C}\right), d_{\mathrm{m}}\left(\boldsymbol{C}\right)\right). \tag{4.156}$$

The material behavior of the tows is assessed on the intermediate configuration. Similar derivations, involving the exploitation of Coleman's method, yield the Schmidt stresses in the reinforcements

$$\bar{\boldsymbol{\Sigma}}^{\mathrm{r}_i} = 2\frac{\partial \Psi^{\mathrm{r}_i}}{\partial \boldsymbol{C}^{\mathrm{e}}} = \boldsymbol{S}^{\mathrm{r}_i}\left(\boldsymbol{C}^{\mathrm{e}}\left(\boldsymbol{G}^{\mathrm{p}}\right), \bar{\boldsymbol{A}}_i^{\mathrm{n}}\left(\boldsymbol{G}^{\mathrm{p}}\right), \boldsymbol{D}^{\pm}\left(\boldsymbol{C}^{\mathrm{e}}\right)\right). \tag{4.157}$$

A *de novo* derivation yields the corresponding moduli

$$\bar{\mathbb{C}}^{\mathrm{r}_i} = 4\frac{\partial^2 \Psi^{\mathrm{r}_i}}{\partial \boldsymbol{C}^{\mathrm{e}2}} = \bar{\mathbb{C}}^{\mathrm{r}_i}\left(\boldsymbol{C}^{\mathrm{e}}\left(\boldsymbol{G}^{\mathrm{p}}\right), \bar{\boldsymbol{A}}_i^{\mathrm{n}}\left(\boldsymbol{G}^{\mathrm{p}}\right), \boldsymbol{D}^{\pm}\left(\boldsymbol{C}^{\mathrm{e}}\right)\right). \tag{4.158}$$

In order to allow the superposition of all stresses, $\bar{\boldsymbol{\Sigma}}^{\mathrm{r}_i}$ and $\bar{\mathbb{C}}^{\mathrm{r}_i}$ are transferred to the Lagrangian setting, applying a *pull-back operation* ($\lhd$). The stresses and moduli in the reinforcements written in the reference configuration then read

$$\boldsymbol{S}^{\mathrm{r}_i} = \boldsymbol{F}^{\mathrm{p}} \lhd \bar{\boldsymbol{\Sigma}}^{\mathrm{r}_i} \qquad \text{and} \qquad \mathbb{C}^{\mathrm{r}_i} = \boldsymbol{F}^{\mathrm{p}} \lhd \bar{\mathbb{C}}^{\mathrm{r}_i}. \tag{4.159}$$

The assembly function f in Eq. 4.153 defines the used homogenization technique and therefore the combination rule of stresses and moduli. For reasons of simplicity, in the present work the assembly rule according to VOIGT [404] is used. This hypothesis serves here as a first guidance, even though the internal microstructure of the weave is not explicitly considered. Other homogenization schemes, taking into account these geometrical particularities, have been proposed by many different authors (cf. Section 2.4.3 for a short review). The homogenization scheme according to Voigt corresponds to the assumption that all phases experience the same uniform strain. As already reported, this assumption is known as the iso-strain assumption in literature. Together with the assumption of the stress-concentration tensor being $\mathbb{A}^{\mathrm{r}_i} = \mathbb{I}$, the Voigt formulation can be recovered from Eq. 2.57, yielding the overall formulation for the stresses and moduli according to

$$\boldsymbol{S} = v^{\mathrm{m}} \boldsymbol{S}^{\mathrm{m}} + \sum_{i=1}^{2} v^{\mathrm{r}_i} \boldsymbol{S}^{\mathrm{r}_i} \tag{4.160}$$

and

$$\mathbb{C} = v^{\mathrm{m}} \mathbb{C}^{\mathrm{m}} + \sum_{i=1}^{2} v^{\mathrm{r}_i} \mathbb{C}^{\mathrm{r}_i} . \tag{4.161}$$

The here proposed framework allows a straightforward implementation of the constitutive equations in Abaqus/Standard, using a user-defined material subroutine (UMAT). After the computation of Eqs. 4.160 and 4.161 defined on the reference configuration, a push-forward operation is necessary in order to obtain the Kirchhoff stresses $\boldsymbol{\tau}$ and the corresponding moduli $\mathbb{c}$. Subsequently, the transformations given in Eqs. 4.146 and 4.147 have to be employed to ensure compatibility with the Abaqus UMAT interface.

5 Simulations

The following section focuses on the compilation of the simulative results, where the structure specified by the multi-scale analysis was also taken into account. The starting point of the simulations was the generation of the microstructures (cf. Section 5.1) needed on the distinct scales. Beginning with a parameter fitting of the basic constituents, the calculations of the deformation and damage behavior of the unidirectional microstructure (cf. Section 5.2) and woven mesostructure (cf. Section 5.3) are presented. Finally, the features of the newly proposed macroscopic material model are demonstrated in Section 5.4. Its behavior with respect to the change of material orientation under deformation as well as the representation of damage evolution is reviewed in the course of exemplary virtual tests.

5.1 Generation of virtual microstructures

For the multiscale approach, two virtual microstructures are needed: the representative volume element characterizing the unidirectionally reinforced composites and the unit cells standing for the woven composite. For an easier handling of the simulation models, a *python* routine was developed in order to embed the routines of the microstructure generation into the preprocessing framework of ABAQUS. The application of case-dependent periodic boundary conditions (PBCs) is described in more detail in Appendix D. For both microstructures, existing generation methods were used. For the creation of topology data of UD-*SRVE*s it therefore is referred to the work of MELRO et al. [259] who developed a three-step methodology to generate UD microstructures. The single steps of one cycle are named by

1. Hard-core step: *Placing* new fibers in the control volume
2. Stirring step: Creating free space by *shaking* the present fibers
3. Compression step: *Compacting* of existing fibers towards the center

and are demonstrated in Fig. 5.1. Described is an iterative process which is repeated several times to achieve the desired fiber volume fractions. The

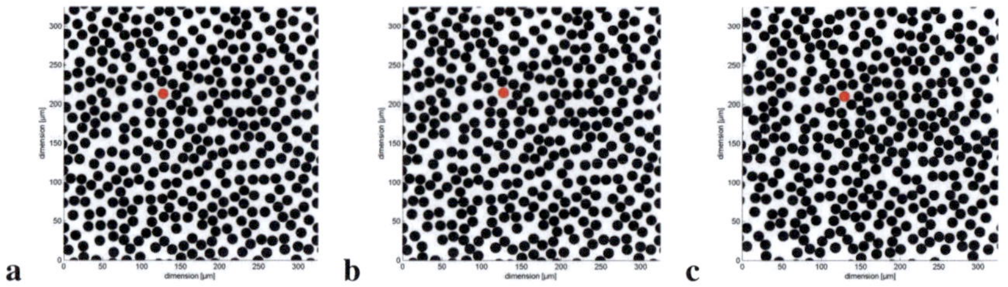

Figure 5.1: The three steps of one cycle of the generation routine by MELRO et al. [259]: (a) Hard-core step, (b) stirring step, and (c) compression step.

movements within one single cycle are rather small but due to the big amount of fibers a very dynamic system emerges. The purpose of steps two and three is the liberation of areas where new fibers could be placed. In such a way it is possible to attain UD cells with a high volume fraction of fibers. Figure 5.2 shows the progress of the routine over several cycles. A further feature for

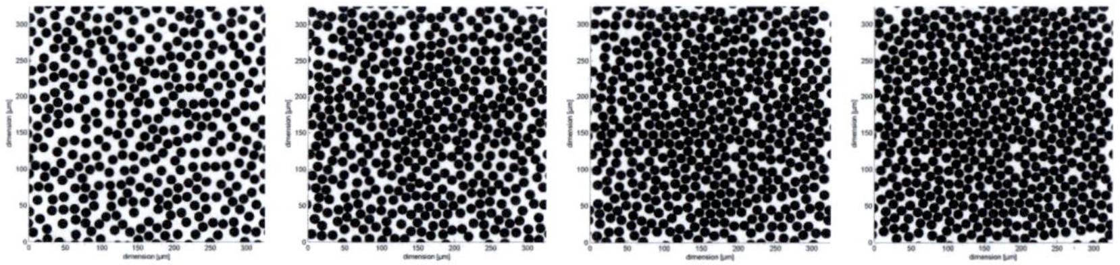

Figure 5.2: Evolving microstructures during the generation process. By repeating the three above-mentioned steps the volume fraction of the control volume increases until the designated target fraction is reached (here $v_f = 63\%$).

the generation of clustered microstructures was developed in the course of the present work. Figure 5.3 visualizes the process steps in order to obtain a clustered geometry. The starting point is a RVE with a high fiber volume fraction ($v_f = 63\%$, cf. Fig. 5.3a). The highlighted green areas in Fig. 5.3b indicate the randomly chosen cluster centers, which are free to overlap. In the present case a target fiber volume fraction of $v_f = 43\%$ was specified. The routine removes fibers from the RVE that are not in the area of the cluster centers until the desired fiber volume content is reached. The red fibers in Fig. 5.3b correspond to the inclusions to be removed. Finally, a control volume with areas of higher density of fibers and such with lower density forms that differs significantly from structures with a more even fiber distribution (cf. Fig. 5.4 and Fig. 5.3c). The routine by MELRO et al. [259] was furthermore

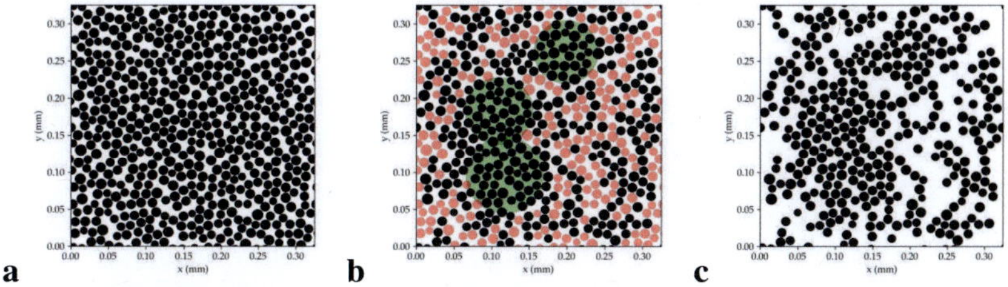

Figure 5.3: Sequence of actions during the generation of clustered microstructures. (a) Original microstructure with a high fiber volume fraction ($v_f = 63\%$). (b) Randomly placed cluster centers are highlighted in green. Fibers that are positioned within these centers stay untouched. The red highlighted fibers belong to the set of fibers that were randomly chosen to be subtracted from the geometry set. The removal process takes place as long as the desired fiber content does not fall below a specified value. (c) Resulting clustered UD-RVE with a volume fraction of $v_f = 43\%$.

extended towards the possibility to use normally distributed fiber diameters (cf. Fig. 5.4). The user can specify a standard deviation in addition to the nominal diameter. The basic principals of the original routine stay unchanged, only the eventual overlap of the fiber volumes has to be checked depending on the present diameters.

Remark. Despite the possibility of using clustered models with varying fiber diameters, these options were not used for the following investigations in order to achieve comparable data records, not being influenced by imperfections on the microstructure. This option builds the basis for future investigations. Carefully note, that due to the special case of the prevailing symmetry of UD materials, the problem of fiber placement in the presented cases is reduced to a two-dimensional one. The resulting output of the microstructure generation is therefore the information about the position of the fiber centers and its diameter as well as the size of the representative volume. A *python* routine, developed throughout this thesis, transfers the created two-dimensional input to ABAQUS, where a three-dimensional model is created by extrusion along the fiber axis. Further preprocessing operations are likewise carried out, such as the material assignment, meshing, and the identification of the sorted element sets on the boundaries of the RVE, which are crucial for a sound definition of the periodic boundary conditions. The connection between fibers and adjacent matrix was

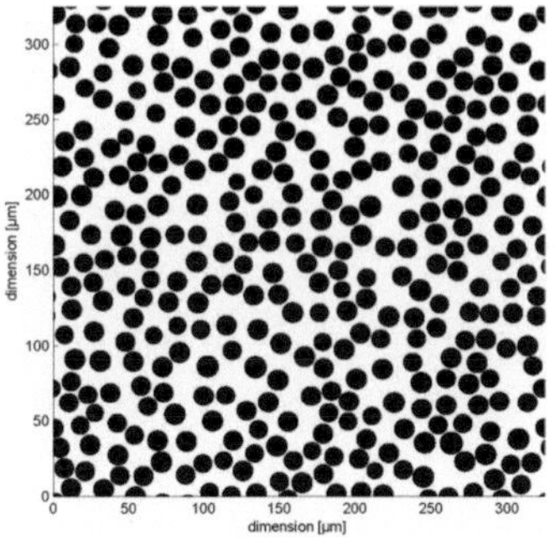

Figure 5.4: UD-SRVE with normally distributed diameters (v_f = 43%).

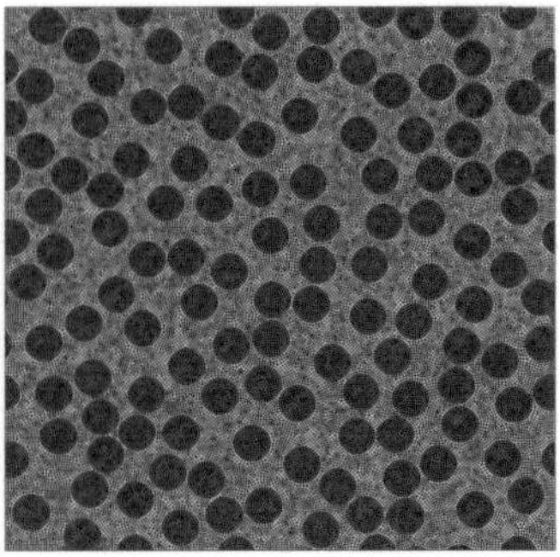

Figure 5.5: Discretized geometry of a UD-SRVE (δ = 30, v_f = 43%, d_f = 13 μm). HEX (*C3D20R*) and WEDGE (*C3D15*) elements were used.

modeled by a conform mesh. The use of cohesive zone elements has been dispensed with. During the experimental investigations, no evidence could be found that failure occurs in the fiber matrix adhesion (cf. Fig. 3.28), supporting this decision. Figure 5.5 shows a fully meshed UD-SRVE geometry. TRIAS et al. [387] performed a thorough study on the necessary size of RVEs, by using several criteria, such as fiber content, effective properties, stress and strain fields including the probability density function (PDF) of occurring stress and strain in the matrix , the Hill condition, and distance distributions. To this end, a parameter $\delta = a/R \geq 50$ was defined for the minimum size of SRVEs regarding the computation of plasticity and failure, defined by the RVE side length a and the radius of the inclusions R. Throughout this thesis, this factor was always respected. A smaller size was selected for the sake of a better representation for the upcoming plots. For the following simulations, the fiber diameter was defined to be $d_\mathrm{f} = 13\,\mu m$. UD-SRVEs for both relevant fiber volume fractions $v_{\mathrm{f},1}$ = 43% and $v_{\mathrm{f},2}$ = 63% were considered.

The virtual unit cells for the impregnated (*wet*) woven composites were created by using the software package *TexGen* (cf. SHERBURN [350]). With this tool, a straight-forward generation of impregnated and dry woven composites is possible. Thus, twill weaves were created considering the geometric dimensions measured during the experimental investigations (cf. Table 3.7). Hereby, the weave was considered to be balanced in terms of both geometry and material properties, i.e. warp and weft yarns have the same properties. The yarns are modeled as lenticular-shaped structures not being in contact with each other. The model is thickened artificially to ensure a meshing of the matrix in the space between two overlapping yarns. The surrounding matrix shares nodes with the yarns, hence setting up a conform mesh, where no other interface is necessary nor defined. Using the integrated meshing tool *TetGen*, a three-

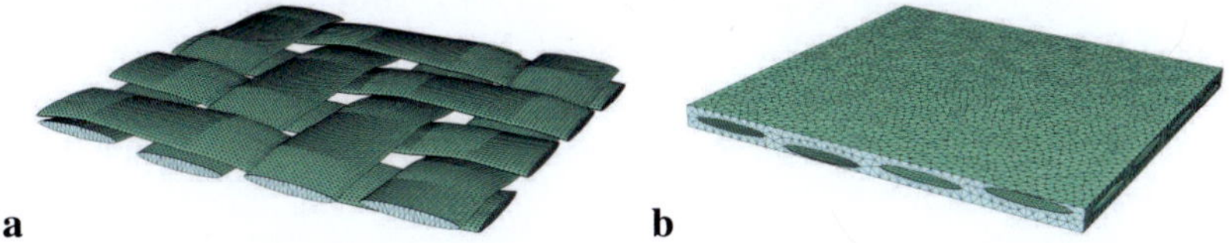

Figure 5.6: (a) Meshed reinforcement structures of a 2x2 twill weave. The single yarns are not in contact with each other but are separated by small gaps to ensure a successful meshing of the complete structure. (b) Meshed body of the surrounding matrix. The matrix is connected to the yarns by a conform mesh, thus sharing nodes on the surface of the yarns. A coarser mesh size is displayed here for the sake of a clearer representation.

dimensional meshed structure was obtained (cf. Fig. 5.6). Please note, that the presented mesh is not considered to be fine enough, but a coarser mesh was chosen here for a better display. Local material orientation, especially the fiber direction, plays an important role for the validity of the following simulations and therefore builds an essential input for the computation. *TexGen*

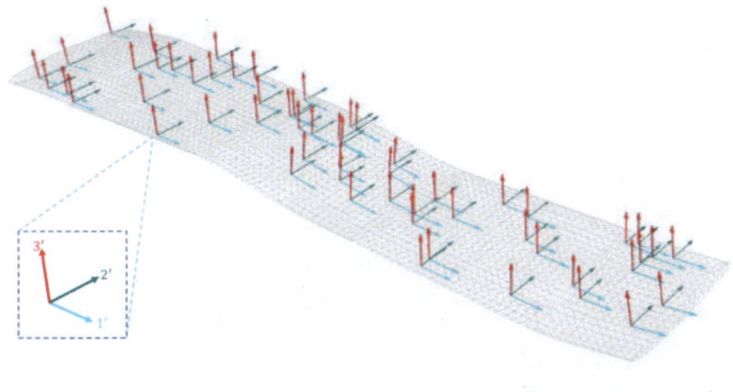

Figure 5.7: Imported material orientation of one single yarn. For each element of the yarns a local material system $\{1', 2', 3'\}$ is defined, where the $1'$-direction corresponds to the fiber axis.

also provides the local material orientation ($\{1', 2', 3'\}$) which can be imported in an ABAQUS computation as it can be seen in Fig. 5.7 for an exemplary single yarn of a 2x2 plain weave. Hereby, the $1'$-direction corresponds to the fiber axis, which follows the undulation of the yarn throughout its length. Hereby, the orientation of the rovings is defined piecewise linearly for each element. Instead of using local coordinate systems, the directional information $A_i \, \forall i \in [1, 3]$ is passed in the material routine as additional parameters. Internally, the results are first calculated with respect to the corresponding orientation system and afterwards transformed onto and saved in the global coordinate system. The use of *TexGen* was embedded in a *python* routine responsible for the completion of further preprocessing steps, necessary for a FE-simulation. Especially the identification of the element sets on the surface of the *unit cell* is taken care of. The user is free to define three-dimensional or in-plane periodic boundary conditions. For the simulation results using woven unit cells presented in this work, in-plane PBCs were used.

5.2 Computations on unidirectional unit cell

One principal idea of using a multiscale approach in order to assess and transfer composite behavior from micro to macro scales is the input of material data on the lowest scale. In the present case the thermoplastic polypropylene matrix and the glass fibers are the basic constituents that build up the basis of the following investigations. In the first step, a parameter identification procedure was carried out. The input for the fitting process was delivered by the experimental campaign presented in Section 3.2.2 and Section 3.2.3. Here, the polymeric matrix was tested under uniaxial tension and compression. The main results are summarized in Fig. 3.3 and Fig. 3.6 consisting of six curves for tension and four curves for compression loading at different loading rates. An elasto-viscoplastic material model including an ultimate damage formulation was presented in Section 4.1. The parameters for that constitutive model were adapted in order to minimize the resulting error during curve fitting. The available software package *LS-Opt* supplied the optimization framework. A polynomial meta model was created and adapted using the sequential response surface method (SRSM). The objective of the optimization was the minimization of the least-square error of the areas underneath the resulting calculated and experimental curves. Furthermore, the identification of the precise position of the failure onset $(\varepsilon_\mathrm{f}(\dot{\varepsilon}), R_\mathrm{m}(\dot{\varepsilon}))$ was an objective. Since the rate dependent parameters for tensile and compressive behavior show a strong connection, all objectives had to be solved in parallel. The resulting set of material parameters for the thermoplastic matrix are summarized in Table 5.1. When comparing the calculated material response and experimentally recorded curves, a good

Table 5.1: Identified material parameters for the thermoplastic matrix material and assumed parameters for the E-Glass fiber material. Parameters with superscript [1] are estimated and such with [2] are measured.

		Neat matrix		E-Glass fiber
		Tension (i=t)	Compression (i=c)	
Elasticity	E (in MPa)	1998.85		72000[1]
	ν (-)	0.42[2]		0.22[1]
Plasticity	ν_{p} (-)	0.42[1]		
Yield onset	$\sigma^0_{i,0}$ (in MPa)	2.62	9.83	
	$\sigma^0_{i,\infty}$ (in MPa)	14.61	21.86	
	$\eta^{\sigma^0_i}$ (-)	39.70	61.92	
	$\alpha^{\sigma^0_i}$ (-)	0.65	1.09	
Yield limit	$\sigma^\infty_{i,0}$ (in MPa)	29.91	44.93	
	$\sigma^\infty_{i,\infty}$ (in MPa)	40.82	74.65	
	$\eta^{\sigma^\infty_i}$ (-)	9.98	4.45	
	$\alpha^{\sigma^\infty_i}$ (-)	0.59	0.49	
Hardening evolution	$\omega_{i,0}$ (-)	70.05	64.98	
	$\omega_{i,\infty}$ (-)	101.72	65.87	
	η^{ω_i} (-)	150.29	150.29	
Damage onset	$X_{i,0}$ (in MPa)	29.86	44.37	2150[1]
	$X_{i,\infty}$ (in MPa)	39.34	73.96	2150[1]
	η^{X_i} (-)	14.06	4.33	
	α^{X_i} (-)	0.63	0.48	
Fracture toughness	K_{Ic} (in MPa/mm$^{\frac{1}{2}}$)	4.50[1]		0.90[1]

agreement is noted and applies both in tension and compression (cf. Figs. 5.8 and 5.10). The results show the basic properties of the implemented material model. All curves start with the same stiffness. The point where the material response becomes non-linear (yield initiation) is dependent on the strain rate, just like the behavior during increasing plasticization. Finally, the stress where the material starts to fail (ultimate failure initiation) varies with the strain rate. Figures 5.9 and 5.11 show a good prediction accuracy when it comes to ultimate

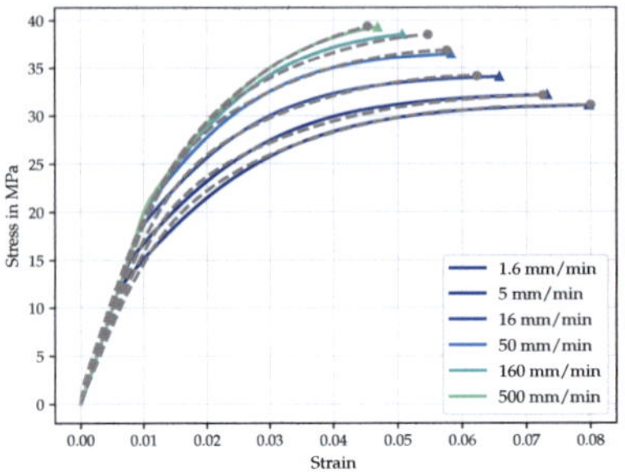

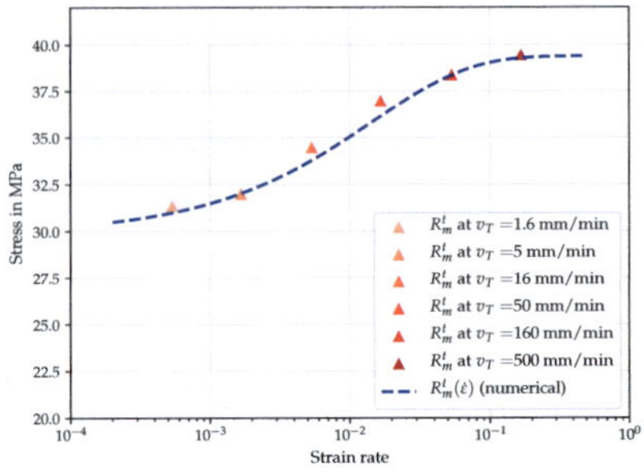

Figure 5.8: Comparison of the material model response (colored, solid lines) and the average experimental results (gray, dashed lines) taken from Fig. 3.3.

Figure 5.9: Comparison of measured ultimate strengths of the matrix material at different strain rates under tension with the model response. Please note, that no direct fitting on this data has been done.

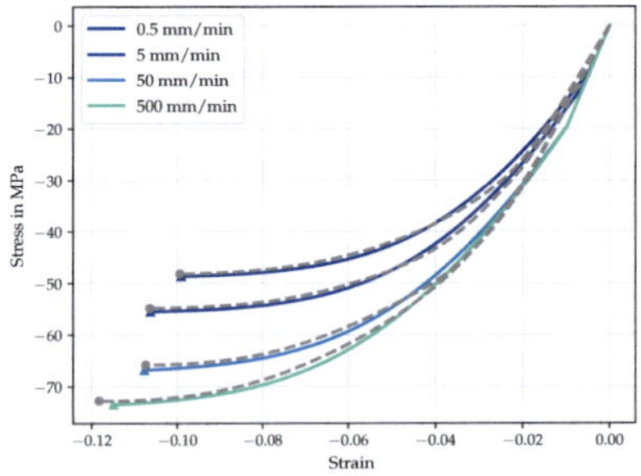

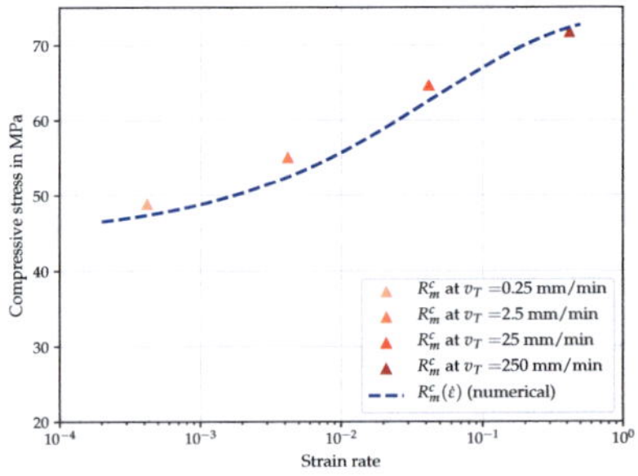

Figure 5.10: Comparison of the material model response (colored, solid lines) and the average experimental results (gray, dashed lines) taken from Fig. 3.6.

Figure 5.11: Comparison of measured ultimate strengths of the matrix material at different strain rates under compression with the model response. Please note, that no direct fitting on this data has been done.

strength, even though this parameter function was not a direct fitting objective but resulted from the curve fitting.

Remark. Since single glass fibers were not available from the material supplier, a detailed experimental investigation was not carried out. Standard material constants for E-glass fibers were taken as summarized in Table 5.1. However the same subroutine as for the matrix material could be used, deactivating the viscoplastic features, rendering a purely elastic model which includes the damage formulation.

5.2.1 Computation of elastic bodies

Having generated the virtual microstructures of the UD-SRVEs, modeled and characterized the isotropic basic constituents, the first step of simulation is the assessment of the elastic properties of the given unidirectionally reinforced microstructures under study. Six calculations corresponding to the six uniaxial load cases are sufficient to determine the full stiffness tensor. The equations of the numerical homogenization presented in Section 2.2.2 and the homogenization integrals in Eq. 2.46 relate the far-field strain and/or stress applied to the RVE with the help of the periodic boundary conditions with the resulting and more complex stress and strain states within the microstructure. Table 5.2 shows the averaged elastic properties for two given volume fractions ($v_{\mathrm{f}} = 43\%$ and $v_{\mathrm{f}} = 63\%$). In total, a number of $n = 10$ microstructures with a size ratio of $\delta = 50$ were used for each volume fraction. Assuming initial orthotropic symmetry, the stiffness tensor can be described by nine independent variables. After computation, the results show that the symmetry can be reduced further, since $\bar{E}_2 \approx \bar{E}_3$, $\bar{\nu}_{12} \approx \bar{\nu}_{13}$, $\bar{G}_{12} \approx \bar{G}_{13}$, and $\bar{G}_{23} \approx \bar{E}_2/(2(1+\bar{\nu}_{23}))$ which are the characteristics of a transversely isotropic material system. The small standard deviations obtained in this study show (cf. Table 5.2), that the calculation of the elastic properties by numerical homogenization is a stable process (using big RVEs) and the random placement of fibers within the microstructure does not play a predominant role at the tested volume fractions. Figure 5.12 shows a three-dimensional representation of the elastic stiffness body of the unidirectionally reinforced material under investigation with a fiber volume fraction of $v_{\mathrm{f}} = 63\%$. Similar forms are obtained with the respective properties of different volume fractions. It becomes clearly visible that one preferred direction is dominating the elastic behavior of the material. Furthermore the characteristic

Table 5.2: Average of computed elastic properties of $n = 10$ different unidirectionally reinforced microstructures with a fiber volume fracture of $v_f = 43\%$ and $v_f = 63\%$. The given nine parameters can be summarized to a reduced set showing the characteristics of a transversely isotropic medium.

	$v_f = 43\%$	$v_f = 63\%$
$\bar{E}_1$ (in MPa)	31923.01 ± 0.25	45924.83 ± 0.29
$\bar{E}_2$ (in MPa)	4972.70 ± 29.92	9724.11 ± 107.76
$\bar{E}_3$ (in MPa)	4966.66 ± 38.56	9702.05 ± 68.94
$\bar{\nu}_{12}$ (-)	0.322 ± 0.001	0.281 ± 0.001
$\bar{\nu}_{13}$ (-)	0.321 ± 0.001	0.280 ± 0.001
$\bar{\nu}_{23}$ (-)	0.595 ± 0.004	0.501 ± 0.003
$\bar{G}_{12}$ (in MPa)	1544.72 ± 9.64	2892.69 ± 26.06
$\bar{G}_{13}$ (in MPa)	1540.34 ± 10.51	2891.69 ± 20.35
$\bar{G}_{23}$ (in MPa)	1557.35 ± 10.42	3164.70 ± 32.74

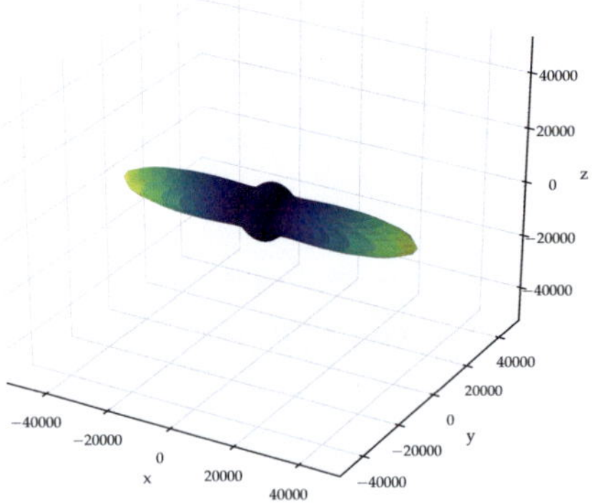

Figure 5.12: Three dimensional representation of the elastic body of the unidirectionally reinforced material with a volume fraction of $v_f = 63\%$. The elastic properties are given in Table 5.2.

of rotational symmetry around the preferred direction (x-axis) is recognized. By projecting the three-dimensional stiffness body onto one symmetry plane,

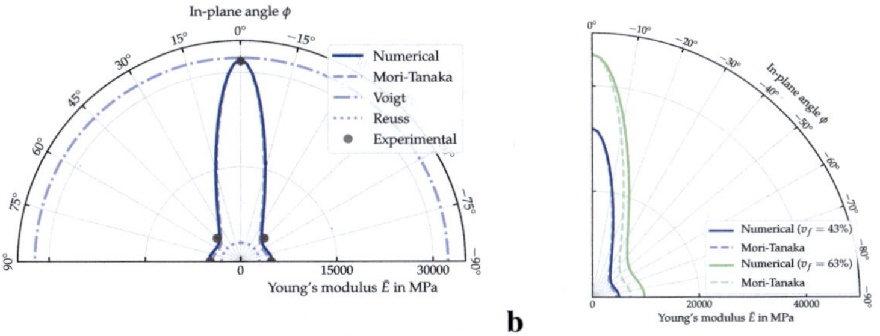

a **b**

Figure 5.13: Projection of the elastic stiffness body from Fig. 5.12 onto polar coordinates, where 0° corresponds to the preferred axis. (a) Comparison of results obtained by selected analytical homogenization techniques and the numerical solution as well as the averaged experimental data from Section 3.3. (b) Comparison of the numerical and Mori-Tanaka (UD) solution at $v_{\mathrm{f}} = 43\%$ and $v_{\mathrm{f}} = 63\%$.

the values of the direction dependent Young's moduli can be directly extracted. Figure 5.13a shows the projected stiffness body of a UD-SRVE with a volume fraction of $v_{\mathrm{f}} = 43\%$. Half symmetry is considered here, and the preferred axis of the composite is aligned with the indicated 0°- direction. Results from selected analytical as well as the numerical solution are depicted. With the numerical solution lying in the interspace of the Voigt and Reuss solution, a physically reasonable outcome can be stated. Please note that the Voigt and Reuss solution is isotropic (circle) since the input materials were assumed to be isotropic. Furthermore experimental data is available for the present volume fraction and is displayed here as an average of the Young's moduli measured at different strain rates (cf. Section 3.3). Comparison of the experimental results with both the numerical and the analytical Mori-Tanaka solution for the special case of UD materials yields a good agreement in terms of longitudinal and transverse Young's moduli. In Fig. 5.13b a comparison of the numerical results using UD-SRVEs and Mori-Tanaka solutions for two different volume fractions is displayed ($v_{\mathrm{f}} = 43\%$ (blue) and $v_{\mathrm{f}} = 63\%$ (green)). The higher number of fibers within the control volume yields an increase of stiffness in both longitudinal and transverse direction. While higher longitudinal stiffness is directly deducable by the increased cross-sectional area of fiber material, the reason for higher transverse stiffness lies in the micromechanical circumstances in the matrix between the fibers. With higher volume fractions the distance of single fibers diminishes and therefore the adjacent matrix is exposed to high multi-axial straining yielding a stiffer composite behavior. While the Mori-Tanaka solution lies in good agreement with the numerical solution for $v_f = 43\%$, the results for $v_f = 63\%$ show a significant deviation especially in the transverse direction. The Mori-Tanaka method assumes a dilute concentration of inclusions and therefore yields an underestimation of the influence of single fibers towards each other. In contrast, by using micro models, the superelevation of strain in the interspace of single fibers is immediately considered. A comparison of results obtained by numerical and a selection of analytical homogenization techniques at different volume fractions is depicted in Fig. 5.14. Both diagrams show the respective Young's modulus over the fiber volume fraction in a semi-logarithmic display. To this end, UD-SRVEs with

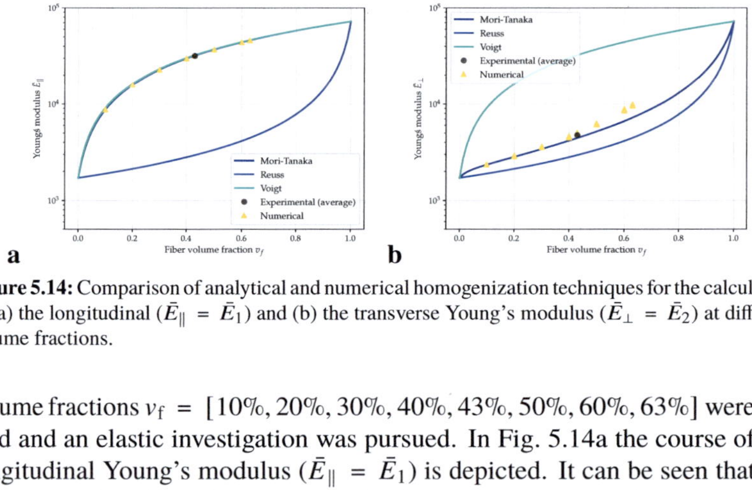

Figure 5.14: Comparison of analytical and numerical homogenization techniques for the calculation of (a) the longitudinal ($\bar{E}_{\parallel} = \bar{E}_1$) and (b) the transverse Young's modulus ($\bar{E}_{\perp} = \bar{E}_2$) at different volume fractions.

volume fractions $v_f = [10\%, 20\%, 30\%, 40\%, 43\%, 50\%, 60\%, 63\%]$ were created and an elastic investigation was pursued. In Fig. 5.14a the course of the longitudinal Young's modulus ($\bar{E}_{\parallel} = \bar{E}_1$) is depicted. It can be seen that the Voigt, Mori-Tanaka and the numerical solution coincide over the whole range of fiber volume fraction, while the Reuss solution yields stiffnesses that are lower due to the assumption of a series connection of the single components. Figure 5.14b depicts the course of the transverse stiffness ($\bar{E}_{\perp} = \bar{E}_2$) over the fiber volume fraction. While the Voigt and Reuss solutions are identical to the ones in Fig. 5.14a, the Mori-Tanaka yields a divergent response, due to the consideration of the internal orientation of reinforcing structures. A similar picture applies to the solutions of the numerical calculations. For volume fractions $v_f \leq 40\%$ the numerical and Mori-Tanaka solution coincide. Exceeding this threshold the effect of an increasing interference of the single fibers plays an ever more important role, resulting in diverging results. The investigations in Section 3.4 showed a volume fraction of $v_f = 63\%$ in the reinforcing yarns. However, only UD material with a volume fraction of $v_f = 43\%$ could be tested (cf. Section 3.3). The studies in this section demonstrated the possibility of the prediction of the elastic properties of the unidirectionally reinforced composite by comparing numerical and experimental results. The assumption applied in the following is that the numerical approach yields realistic elastic constants for the material at $v_f = 63\%$ (cf. Table 5.2) which will be used throughout this thesis.

5.2.2 Plasticity and damage evolution

The material parameters determined in Section 5.2 were used to simulate six different representative loading cases described in the following. Therefore, UD-SRVEs with a side ratio of $\delta = 50$ were created. The plastic and damage evolution in the matrix material were solved simultaneously and are depicted in the figures below for a fiber volume fraction of $v_f = 43\%$. For the sake of a clear representation, RVEs with a side ratio of $\delta = 30$ are displayed here and the fibers where assumed to be purely elastic for all presented cases except for the case of longitudinal tension. Carefully note, that having implemented the general set of periodic boundary conditions, any arbitrary loading scenario can be applied on the RVE and the results here serve solely as a small selection.

5.2.2.1 Transversal tension

Figure 5.15 shows the results of an applied horizontal transversal tension load to a UD-SRVE with a volume fraction of $v_f = 43\%$ (cf. Fig. 5.15a) and $v_f = 63\%$ (cf. Fig. 5.15b) in five consecutive steps (1-5) respectively. The top row represents the damage progression and the one at the bottom row shows the development of the accumulated plastic strain. The first figure (1) shows the RVE in its pristine state. In (2) a damage initiation in the material is observable where the stress state locally exceeds the failure criterion. This process is preferably triggered in the space between two fibers situated in close vicinity to each other and thus the tensile hydrostatic stresses are high. From Fig. 5.16 it can be seen that this process starts around the 50% threshold of the ultimate strength. By increasing the external load, the existing damaged regions grow, new failure spots are introduced and existing cracks coalesce (cf. (3)). Comparing Fig. 5.15a and b at stage (3) it can be stated, that RVEs with a higher volume fraction show considerably more locations of damage initiation due to the smaller distances in-between fibers and thus the resulting higher shares of tensile hydrostatic stress. By decreasing the undamaged volume, a further loss of stiffness is observed, yielding non-linearity of the stress-strain curve (cf. Fig. 5.16). A localized band of damaged elements starts to evolve through the RVE (cf. (4)) which is oriented perpendicular to the external loading direction. Carefully note, that other existing cracks stop to grow and stay constant eventually. Subsequently, the section of localized cracks grow further and finally coalesce to a transverse crack throughout the RVE. The described behavior of the material in the simulation above lies in good agreement with the reported failure behavior of UD composites in Section 2.3.2 under transverse tensile load. The plastic evolution inside the RVE behaves similar to the development of the damage variable. Initiating plastic flow occurs in-between the fibers and progresses from there. As soon as cracks begin to grow and to coalesce, the areas of high plastic strain rates are located near the crack tips. Keep in mind for the following observations that as soon as damage evolution is triggered, further plastic evolution is deactivated in the corresponding element.

Remark. Even though it is talked about *cracks* at this point here, it must be kept in mind, that this is synonymous with a band of elements with a highly reduced stiffness rather than discrete cracks in the sense of material separation. Additionally to the above mentioned qualitative simulations where the focus lied on the correct reproduction of the damage mechanism, the homogenized stress-strain curves performed at three strain rates under investigation were compared to the corresponding experimental results (cf. Section 3.3). Figure 5.17 shows the results in terms of stress-strain curves for a UD-SRVE with a volume fraction of $v_{\mathrm{f}} = 43\%$ objected to a transverse tensile load at three different loading rates, (a) 1.25 mm/min, (b) 12.5 mm/min, and (c) 125 mm/min. These correspond to the loading rates applied during the experimental assessment, allowing therefore a direct comparison. Keeping in mind the viscoplastic behavior of the matrix, differences in the material response are observed. At the lowest strain rate (cf. Fig. 5.17a), the plasticity in the matrix dominates the

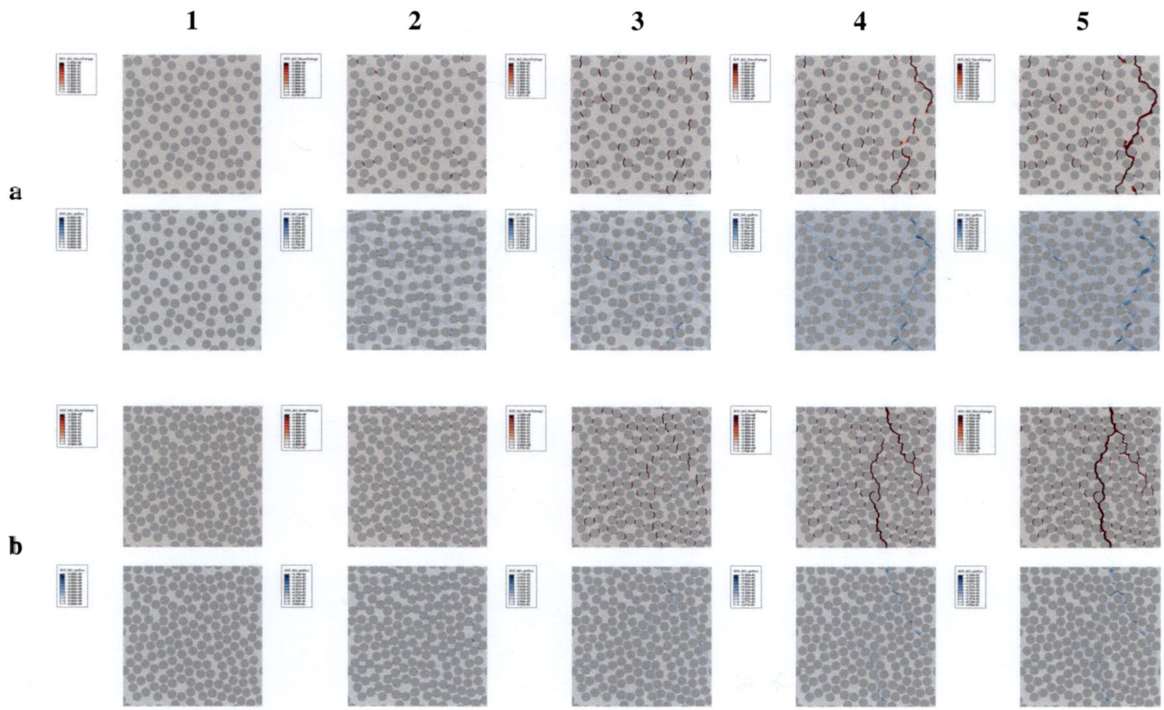

Figure 5.15: Evolution of damage and equivalent plastic strain for RVEs with fiber volume fractions of (a) $v_f = 43\%$ and (b) $v_f = 63\%$ due to a transverse tensile load. The images show the different stages of the RVE: pristine state (1), initiation, grow and coalescence of existing cracks (2) and (3), starting localization of a macroscopic crack (4) and a fully damaged RVE (5).

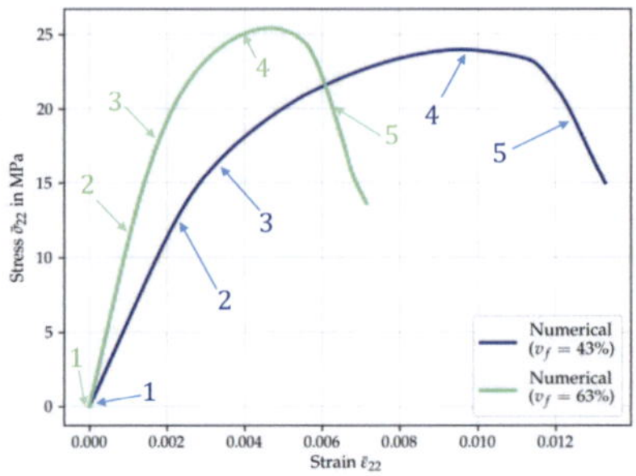

Figure 5.16: Resulting stress-strain diagrams for transverse tensile loading of UD-SRVEs with volume fractions of $v_f = 43\%$ and $v_f = 63\%$. The numbers indicated correspond to the failure stages depicted in Fig. 5.15: (1) describes the pristine state, (2) is the point of beginning damage initiation, (3) and (4) show the effect of increasing crack lengths and coalescing cracks, and (5) indicates the point in the loading history where the resulting crack traverses the whole RVE.

composite behavior in such a way, that the matrix yields already early on and with the formation of large plastic strain. To this end, a failure strain results, overestimating the one seen in the experiments. The failure strength is predicted in good agreement with the experimental results. In Fig. 5.17b the strain rate influence in the numerical model becomes visible. Here, the UD-SRVE was loaded by a medium strain rate. Compared to before, the composite material shows a stiffer response, a lower degree of non-linearity which is associated to a slower plastic evolution, and a lower failure strain. The form of the numerical and experimental curves align in good agreement and also the failure onset is predicted well. By increasing the strain rate further as shown in Fig. 5.17c the material response takes yet another form. Also here, the numerical response is in good agreement with the experimental results. However, the material

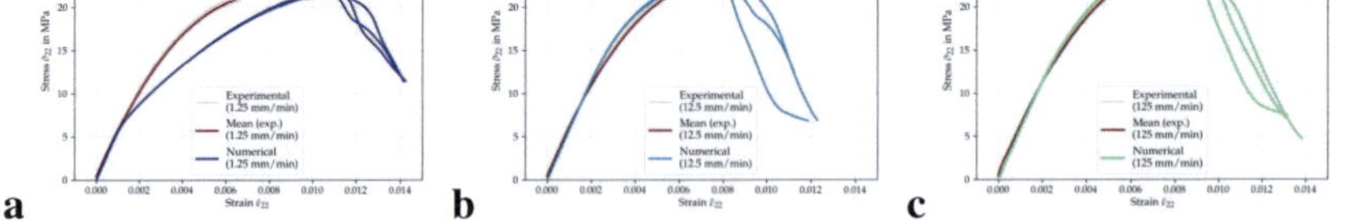

Figure 5.17: Comparison of numerically and experimentally obtained stress-strain curves for loading rates of (a) 1.25 mm/min, (b) 12.5 mm/min, and (c) 125 mm/min under transverse tension. For this investigation a UD-SRVE with a fiber volume fraction of $v_f = 43\%$ was used.

strength is slightly overestimated. Summarizing the above mentioned results it can be stated, that the visco-plastic constitutive model used for the matrix is capable of transferring rate dependent effects to the composite level. However, using the model parameters determined from specimens of pure matrix (cf. Section 5.2) the model overestimates the rate dependency to some degree compared to experimental results obtained in Section 3.3, where only a small rate influence was observed. The material strengths are predicted sufficiently well for the presented case here.

5.2.2.2 Transversal compression

Figure 5.18 shows the result from an UD-SRVE subjected to transversal compression, which is applied horizontally here. Figure 5.18a represents the spatial distribution of the damage variable d_{m} of the matrix, whereas Fig. 5.18b corresponds to the equivalent plastic strains $\varepsilon^{\mathrm{p}}_{\mathrm{acc}}$. The regions with a higher

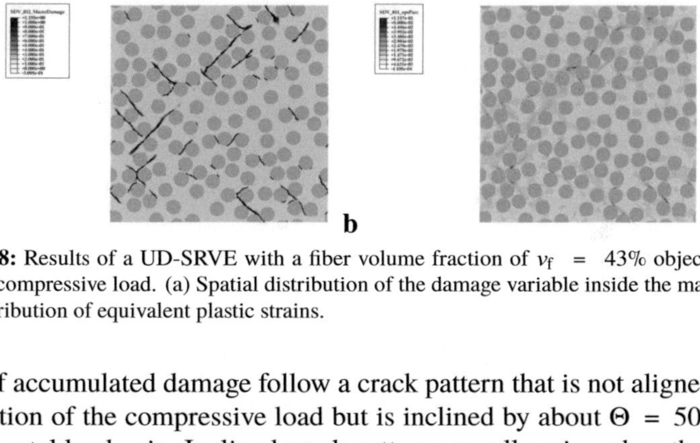

a b

Figure 5.18: Results of a UD-SRVE with a fiber volume fraction of $v_{\mathrm{f}} = 43\%$ objected to a transverse compressive load. (a) Spatial distribution of the damage variable inside the matrix, (b) spatial distribution of equivalent plastic strains.

degree of accumulated damage follow a crack pattern that is not aligned with the direction of the compressive load but is inclined by about $\Theta = 50°$ with the horizontal load axis. Inclined crack patterns usually arise when the shear strength is exceeded. Similar results where observed and reported in the experimental section of this thesis (cf. Section 3.3) as well as in the literature (cf. Section 2.3). Carefully note, that the crack growth is accompanied by the formation of plasticized bands with locally large plastic strains ($\varepsilon^{\mathrm{p}}_{\mathrm{acc}} > 10\%$). Again, the high plastic strains form preferably in front of the crack tip, once it occurs. Figure 5.19 shows the stress-strain curves of the UD-SRVEs subjected to three different loading rates after homogenization. The influence of the rate

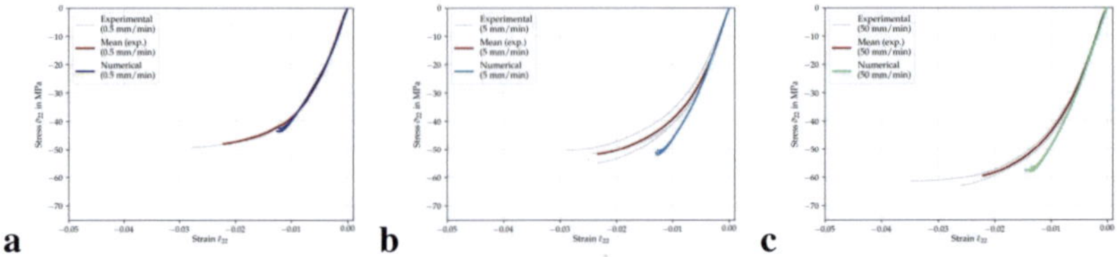

Figure 5.19: Comparison of numerically and experimentally obtained stress-strain curves for loading rates of (a) 0.5 mm/min, (b) 5 mm/min, and (c) 50 mm/min under transverse compression. For this investigation a UD-SRVE with a fiber volume fraction of $v_\mathrm{f} = 43\%$ was used.

dependency of the constitutive model is observable by the increasing strength values for higher load velocities rather than the degree of non-linearity of the material responses as it was the case in the previous example. Structural failure is identified by an abrupt change in the stress-strain curve and only a small scatter in the material strength can be observed. Once initiated, crack propagation evolves very fast. The formulation of damage progression as implemented in the present case has been calibrated for a uniaxial tensile case (mode I) therefore overestimating the pace of damage evolution. To this end, the predicted failure strengths are slightly lower than the experimentally recorded ones. With the damage variable growing too rapidly inside the RVE, the resulting failure strains do not reach the same extent as in the experiments.

5.2.2.3 Longitudinal tension

Figure 5.20 shows the stress-strain diagrams of UD-RVEs loaded under longitudinal tension at different strain rates. The material response is perfectly linear

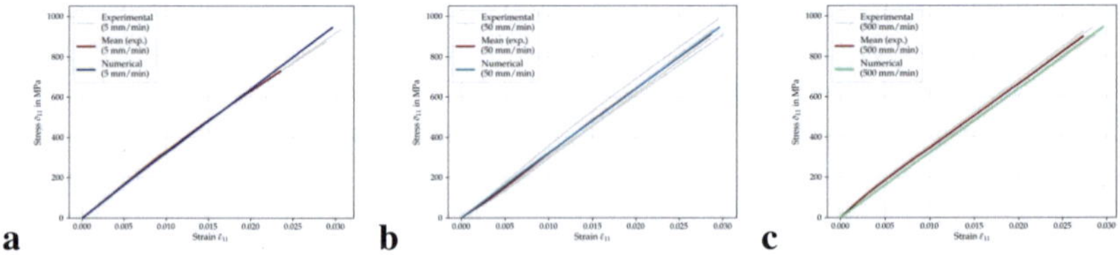

Figure 5.20: Comparison of numerically and experimentally obtained stress-strain curves for loading rates of (a) 5 mm/min, (b) 50 mm/min, and (c) 500 mm/min under longitudinal tension. UD-SRVEs with a fiber volume fraction of $v_\mathrm{f} = 43\%$ were used.

up to failure and no significant difference between the single tests in terms of a strain-rate effect can be found. This behavior is however not surprising, since the material response here is dominated by the fibers which has not been modeled to be rate dependent. Once the stresses in the fibers reach the local strength, damage is initiated. As soon as the stiffness has been reduced to such an extent that the matrix is reaching a critical loading state, damage also grows within the matrix. Interestingly, the matrix damage grows not entirely in the plane perpendicular to the macroscopic load but also along the fiber axis. This phenomenon is observed also in literature and is finally referred to as fiber pullout. Figure 5.21 shows the spatial damage distribution of a quasi fully damaged SRVE due to a longitudinal tensile load for the fiber and matrix regions respectively.

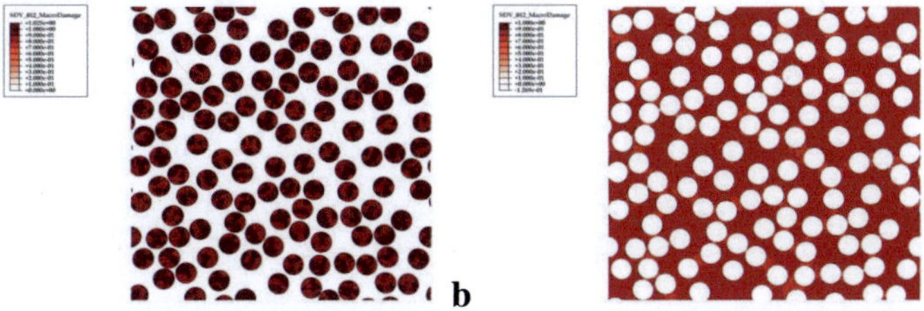

a b

Figure 5.21: UD-SRVE with a fiber volume fraction of $v_f = 43\%$ subbjected to longitudinal tension. Spatial distribution of the damage variable (a) in the fibers and (b) in the matrix.

5.2.2.4 Transversal shear

Figure 5.22 shows the results of a RVE loaded by a symmetrically applied pure ambilateral transversal shear load. Figure 5.22a hereby represents the resulting damage pattern within the microstructure. Both horizontal and vertical cracks can be observed. The regions of localized plastic strains coincide with the ones of high damaged material (cf. Fig. 5.22b). The resulting stress-strain curve after homogenization is depicted in blue in Fig. 5.22c. Furthermore, an equivalent investigation was done for a UD-SRVE with a fiber volume fraction of $v_f = 63\%$. The homogenized stress-strain curve is displayed as the green dashed line. Confronting the two curves, the effect of the higher volume fraction is visible. Apart from the increased stiffness, the composite with the higher volume fraction yields a slightly higher ultimate failure strength. Due to the lower matrix share and the smaller associated plasticity-related effects, a lower strain at failure is observed. Moreover, since the deformation is matrix dominated, a lower degree of non-linearity manifests.

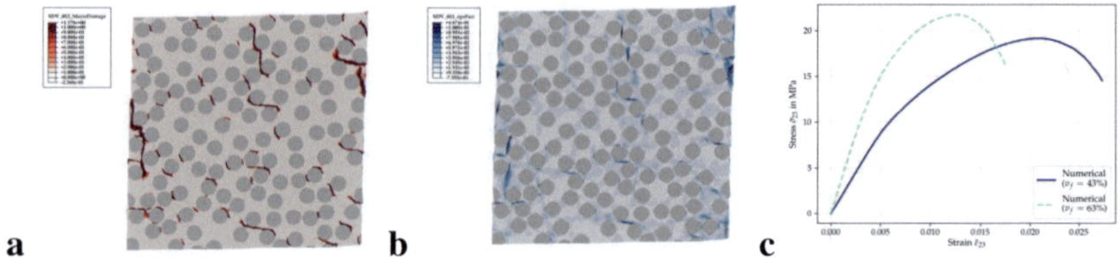

Figure 5.22: Results of a UD-SRVE with a fiber volume fraction of $v_{\mathrm{f}} = 43\%$ objected to a pure transverse shear load. Hereby, the load was applied symmetrically. (a) Spatial distribution of the damage variable inside the matrix, (b) spatial distribution of equivalent plastic strains, and (c) comparison of two resulting stress-strain curves after homogenization for the fiber volume fractions $v_{\mathrm{f}} = 43\%$ and $v_{\mathrm{f}} = 63\%$.

5.2.2.5 Longitudinal shear

Figure 5.23 shows the results from an applied longitudinal shear load to a UD-SRVE where a and b correspond to the spatial distribution of the damage variable and the equivalent plastic strain respectively. The load was applied in

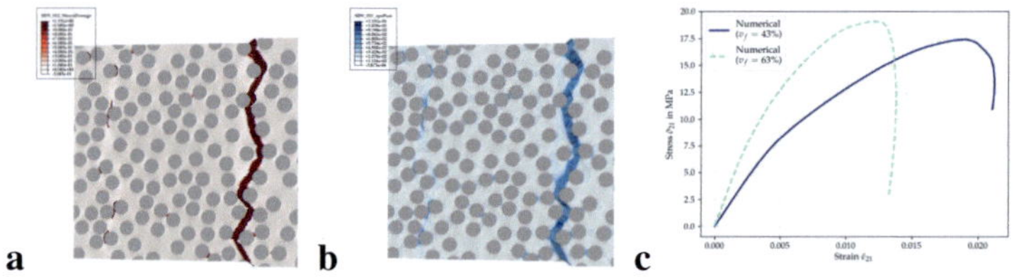

Figure 5.23: Results of a UD-SRVE with a fiber volume fraction of $v_{\mathrm{f}} = 43\%$ objected to a pure longitudinal shear load. (a) Spatial distribution of the damage variable inside the matrix, (b) spatial distribution of equivalent plastic strains, and (c) comparison of two resulting stress-strain curves after homogenization for the fiber volume fractions $v_{\mathrm{f}} = 43\%$ and $v_{\mathrm{f}} = 63\%$.

such a manner that the left edge of the RVE was moved into the visible plane along the fiber direction. This displacement resulted in a sharp crack, i.e. a narrow band of localized damage and plastic strains in the matrix material. The rest of the matrix material remained largely intact and only a few areas show slight damage. Again, the plastic hot-spots coincide with those of damage.

Homogenization of the local stresses and strains within the UD-SRVE yields the resulting stress-strain curves depicted in Fig. 5.23c for the relevant fiber volume fractions $v_f = 43\%$ and $v_f = 63\%$. A direct comparison of the two obtained curves shows the same tendencies as observed in the previous case. Again, the higher volume fraction leads to a slightly higher material strength as well as a reduced strain at failure. Due to the less pronounced matrix related effects a reduced non-linearity is observed.

5.2.2.6 Transversal tension and longitudinal shear

The main advantage of the virtual material characterization is the possibility of applying a wide range of arbitrary loading scenarios, also those which are hardly to achieve experimentally. The present loading case is an example for this. A UD-SRVE was loaded simultaneously by a horizontal transversal tension and a smaller amount of longitudinal shear. The results are demonstrated in Fig. 5.24 as the spatial distribution of the damage variable as well as the equivalent plastic strain. As expected and described by Puck [319], the fracture plane forms perpendicular to the tensile load. Plastic strain is mainly observed in the areas of damage evolution. Homogenization of the local stress and strain states

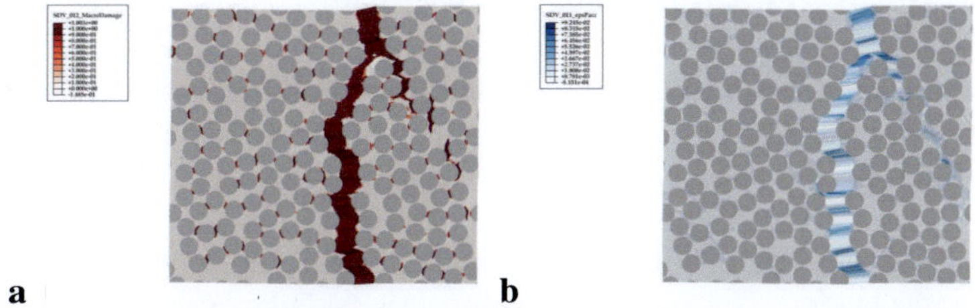

Figure 5.24: Results of a UD-SRVE with a fiber volume fraction of $v_f = 43\%$ objected to a transverse tensile load superimposed by a longitudinal shear load. (a) Spatial distribution of the damage variable inside the matrix, (b) spatial distribution of equivalent plastic strains.

yields the overall stress-strain curve of the composite. Figure 5.25a shows the resulting curve for the transverse stress $\bar{\sigma}_{22}$ versus the transverse strain $\bar{\varepsilon}_{22}$, whereas Fig. 5.25b depicts the material response in terms of the shear stress $\bar{\sigma}_{21}$ and the corresponding shear strain $\bar{\varepsilon}_{21}$. Please note the differently scaled ordinate axis of both diagrams. Again the investigation was performed for both relevant fiber volume fractions $v_f = 43\%$ and $v_f = 63\%$. Analyzing

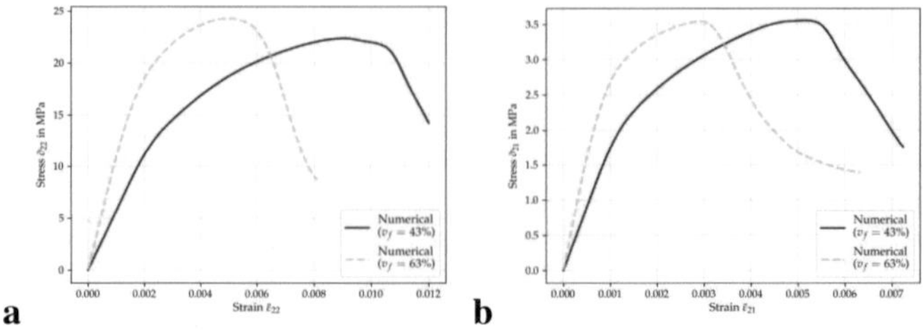

Figure 5.25: Homogenized stress-strain curves for UD-SRVE microstructures with a fiber volume fraction of $v_\mathrm{f} = 43\%$ and $v_\mathrm{f} = 63\%$. (a) Transverse stress versus transverse strain ($\bar{\sigma}_{22}$-$\bar{\varepsilon}_{22}$), (b) Longitudinal shear stress versus longitudinal shear strain ($\bar{\sigma}_{21}$-$\bar{\varepsilon}_{21}$).

Fig. 5.25a, a similar form of the curves as for the pure transverse tensile load calculated previously can be stated. However, due to the additional load by the shear deformation, the strength values are slightly reduced. The interaction of transverse tensile and longitudinal shear loads plays an important role for the parametrization of the homogenized failure model of the UD material recently presented in Section 4.2. The resulting effects are further investigated in Section 5.2.3.

5.2.3 Computation of failure envelope

In Section 4.2, a constitutive model for unidirectionally reinforced fiber composites was presented taking into account the most prominent damage mechanisms that are observed experimentally. The appearing failure modes were classified in two species, along the fiber axis and transversally to it. While the material strengths for the first can be obtained directly from (virtual) characterization, the latter considered the three-dimensional failure criterion introduced by Puck [319]. To this end, a failure envelope (cf. Table 4.2) described by the material strengths $R_\perp^\pm$ and $R_{\perp\parallel}$ as well as the model parameters $p_{\perp\parallel}^\pm$, giving the inclination of the failure surface at $\sigma_\perp = 0$, has been implemented. One purpose of the undertaken computations was the virtual characterization of the chosen criterion. In Section 5.2.2 some examples for possible loading scenarios were presented. From there it is a straightforward operation to extract the set of material strengths as the highest appearing stresses throughout the analysis. Transversal tension superimposed by a longitudinal shear at various in different relations defines the failure envelope defined by Puck [319]. The computations were conducted using $n = 10$ different realizations of UD-SRVE microstructures with a fiber volume fraction of $v_f = 43\%$ and $v_f = 63\%$. In order to discretize the failure surface in a sufficient manner, $m = 21$ different ratios of transversal tensile/compressive and longitudinal shears were defined. Figure 5.26a shows the raw results obtained from this analysis in $\tau_{\perp\|}$-$\sigma_\perp$ di-

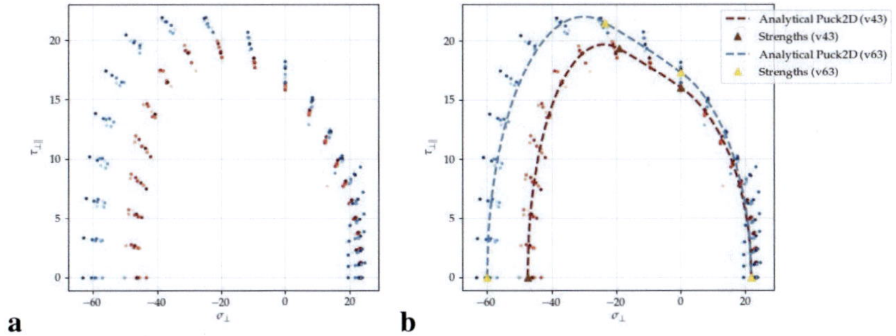

Figure 5.26: Representation of the computation results for different load ratios in a $\tau_{\perp\|}$-$\sigma_\perp$ diagram for fiber volume fractions $v_f = 43\%$ and $v_f = 63\%$. The failure envelope was discretized by $m = 10$ points, each representing a certain load ratio. For each support point $n = 10$ realization were analyzed. (a) Display of the raw results, (b) Display of the fitted failure criterion introduced by Puck [319] and implemented in Section 4.2. The parameters are summarized in Table 5.3.

agram. The results in red correspond to the UD-SRVEs with a fiber volume fraction of $v_f = 43\%$ whereas the blue results represent UD-SRVEs with $v_f = 63\%$. The $m = 21$ single sets due to discretization as well as the scatter of the material strength for each loading ratio due to the different microstructure are clearly distinguishable. The form of the obtained failure envelope lies in good agreement with the model of Puck [319]. Figure 5.26b displays moreover the analytical formulation of Puck's failure criterion for both fiber volume fractions. The curve was obtained by fitting the descriptive parameters with the help of a least square error method. The material strengths are marked as triangles. The resulting strength parameters are given in Table 5.3 together with the material strengths for transverse shear and longitudinal loads. Please note, that the material strength for longitudinal compressive loads could not be investigated by using the UD-SRVEs in the present form. Failure in fiber direction under compressive is triggered by small misalignments and concomitant buckling effects of single fibers which was not considered in the present model. Based on the experimental results presented in Section 3.3, for the following investigations a longitudinal compressive failure strength was assumed that is three times lower than the tensile strength. A detailed examination of the related effects must be the subject of future work.

Table 5.3: Computed strength properties for unidirectionally glass-fiber reinforced composites with a thermoplastic polypropylene matrix. The values were determined for two fiber volume fractions ($v_f = 43\%$ and $v_f = 63\%$) with the help of $n = 10$ realizations with a side ratio of $\delta = 50$. The values marked with the superscript [1] are estimates.

	$v_f = 43\%$	$v_f = 63\%$
$R_{\parallel}^{+}$	989.93	1361.63
$R_{\parallel}^{-}$	330[1]	460[1]
$R_{\perp\perp}$	18.41	21.78
$R_{\perp}^{+}$	22.01	22.12
$R_{\perp}^{-}$	47.64	59.87
$R_{\perp\parallel}$	16.12	17.31
$p_{\perp\parallel}^{+}$	0.19	0.20
$p_{\perp\parallel}^{-}$	0.19	0.20

5.3 Computations on weave reinforced unit cell

5.3.1 Basic features of the homogenized UD model

The work at hand introduces a three-dimensional constitutive model for the prediction of damage onset and growth in unidirectional reinforced plastics, corresponding to the impregnated rovings in the weaves (cf. Section 4.2). Taking into account large deformations, stress-based failure criteria and damage evolution laws are introduced. The proposed model was implemented into the commercial finite element code ABAQUS using the implicit user material subroutine *UMAT*. The capability of taking into account both the tension/-compression anisotropy in damage evolution and the resulting damage effects with respect to different loading directions is demonstrated in the following by means of selected numerical use cases, involving single element simulation for a fictitious material. Special attention was paid to the transversal failure modes, while the longitudinal damage evolution behaves straightforward and is known from many existing failure models. As a first example, a tensile transverse loading in 2-direction and a subsequent unloading/reloading scenario is investigated. The resulting loading cycle is shown in Fig. 5.27. Figure 5.27a shows the stress-strain relation, whereas Fig. 5.27b depicts the evolution of the damage variable d_2^+ that is associated to the loading direction and the normalized deformation over time. It can be seen that the proposed model predicts damage

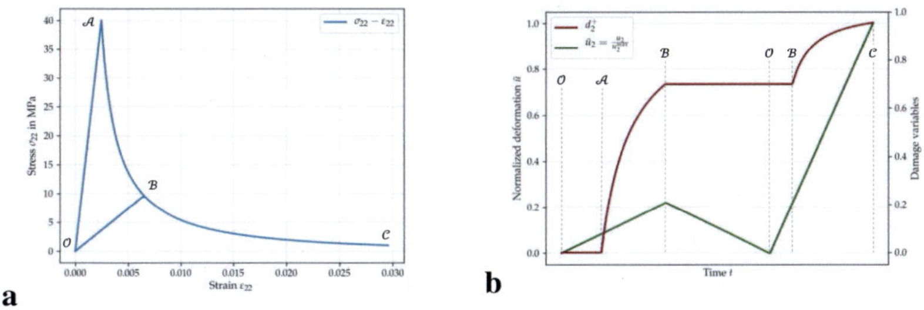

Figure 5.27: (a) Stress-strain response for a loading/unloading/loading ($\mathcal{O}$-$\mathcal{A}$-$\mathcal{B}$/$\mathcal{B}$-$\mathcal{O}$/$\mathcal{O}$-$\mathcal{B}$-$\mathcal{C}$) scenario in transverse direction of a single element. (b) Representation of the normalized deformation $\bar{u}_2$ and the corresponding damage evolution for d_2^+ over time.

onset in loading direction without any previous inelastic material behavior ($\mathcal{O}$-$\mathcal{A}$). Exceeding the transverse tensile stiffness $R_\perp^+$, stiffness degradation follows due to damage evolution ($\mathcal{A}$-$\mathcal{B}$). Carefully note, that only the damage variable d_2^+ along the loading direction grows. All other damage variables ($d_1^\pm$, d_2^-, $d_3^\pm$) stay zero. This behavior changes the symmetry class of the stiffness tensor. During unloading, the degraded stiffness becomes visible by a reduced secant ($\mathcal{B}$-$\mathcal{O}$). However, the calculated damage variables stay constant (cf. Fig. 5.27b), which is a requirement for thermodynamic consistency. By reloading, the material sees a reduced stiffness since the model represents a crack opening process of a preexisting crack ($\mathcal{O}$-$\mathcal{B}$). Exceeding the predamaged strength in $\mathcal{B}$, damage evolution for d_2^+ progresses further (Fig. 5.27b) and a continuously reduced stress response ($\mathcal{B}$-$\mathcal{C}$). The second example examines the model response under compressive load reversal and a subsequent tensile load. The loading cycle is depicted in Fig. 5.28, where Fig. 5.28a shows the resulting stress-strain relationship, while Fig. 5.28b monitors the evolution of damage and presents the normalized deformation $\bar{u}_2$. At first, the material does not see any degradation of stiffness ($\mathcal{O}$-$\mathcal{A}$). As soon as the stress reaches the material strength for compression $R_\perp^-$ (at $\mathcal{A}$), damage grows ($\mathcal{A}$-$\mathcal{B}$). In contrast to the first example, damage evolves now not only in the direction of external load,

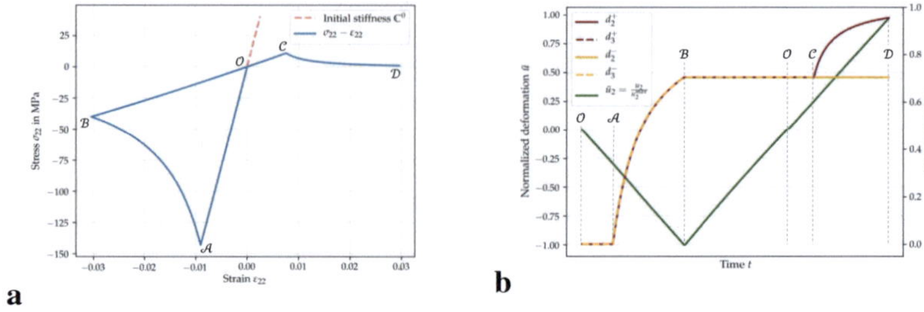

Figure 5.28: (a) Stress-strain response for a compressive loading/unloading (O-$\mathcal{A}$-$\mathcal{B}$/$\mathcal{B}$-O) followed by a tensile loading scenario (O-C-$\mathcal{D}$) in transverse direction of a single element. (b) Representation of the normalized deformation $\bar{u}_2$ and the corresponding damage evolution for d_2^- and d_2^+ over time.

but, due to the inclination of the damage system by Θ, also in the perpendicular 3-direction. In the present case, $\Theta = 45°$ yields a symmetric evolution of transverse damage, $d_2^+ = d_3^+ = d_2^- = d_3^-$ up to point $\mathcal{B}$. The underlying model representation yields that the crack emerging due to external load is visible throughout the transverse isotropic plane. Note, that damage evolution in the various directions is highly dependent on the occurring orientation of the crack system. Once more, the evolution of damage changes the symmetry class of the stiffness tensor. During unloading ($\mathcal{B}$-O) the crack closes and the calculated damage state remains unchanged. Due to the introduction of the previously introduced damage ($d_2^\pm > 0$), the stiffness under a subsequent tensile load stays reduced (O-C). The model represents an opening process of a preexisting crack. Up to C the material behaves linearly. Once exceeding the remaining strength, continued damage progression results in a further decrease of the stiffness, since the damage variable d_2^+ keep growing (cf. Fig. 5.28b). The opposite case is discussed next. In the third example, a tensile transverse loading/unloading scenario is followed by a compressive load, both in 2-direction. Figure 5.29a shows the resulting stress-strain relation, whereas Fig. 5.29b depicts the evolution of the damage variables $d_2^\pm$ and $d_3^\pm$ and the normalized deformation $\bar{u}_2$ over time. The tensile loading cycle (O-$\mathcal{A}$-$\mathcal{B}$-O) shows a similar behavior as in the first example when it comes to the material response. In terms of the model representation, a crack opens perpendicular to the loading direction, decreasing the stiffness under a tensile load. After unloading, a compressive load is applied upon the single element. The model representation now states that the preexisting crack is not only closed but the compressive load results moreover in an active closing of the crack flanks. Therefore, no reduced stiffness is observed

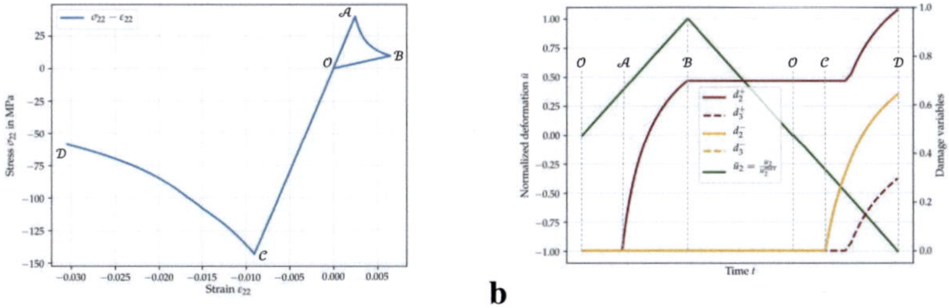

a b

Figure 5.29: (a) Stress-strain response for a transverse tensile loading/unloading ($\mathcal{O}$-$\mathcal{A}$-$\mathcal{B}$/$\mathcal{B}$-$\mathcal{O}$) followed by a transverse compressive loading scenario ($\mathcal{O}$-$\mathcal{C}$-$\mathcal{D}$) in 2-direction of a single element. (b) Representation of the normalized deformation $\bar{u}_2$ and the corresponding damage evolution for d_2^+, d_3^+, d_2^-, and d_3^- over time.

at first ($\mathcal{O}$-$\mathcal{C}$). Once exceeding the compressive strength $R_\perp^-$, further damage is introduced representing shear damage ($\mathcal{C}$-$\mathcal{D}$, cf. Fig. 5.29b). So far only loads in one load direction were presented. The forth example examines the model response under subsequent transverse tensile loads in two perpendicular directions. The loading cycle is depicted in Fig. 5.30, where Fig. 5.30a shows the resulting stress-strain relationship, while Fig. 5.30b represents the evolution of damage and presents the normalized deformation $\bar{u}_2$ and $\bar{u}_3$. Similar to the

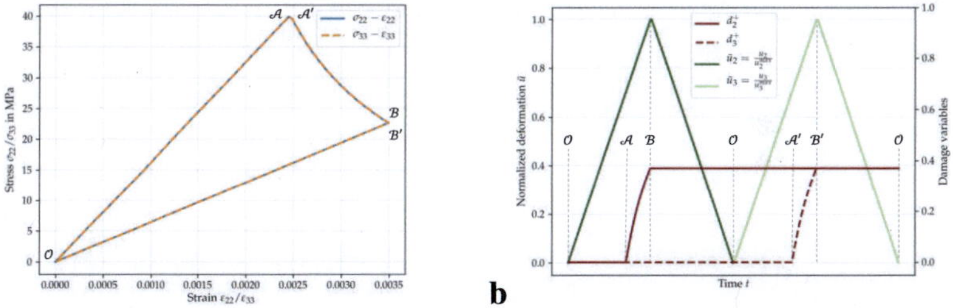

a b

Figure 5.30: (a) Stress-strain response for a transverse tensile loading/unloading ($\mathcal{O}$-$\mathcal{A}$-$\mathcal{B}$/$\mathcal{B}$-$\mathcal{O}$) in 2-direction followed by a transverse tensile loading scenario ($\mathcal{O}$-$\mathcal{A}'$-$\mathcal{B}'$/$\mathcal{B}'$-$\mathcal{O}$) in 3-direction of a single element. (b) Representation of the normalized deformation $\bar{u}_2$ and $\bar{u}_3$ and the corresponding damage evolution for d_2^+ and d_3^+ over time.

first example, a linear increase is calculated until the material strength $R_\perp^+$ is reached ($\mathcal{O}$-$\mathcal{A}$). Exceeding the material strength, damage evolution is initiated, resulting in a increasingly reduced stiffness ($\mathcal{A}$-$\mathcal{B}$). The reduced stiffness becomes visible during the unloading path, where the damage variables stay constant. The stress-strain response is identical for both loading directions. The model concept states that the two loads in the 2- and 3-direction cause a crack in each case. Since both cracks are oriented perpendicularly, they do not affect each other, which can be seen in the unchanged initial stiffness.

5.3.2 Computation of elastic bodies

Having generated the virtual microstructures of the woven unit cells respecting the geometrical parameters given in Table 3.7, modeled and parametrized the UD model as well as the isotropic matrix, the elastic properties of the weave-reinforced mesoscopic structure under study can be assessed. Carefully note, that the height of the unit cell had to be adjusted slightly in order to allow a meshing in the interspace of two crossing rovings. Equivalent to before, six calculations corresponding to the six uniaxial load cases were performed in order to determine the full stiffness tensor. Again, the equations of the numerical homogenization presented in Section 2.2.2 and the homogenization integrals in Eq. 2.46 relate the far-field strain and/or stress applied to the unit cell with the help of the in-plane periodic boundary conditions with the resulting and more complex stress and strain states within the microstructure. Table 5.4 shows the resulting elastic properties of the given virtual weave. Assuming initial orthotropic symmetry, the stiffness tensor can be described by nine independent variables. After computation, the results show that the orthotropic symmetry class is well represented. Figure 5.31a shows a three-dimensional representation of the elastic stiffness body of the weave-reinforced material under investigation in terms of Young's modulus. It is clearly visible that two preferred direction are dominating the elastic behavior of the material corresponding with the preferred directions of the inter-woven rovings. Orthotropy is defined by an invariance to all rotations about 180° around the orthotropic axes which is reflected well by the given elastic body. By projecting the three-dimensional stiffness body onto the symmetry plane aligning with the x-y plane in Fig. 5.31a, the values of the direction dependent Young's moduli can be directly extracted. Figure 5.31b shows the projected stiffness body of the virtual weave. Half symmetry is considered here and the preferred axis of the composite is aligned with the indicated 0°- and 90°-direction respectively. Furthermore, experimental data is available and is displayed here as an average of the Young's moduli measured at different strain rates (cf. Section 3.4). Comparison of the experimental results

Table 5.4: Obtained stiffness parameters by volumetric homogenization of a woven unit cell. The geometrical parameters were taken from Table 3.7.

Elastic properties	
$\bar{E}_1$ (in MPa)	16088.29
$\bar{E}_2$ (in MPa)	15908.79
$\bar{E}_3$ (in MPa)	6271.47
$\bar{\nu}_{12}$ (-)	0.130
$\bar{\nu}_{13}$ (-)	0.501
$\bar{\nu}_{23}$ (-)	0.501
$\bar{G}_{12}$ (in MPa)	1280.42
$\bar{G}_{13}$ (in MPa)	1289.96
$\bar{G}_{23}$ (in MPa)	1554.49

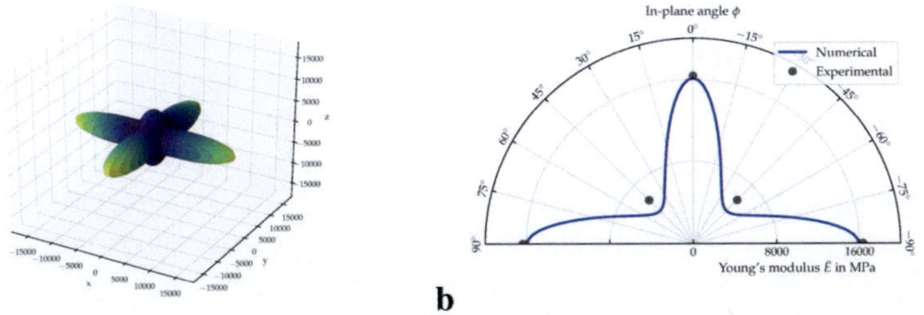

a **b**

Figure 5.31: (a) Display of the three-dimensional stiffness body computed from six uniaxial load cases using a weave-reinforced unit cell. (b) Projection of the three-dimensional stiffness body in the symmetry plane containing the reinforcing rovings (x-y plane from (a)). Half-symmetry is considered here. Moreover the stiffness results from the experiments presented in Section 3.4 are depicted.

with the numerical solution yields a good agreement in terms of Young's modulus along the preferred axis, even though a slightly smaller value is obtained. Lower values are also obtained in the 45°-direction which can be compared to the results of the ±45° off-axis experiments. In both cases, the cause of the lower stiffness is the artificial additional matrix material in the interspace of overlaid rovings. For the longitudinal case this yields a higher sectional matrix share, for the latter the additional matrix allows an easier rotation of the reinforcement structures with the load direction.

5.3.3 Plasticity and damage evolution

For the following load cases the in-plane formulation of the periodic boundary conditions (cf. Appendix D) were used. Here, periodicity is assumed only in the plane of reinforcements and does not apply in the perpendicular z-direction, which is defined in the thickness of the ply. Four different loading scenarios are considered: uniaxial tension, in-plane shear, biaxial tension along the rovings, and a uniaxial tension case that superposed by an in-plane shear.

5.3.3.1 Uniaxial tension

Figure 5.32 shows the stress-strain ($\bar{\sigma}_{11}$-$\bar{\varepsilon}_{11}$) curve of a virtual uniaxial tension test obtained by volumetric homogenization. The load was applied along the reinforcement axis (horizontal 1-direction). The initial stiffness of the numerical solution is just slightly lower than the experimental results which agrees with the previous investigations (cf. Fig. 5.31b). The numbers indicated correspond to five consecutive loading stages (1-5) depicted in Fig. 5.33. Hereby, the top row represents the evolution of the equivalent plastic strain $\varepsilon^{\mathrm{p}}_{\mathrm{acc}}$ while the second shows the isotropic damage variable d_{m} in the surrounding matrix. The last two rows concentrate on the reinforcement structures. Here, the longitudinal damage variable d_1^+ is displayed in black whereas the transverse damage variable d_2^+ is displayed in orange. The first column (1) shows the unit cell in its pristine state. In (2) a initiation of transverse damage in the rovings oriented perpendicular to the loading direction is observable where the stress states locally exceeds the implemented failure criterion. Predestined locations of damage onset are the crimp regions. From Fig. 5.32 it can be seen that this process starts early on compared to the point of ultimate failure.

By increasing the external load, the existing damaged regions grow across the whole perpendicular rovings. Decreasing the undamaged volume yields an increasing loss of stiffness (cf. Fig. 5.32). By decreasing the transverse

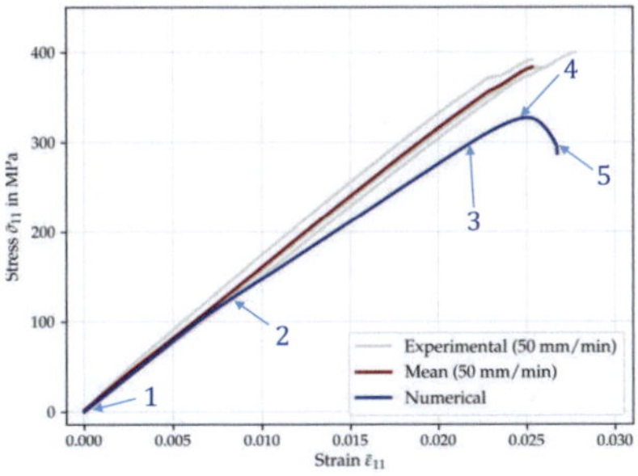

Figure 5.32: Resulting stress-strain diagram ($\bar{\sigma}_{11}$-$\bar{\varepsilon}_{11}$) for a transverse tensile load in horizontal direction of a twill weave reinforced unit cell. The indicated numbers correspond to the failure stages depicted in Fig. 5.33: (1) describes the pristine state, (2) is the point of beginning transverse damage initiation, (3) shows increasing transverse damage progression accompanied by plastic evolution in the adjacent matrix, (4) additional initiation of longitudinal damage in the loaded rovings, and (5) corresponds to a fully damaged unit cell.

stiffness of the damaged rovings, the adjacent matrix sees increasingly higher deformation, resulting in the formation of plastic regions. Carefully note, that the hot spots of the plastic evolution coincide with those of transverse damage in the rovings. Due to the relatively small extent of plastic strains, this effect attributes no additional non-linearity to the stress-strain curve at this stage (cf. Fig. 5.32, (3)). When the local strength of the rovings in the preferred axis is exceeded, fiber damage starts to arise (cf. Fig. 5.33, (4)), by building cracks that are oriented perpendicular to the loading axis. Again, the crimp regions are the critical locations for damage onset. A localized band of damaged elements starts to form in the rovings aligned with the loading direction by further increasing the external load. This behavior is associated with a drastic drop in the stress-strain curve (cf. Fig. 5.32, (4-5)) since the principal load carrier are significantly weakened. At stage (5), all horizontally aligned rovings show a crack over their complete width associated with a significant stiffness drop in the composite response. This is leading to a higher loading of the adjacent matrix around the cracks, resulting in high plastic straining and the onset of matrix damage.

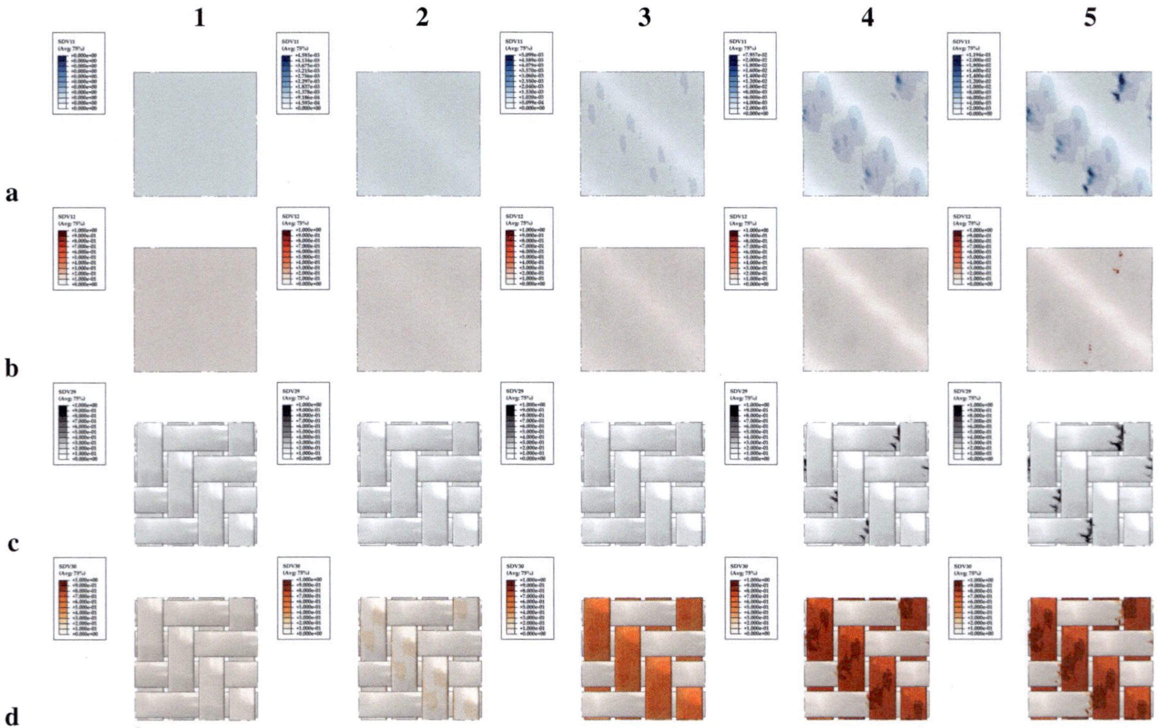

Figure 5.33: Plastic (a) and isotropic damage variable in the matrix (b), and longitudinal (c) and transversal damage variables (d) in the rovings under horizontal tensile load. Stages: pristine state (1), initiation of transverse damage in the rovings (2), grow of transverse damage and plastic evolution in the matrix (3), starting longitudinal damage in the horizontal rovings (4), and fully damaged unit cell (5).

5.3.3.2 In-plane shear

The following examples focuses on a pure ambilateral in-plane shear load case. Fig. 5.34 shows the evolution of the in-plane shear stress $\bar{\sigma}_{21}$ versus the far-field strain $\bar{\varepsilon}_{21}$ obtained by volumetric homogenization that is compared to the corresponding experimental results. As already observed in Fig. 5.31b,

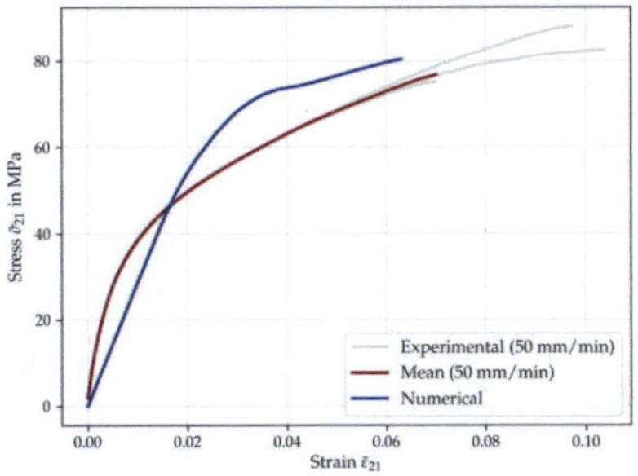

Figure 5.34: Resulting stress-strain diagram ($\bar{\sigma}_{21}$-$\bar{\varepsilon}_{21}$) for an in-plane shear load of a twill weave reinforced unit cell obtained by volumetric homogenization. Additionally, the experimental results for the given load case are presented again for comparison (cf. Section 3.3).

a lower stiffness is computed in the in-plane shear case. The following non-linearity is caused mainly by plastic effects in the matrix, superimposed with transverse damage effects in the reinforcement structures.Figure 5.35 represents the resulting damage pattern and plastic evolution within the mesostructure. Transverse damage caused by the present shear load tends to concentrate in the crimp area of the interwoven rovings, where the yarns cross each other and from there propagate across the yarns. Damage in the longitudinal direction of the rovings has not been initiated. In conclusion it can be stated, that under the given shear load the narrow areas in the interspace of the rovings show high stress concentrations due to the different shear moduli of the single components. Therefore, plastic strains arise in those locations. The mentioned effects explain the non-linear progression of the stress-strain curve in Fig. 5.34.

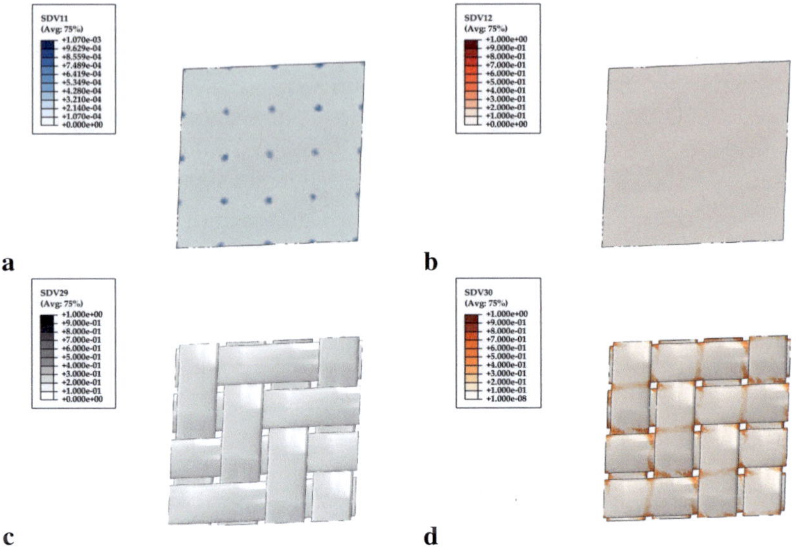

Figure 5.35: Results of twill weave reinforced unit cell loaded by an in-plane shear load. (a) Spatial distribution of the accumulated plastic strain, (b) isotropic matrix damage variable, (c) longitudinal damage, and (d) transversal damage variable in the rovings.

5.3.3.3 Biaxial tension

Figure 5.36 shows the results of an unit cell loaded by a two-sided in-plane tension load after volumetric homogenization. As it can be seen from the diagram, a quasi linear curve for both tensions ($\bar{\sigma}_{11}$ and $\bar{\sigma}_{22}$) can be seen up to ultimate failure. A slight non-linear behavior is however reported due to the occurring transverse damage in all rovings (cf. Fig. 5.37). Hereby, the damage in the horizontal rovings is a result of the tension in the vertical direction and vice versa. At the point of ultimate failure, the vertical rovings abruptly develop a crack band due to the evolution of longitudinal damage leading to a drop in the stress response. This is accompanied by plastic evolution in the matrix in close vicinity of the occurring cracks. The stresses in the horizontal rovings however, continue to rise and are not affected significantly by the damage effects in their counterparts. Furthermore, comparing Fig. 5.32 for the uniaxial tensile case and Fig. 5.36, a slightly lower ultimate strength together with a lower failure strain can be stated for the superimposed case.

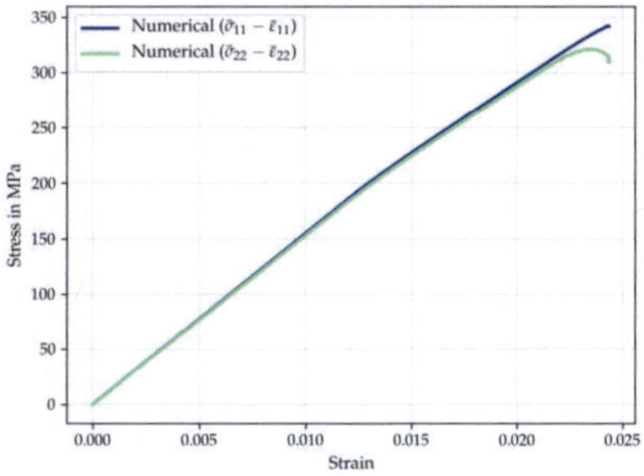

Figure 5.36: Resulting stress-strain diagram ($\bar{\sigma}_{11}$-$\bar{\varepsilon}_{11}$ and $\bar{\sigma}_{22}$-$\bar{\varepsilon}_{22}$) for a biaxial tensional load of a twill weave reinforced unit cell obtained by volumetric homogenization.

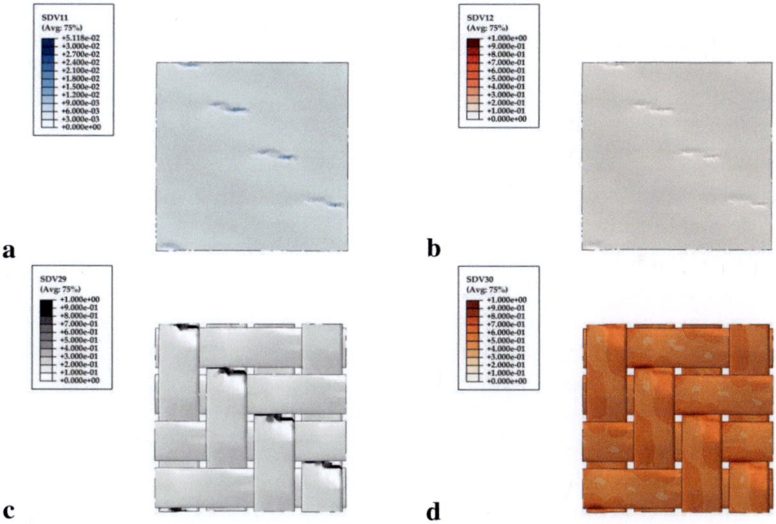

Figure 5.37: Results of twill weave reinforced unit cell loaded by a biaxial tensile load. (a) Spatial distribution of the accumulated plastic strain, (b) isotropic matrix damage variable, (c) longitudinal damage, and (d) transversal damage variable in the rovings.

5.3.3.4 Uniaxial tension and in-plane shear

Here again, the possibility of applying a superimposed load was used. Figure 5.38 shows the resulting stress-strain diagram for a combined horizontal tensile ($\bar{\sigma}_{11}$-$\bar{\varepsilon}_{11}$) and a smaller in-plane shear load ($\bar{\sigma}_{21}$-$\bar{\varepsilon}_{21}$). Figure 5.39 shows the corresponding failure pattern and plastic evolution in the matrix and roving structures respectively. The observed non-linearity in Fig. 5.38 prior to

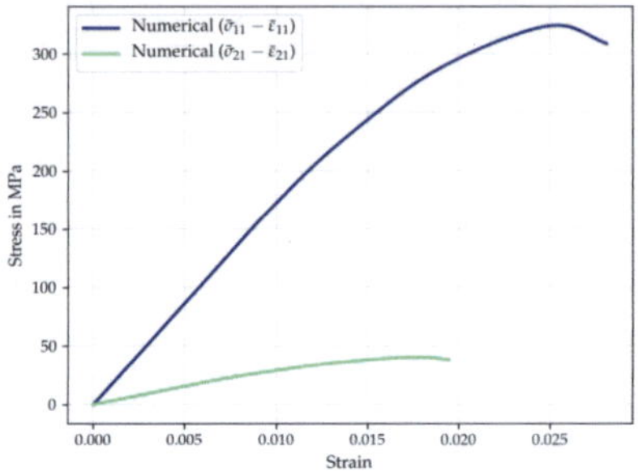

Figure 5.38: Resulting stress-strain diagram ($\bar{\sigma}_{11}$-$\bar{\varepsilon}_{11}$ and $\bar{\sigma}_{21}$-$\bar{\varepsilon}_{21}$) for a horizontal tensional load and a superimposed in-plane shear load of a twill weave reinforced unit cell obtained by volumetric homogenization.

ultimate damage can be traced back to the evolving transverse damage in the vertical rovings together with a plastic evolution in the matrix (cf. Fig. 5.39a and d) mainly provoked by the shear load. The point of ultimate failure is indicated by a stress drop caused by a sudden reduction of element stiffness in vertical crack bands in the load carrying rovings. Once the stiffness locally is reduced by a great extent the surrounding material gets locally objected to a higher degree of deformation due to stress redistribution. As a result plastic bands and finally matrix damage evolves in these hot spots. For the present case, the current strength ($R_{\|,1}$ = 317 MPa) is slightly lower than the strength under uniaxial tension ($R_{\|,1}$ = 330 MPa). This allows the conclusion that a influence of the longitudinal strength on superimposed shear loads exists. The extent of this dependency is to clarify in future work.

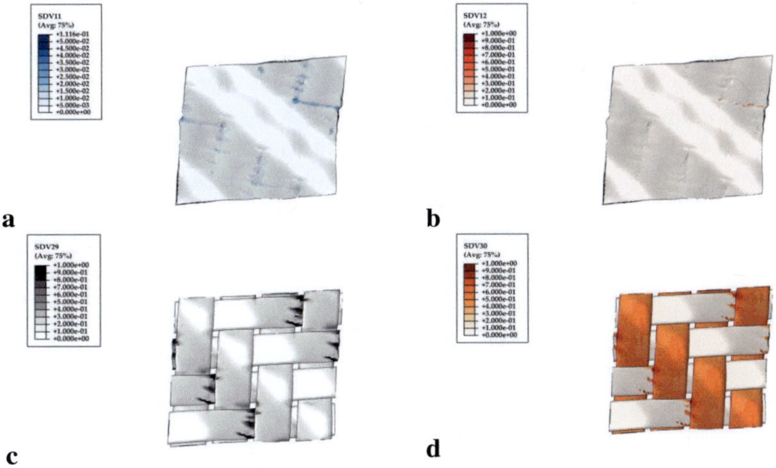

Figure 5.39: Results of twill weave reinforced unit cell loaded by a horizontal tensile load in combination with a superimposed in-plane shear load. (a) Spatial distribution of the accumulated plastic strain, (b) isotropic matrix damage variable, (c) longitudinal damage, and (d) transversal damage variable in the rovings.

5.4 Computations for macroscopic woven composites

5.4.1 Computation of elastic bodies

The constitutive model for weave reinforced composites (*HomWeave*) does not take the overall elastic properties of the composite, but the stiffness parameters of the single components for rovings and matrix. For the thermoplastic matrix, a parameter identification based on experimental results was undertaken, presented in Section 5.2. The UD structures are represented by an initially transverse isotropic material model (cf. Section 4.2) and the resulting elastic properties are summarized in Table 5.2. The fiber volume fractions plays an important role for the following investigations. The roving structures in the present weave are characterized by a local fiber volume fraction of $v_\mathrm{f} = 63\%$. The overall fiber volume fraction in the weave was determined to be $v_\mathrm{tow} = 74.6\%$ (cf. Table 3.7). This results in the single volume fractions of the homogenized rovings of $c^{\mathrm{r}_1} = c^{\mathrm{r}_2} = 0.746/2$ where a balanced weave is assumed. The volume fraction of the matrix is determined by Eq. 4.154 and was defined as $c^{\mathrm{m}} = 0.254$. Having the computational model implemented, the resulting stiffness body can now be calculated by means of six calculations corresponding to the six uniaxial load cases. After volumetric homogenization, the full stiffness tensor is obtained. Table 5.5 shows the averaged elastic properties for a composite. Assuming an initial orthotropic symmetry, the stiffness tensor can be described by nine independent variables. Figure 5.40a

Table 5.5: Obtained stiffness parameters computed by the homogenized weave model (*HomWeave*). The parameters concerning the volume fraction were taken from Table 3.7. The elastic properties of the yarns were obtained by volumetric homogenization (cf. Table 5.4).

Elastic properties	
$\bar{E}_1$ (in MPa)	21562.82
$\bar{E}_2$ (in MPa)	21562.82
$\bar{E}_3$ (in MPa)	9246.24
$\bar{\nu}_{12}$ (-)	0.111
$\bar{\nu}_{13}$ (-)	0.449
$\bar{\nu}_{23}$ (-)	0.45
$\bar{G}_{12}$ (in MPa)	2336.31
$\bar{G}_{13}$ (in MPa)	2336.31
$\bar{G}_{23}$ (in MPa)	2467.43

shows the resulting three-dimensional representation of the elastic stiffness body of the homogenized weave model under investigation. Again, the two preferred directions in terms of stiffness corresponding with the weft and warp directions become clearly visible. Furthermore, the characteristic symmetry planes are observed. Projecting the three-dimensional stiffness body onto the z-plane, the values of the direction dependent Young's moduli can be directly extracted. Figure 5.40b shows the projected stiffness body considering half symmetry in the x-plane. The preferred axes of the composite are aligned with the indicated 0°- and 90°- direction. Additionally to the results of the material model under investigation, the experimental results as well as the numerical solution obtained by volumetric homogenization of the woven meso structure is depicted here again. As expected for a material model using the homogenization technique according to Voigt [404], the material responds stiffer due to the assumption made in such models. By neglecting the structural conditions on the meso level and instead presuming a simple parallel connection of all constituents the overall elastic properties are overestimated (ca. 30% higher stiffness in the preferred directions compared to the experimental results).

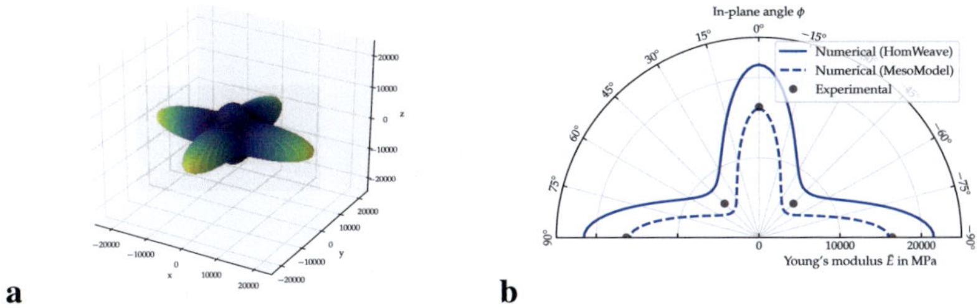

Figure 5.40: (a) Three-dimensional stiffness body computed from six uniaxial load cases using the homogenized three-dimensional weave model. (b) Projection of the three-dimensional stiffness body in the symmetry plane containing the reinforcing rovings (x-y plane from (a)). Half-symmetry is considered here. Moreover the stiffness results from the experiments presented in Section 3.4 as well as the results from the mesomodel are depicted.

5.4.2 Effective composite response

In the following, first results obtained by the simulation of a single element using the newly introduced constitutive model for weave reinforced composites are presented. The first example deals with the case of an 45° off-axis shear load, neglecting damage effects. The purpose of this loading scenario is therefore to show the possibility regarding the reproduction of the kinematical reorientation of the fibers due to an external load. Figure 5.41 depicts the computed stress response (σ_{21} versus ε_{21}) together with the corresponding experimental results. The two preferred axis initially are perpendicular to each other ($\gamma_0 = \measuredangle(\boldsymbol{A}_1, \boldsymbol{A}_2) = 90°$) but are rotated by $\varphi = 45°$ with respect to the loading direction. A higher stiffness in the numerical solution compared to the experiments is obtained. This behavior reproduces the results already observed in Fig. 5.40b. The following non-linearity of the stress-strain curve results from the evolution of plastic strain in the matrix material. This is accompanied by a change in the angle between the two preferred directions as depicted in Fig. 5.42. Here, the angle $\gamma = \measuredangle(\boldsymbol{a}_1, \boldsymbol{a}_2)$ is plotted over time t. With an increasing extent of deformation the angle is steadily decreased. To this end, a stiffer material response results over time, since the reinforcement structures are rotated in load

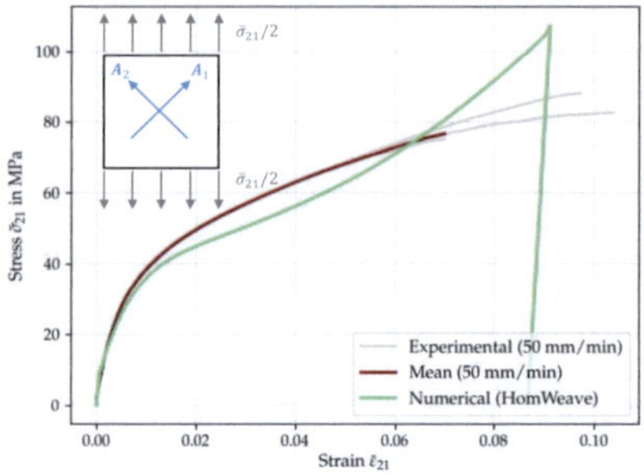

Figure 5.41: Resulting stress-strain diagram ($\bar{\sigma}_{21}$-$\bar{\varepsilon}_{21}$) for an in-plane shear load of a twill weave reinforced unit cell computed by the omogenized material model for weave reinforced composites (*HomWeave*). Additionally, the experimental results for the given load case are presented for comparison (cf. Section 3.3).

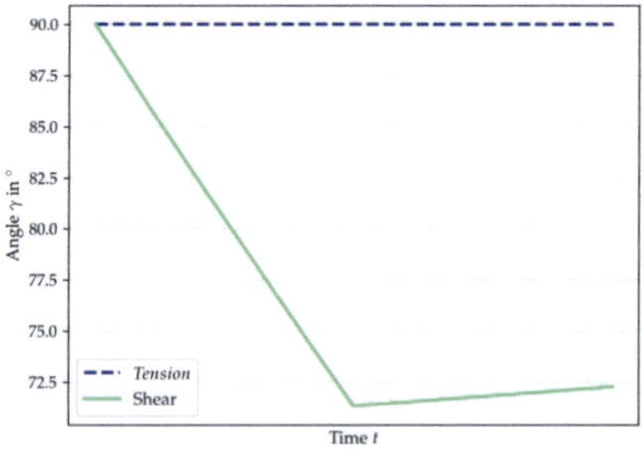

Figure 5.42: Inner angle $\gamma = \measuredangle(\boldsymbol{a}_1, \boldsymbol{a}_2)$ between the reinforcing fibers for the two presented examples over time t. Shear load provokes a reorientation of the reinforcements that is characterized by a change of the inner angle. Upon unloading, the reversible parts of the deformation decrease, leaving behind a remaining reorientation, characterizing a new stress-free state. During the uniaxial load aligned with a preferred direction no change of the angle γ is observed.

directions and therefore make up an increasingly larger proportion of the total stiffness of the composite. Upon unloading, the elastic parts of deformation in the matrix and the reinforcing fibers spring back, while plastic strains remain in the matrix. These prevent a complete back rotation of the roving. This effect can be observed by the increased slope of the stress-strain curve during unloading in Fig. 5.41 and the final angle $\gamma \approx 72.5° < 90°$ in Fig. 5.42. Hence, the introduced plastic strain in the matrix create a changed stress-free state of equilibrium in the composite. In contrast to this is the deformation behavior of a composite, which is loaded in its preferred direction. Figure 5.43 shows the computed $\bar{\sigma}_{11}$-$\bar{\varepsilon}_{11}$ diagram. The load was applied along the reinforcement axes

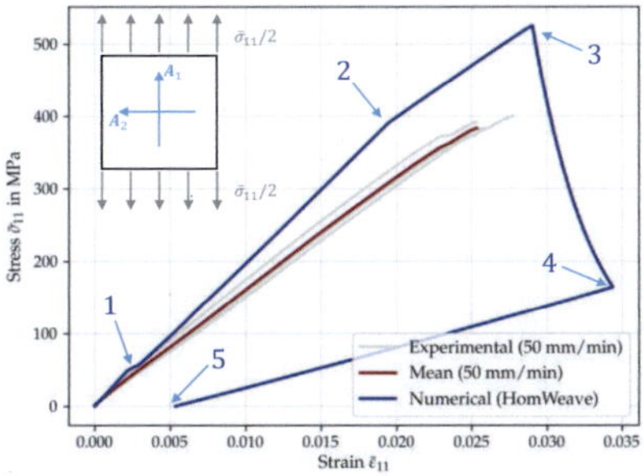

Figure 5.43: Resulting stress-strain diagram ($\bar{\sigma}_{11}$-$\bar{\varepsilon}_{11}$) for a uniaxial tensile load in a preferred direction ($\boldsymbol{A}_1$) of a twill weave reinforced unit cell modeled by the homogenized weave model (*HomWeave*). The indicated numbers correspond to the following events: (1) transverse damage initiation in the roving with the direction $\boldsymbol{A}_2$, (2) plastic onset in the matrix, (3) longitudinal damage initiation in $\boldsymbol{A}_1$, (4) point of beginning unloading, (5) unloaded state, revealing the permanent plastic strains.

with the direction $\boldsymbol{A}_1$. The initial stiffness of the numerical solution is higher than the experimental results which agrees with the previous investigations (cf. Fig. 5.40b). The numbers indicated correspond to five consecutive loading stages (1-5). Already early on, transverse damage in the direction perpendicular to the loading ($\boldsymbol{A}_2$) is activated. With the composite stiffness being a function of all constituents, this results directly in a reduced stiffness in the stress response. The superposition of the three material models treats the matrix material as independent, solely coupled to the reinforcing rovings. To this end, when a sufficient degree of deformation is reached, plastic strains occur in the matrix. The kink in the stress-strain curve at (2) corresponds to the onset of plastic evolution in the matrix that continues until the unloading point in (4). At (3) the ultimate strength in longitudinal direction is reached in $\boldsymbol{A}_1$ yielding an abrupt drop in the stress-strain curve up to (4) as the stiffness of the major load carrying structure is reduced significantly by the evolving damage parameter d_1^+. (4) denotes the point of unloading. Until here, no damage in the matrix d_{m} was observed. The higher calculated ultimate stresses of the composite result again due to the VOIGT-homogenization. After unloading a permanent deformation remains (5), revealing the plastic strains evolved during the loading cycle.

6 Conclusion

This thesis deals with the multiscale investigation of a weave reinforced composite with a thermoplastic polypropylene matrix. To this end, three different subscales were examined throughout this work, which have been identified by an optical analysis: (a) matrix and fibers as the basic constituents, (b) unidirectionally reinforced microstructures, and (c) the full composite of woven fiber structures in thermoplastic matrix. For all three geometrical scales an experimental and numerical treatment is proposed, which were developed in close coordination. The common thread of the present work was to introduce a closed multiscale chain in order to computationally predict the material behavior of the macroscopic composite in terms of deformation and failure. The first part of the presented work discusses the experimental assessment of the different levels (a)-(c) of the material system at hand (cf. Chapter 3). Carefully note, that experimental testing gives an essential input for the subsequent modeling activities, not only for characterization purposes but also for the validation of the model response. Therefore, it is of great importance to use consistent material systems throughout the multiscale approach. In the present case, this requirement could not be strictly adhered to since the unidirectionally reinforced composites could not be purchased from the supplier of the organo sheets. Furthermore, it was not possible to test the material behavior of the fibers individually. The main idea of the present work was to use fitted data as an input only on the lowest scale, i.e. matrix (and fibers). For the matrix, both tensile and compressive tests were performed. Considering different loading rates, the rate dependency upon elastic moduli, plastic properties, and material strength was assessed. The goal of the experimental analysis of the relevant composites (UD and woven) was slightly different. For the later simulation of the UD microstructure and the woven mesostructure it was of great importance to respect the spatial conditions within. For this reason, the structures were initially measured in terms of geometry. The focus of mechanical testing was on determining the anisotropy influence on the deformation behavior of the composites and comparing the observed failure modes with those known from literature. The obtained stress-strain diagrams were used to validate the model response. Chapter 4 lays the theoretical framework of the corresponding computational material models. Three constitutive models were introduced in the course of this work, all formulated in the framework of finite strains. The theoretical basis of this framework was introduced and a geometric interpretation of the basic kinematic equations was presented. Besides the prediction of the deformation, a main focus was to respect material non-linearities, such as plasticity in the matrix material and damage evolution in all underlying material systems. To ensure a thermodynamically consistent behavior, the provided models are derived from the free energy and the dissipation functions.

Starting on the lowest level of basic constituents, the matrix material was characterized with regard to its viscoplastic behavior in combination with rate-dependent failure using standard tests during the experimental campaign. The information obtained here builds the fundamental basis for the identification of the relevant material parameters and hence the starting point of all subsequent investigations. An isotropic elasto-viscoplastic constitutive material model including a formulation for progressive ultimate damage was defined. In a first step, a standard plasticity model is introduced, using an additive elastic-plastic strain split in the framework of small strains. By including a stress-based failure criterion and adding formulations for damage evolution, the model was enabled to predict progressive failure. The problem of mesh-dependent solutions during calculations was addressed by implementing the crack band model, incorporating a characteristic length scale parameter regularizing the dissipation energy of the localizing finite element. Finally, the existing model was extended towards a rate-dependent formulation for plasticity and damage onset. By using purely geometrical operations on the strain tensor, the model was then transferred to a finite strain framework by means of logarithmic strains. The consideration of viscoelasticity was neglected at this point but should be subject of future work. Using the experimental results for the matrix material, the model parameters were adapted. It has to be emphasized, that this is the only time in the present work when a parameter fitting was carried out. The adaptation results showed a good ability of the presented material model to predict the behavior of the thermoplastic matrix both in terms of viscoplasticity and rate-dependent damage onset. Carefully note, that the given model was also used for the fibers with the characteristic of deactivated plasticity and rate-independent damage behavior. With the help of numerical micromechanical analyses a better understanding of the constitutive behavior of composite materials can be achieved. Conforming with the multiscale strategy, such analyses were performed on two geometrical levels. In a first step, UD-SRVEs were generated using an existing method. This routine was then extended towards the consideration of variable fiber diameters as well as the feature of clustered microstructures. Secondly, impregnated woven unit cells were generated based on the determined geometric data. Implementing the given three-dimensional and in-plane periodic boundary conditions, the created microstructures can be loaded in any desired load combination, enabling the assessment of their deformation and failure behavior. With the geometric models of the microstructure generated and the constitutive model for matrix and fibers implemented, an assessment of the unidirectionally fiber reinforced RVEs was performed for two fiber volume fractions. The results regarding elasticity and a selection of load cases with focus on resulting plasticity and damage were presented. Volumetric homogenization of the calculated stress and strain fields enabled the comparison with known experimental investigations. Overall, a good forecasting quality of the predicted solutions and the experimentally obtained results in terms of the transverse isotropic elastic properties can be stated. The failure behavior of UD-SRVEs under transversal loading was investigated in detail and is very consistent with the failure behavior known from literature. However, it should be noted that the matrix of the tested material does not match the one of the adjustment and therefore may explain occurring slight deviations in the forecast of too high plastic contributions and therefore overestimated strength when loaded in the matrix dominated direction. Deviating crystallization states in the matrix can be achieved due to the fiber input, which can lead to a change in material behavior. The usage of the presented methodology allows the definition of a vast number of possible superimposed loads whereby most of them cannot be achieved experimentally. However, one load case has to be already a priori excluded from this definition, the longitudinal compressive load. In reality, local instabilities inside the material initially lead to defects in the matrix, which allows local buckling of the fibers. In the following, this leads to total failure of the whole material. Neglecting an initial undulation of the fibers, this effect can therefore not be reproduced computationally with the described method. As an interface to the next higher geometric scale, a failure envelope was calculated and approximated by the Puck criterion. Hereby the influence of fiber stresses on the transverse failure behavior was neglected and has to be numerically investigated in future work. The required parameters are included in the presented constitutive model for unidirectionally reinforced plastics.

A homogenized material model considering anisotropic damage initiation and progression is necessary to predict the behavior of the impregnated rovings within the woven composite. This model is concerned with an initially transverse isotropic formulation of elasticity in the framework of large strains. After the deduction of the material stiffness by means of the free energy function, a methodology to cover anisotropic damage initiation and progression is presented. Hereby the most important failure modes identified in literature have been implemented. These can be grouped in fiber failure and matrix related failure modes that are effective in perpendicular directions. In both cases a unilateral damage behavior is defined allowing the distinction of tensile and compressive failure and moreover takes into account the interacting effects of damage introduced by tensile or compressive loads. Generally, as soon as damage is introduced, the symmetry class of the material changes. Mesh sensitivity is accounted for by means of the crack band model as before. A known general framework of elastic degradation and damage at small strains was extended towards a formulation of large strains in a logarithmic setting. The kinematic features in terms of deformation, damage onset, and progression were clarified using selected exemplary load cases, showing the model response of single element tests. Virtual assessments of the deformation and failure behavior of the woven fiber reinforced composite were performed in order to show the capabilities of the geometrical as well as the constitutive models. However, the purpose of this demonstrations was not the thorough study of exact damage mechanisms but rather a feasibility study. An intensified study of the micromechanical processes in the wet weaves and a comparison with experiments or findings from literature is recommended.

The last modeling step closes the multiscale chain proposed in this work. Considering a finite change of reinforcement direction due to external loads, a composite formulation of a constitutive model is presented taking into account the aforementioned material models. Hereby, an intermediate stress-free state is established that originates from reversible and irreversible contributions of strains. The main idea of the proposed model is the superposition of two angled preferred directions, which are coupled by the isotropic matrix by means of the analytical homogenization. For reasons of simplicity in the present case the homogenization rule of Voigt [404] was chosen. Using the already introduced constitutive models, it was possible to incorporate the mechanism-based damage formulations for both the matrix as well as the UD material. Thermodynamic consistency is ensured by the separate derivation of the damage evolution respecting the dissipation inequality. The approach of numerical implementation of an analytical homogenization technique with damage formulations in the individual phases in combination with the large strain formulation has not yet been discussed in literature. However, the constitutive equations of the presented model are derived without a sufficient testing of the model response under different load cases. Only two examples showing the kinematic behavior in the event of a load-related change in the preferred direction and the events during damage evolution were presented. The charm of such a simple formulation goes hand in hand with a few drawbacks. By ignoring the inner structure of the weave to a large extent, i.e. the crimp regions and the waviness of the yarns herein, and assuming a simple superposition in the form of a parallel connection instead, an important stress concentration region is ignored. Hence, the results from these models tend to overestimate the elastic moduli and do not reflect relevant local stress conditions which are important for the correct prediction of damage initiation. Furthermore, an asymmetry of stiffness under tension and compression is observed in reality, since the undulated rovings buckle easily under compression whereas they get pulled straight under tensile load. This behavior is not considered in the presented model. Reviewing literature in this regard show alternative homogenization techniques that take into consideration geometrical conditions in the microstructure. The Mori-Tanaka model for woven composites of GOMMERS et al. [127] is to be named here as an example for a more sophisticated model, however formulated in the small strain framework. The extent to which a change of the homogenization technique is successful must be investigated in future work.

A

Detailed experimental data

A.1 Polypropylene matrix

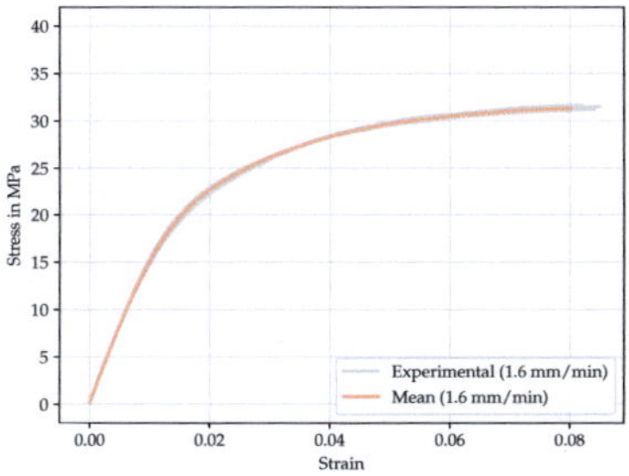

	$E^{\mathrm{t}}_{\mathrm{m}}$ (MPa)	$R^{\mathrm{t}}_{\mathrm{m}}$ (MPa)	$\varepsilon^{\mathrm{t}}_{\mathrm{F}}$ (%)
	1670.40	31.22	8.41
	1641.01	31.51	8.50
	1635.62	31.65	8.21
	1646.67	31.14	8.01
	1624.80	31.05	8.04
μ	1643.70	31.31	8.22
s	16.96	0.22	0.20

Figure A.1: Stress-strain relation and determined parameters of quasi-static tension tests of pure matrix at $v_{\mathrm{T}} = 1.6\,\mathrm{mm/min}$.

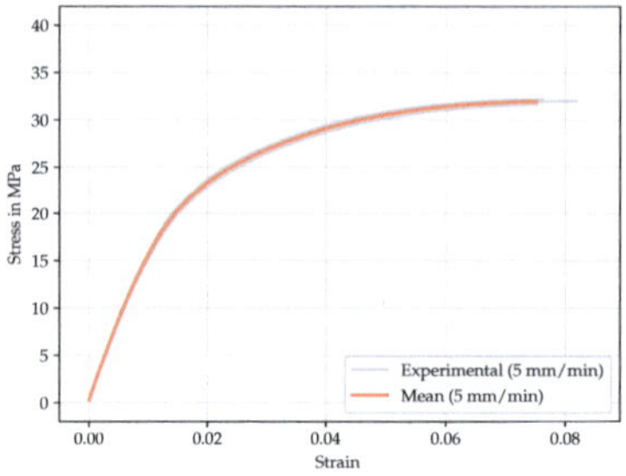

	E_m^t (MPa)	R_m^t (MPa)	ε_F^t (%)
	1661.60	31.64	7.51
	1760.71	32.18	7.61
	1666.86	32.01	8.17
μ	1696.39	31.94	7.77
s	55.76	0.23	0.29

Figure A.2: Stress-strain relation and determined parameters of quasi-static tension tests of pure matrix at v_T = 5 mm/min.

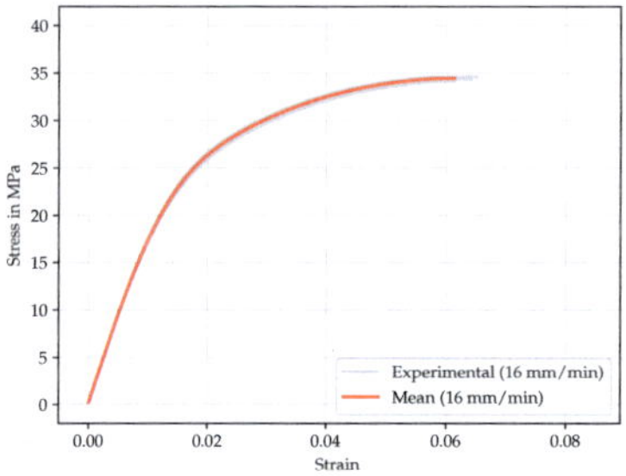

	E_m^t (MPa)	R_m^t (MPa)	ε_F^t (%)
	1887.88	34.32	6.41
	1882.36	34.57	6.52
	1903.70	34.59	6.57
	1933.37	34.49	6.30
	1931.29	34.17	6.12
μ	1907.72	34.43	6.36
s	23.80	0.16	0.14

Figure A.3: Stress-strain relation and determined parameters of quasi-static tension tests of pure matrix at v_T = 16 mm/min.

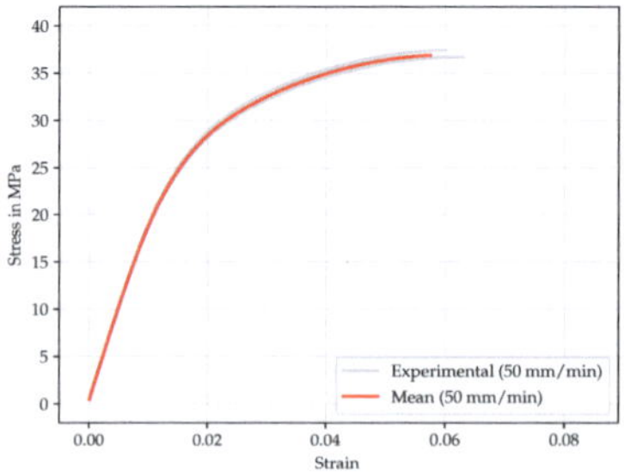

	E_m^t (MPa)	R_m^t (MPa)	ε_F^t (%)
	2031.90	37.43	5.99
	1989.87	36.69	6.28
	1989.40	36.72	6.03
	1946.91	36.87	5.74
μ	1989.52	36.93	6.01
s	43.70	0.29	0.19

Figure A.4: Stress-strain relation and determined parameters of quasi-static tension tests of pure matrix at v_T = 50 mm/min.

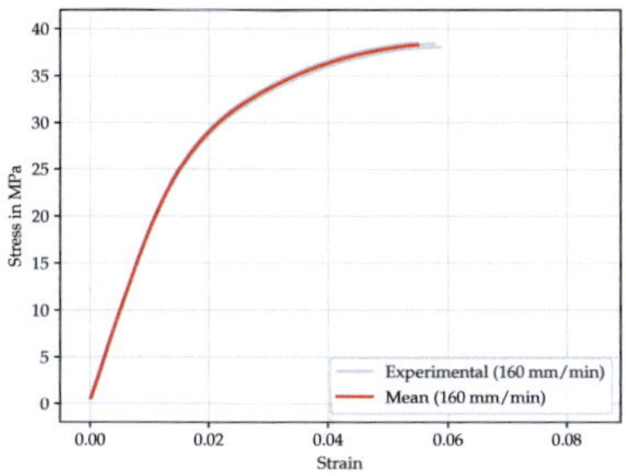

	$E_{\mathrm{m}}^{\mathrm{t}}$ (MPa)	$R_{\mathrm{m}}^{\mathrm{t}}$ (MPa)	$\varepsilon_{\mathrm{F}}^{\mathrm{t}}$ (%)
	1858.31	38.41	5.77
	1925.03	38.04	5.86
	1920.44	38.52	5.49
μ	1901.26	38.32	5.71
s	37.27	0.21	0.16

Figure A.5: Stress-strain relation and determined parameters of quasi-static tension tests of pure matrix at $v_{\mathrm{T}} = 160\,\mathrm{mm/min}$.

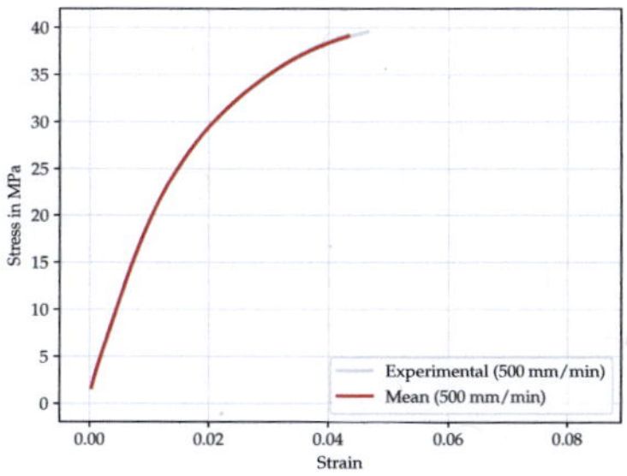

	$E_{\mathrm{m}}^{\mathrm{t}}$ (MPa)	$R_{\mathrm{m}}^{\mathrm{t}}$ (MPa)	$\varepsilon_{\mathrm{F}}^{\mathrm{t}}$ (%)
	2055.26	39.54	4.65
	2042.56	39.51	4.61
	2072.88	39.29	4.45
	1988.80	39.26	4.32
μ	2039.88	39.39	4.51
s	36.26	0.13	0.13

Figure A.6: Stress-strain relation and determined parameters of quasi-static tension tests of pure matrix at $v_{\mathrm{T}} = 500\,\mathrm{mm/min}$.

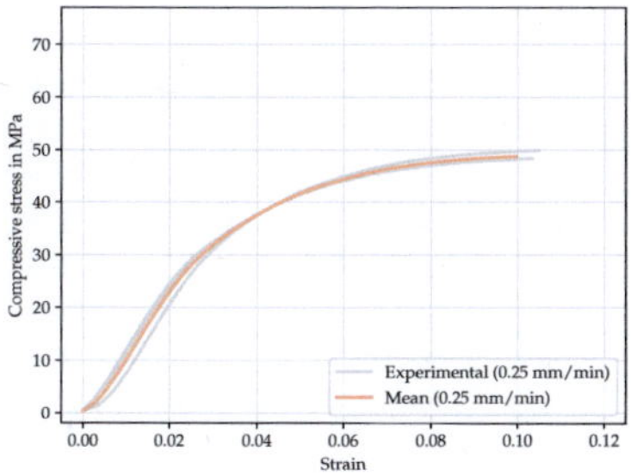

	E_{c} (MPa)	$R_{\mathrm{m}}^{\mathrm{c}}$ (MPa)	$\varepsilon_{\mathrm{F}}^{\mathrm{c}}$ (%)
	1314.30	49.92	10.51
	1163.97	48.39	10.33
	1223.21	48.19	9.94
μ	1233.83	48.83	10.26
s	61.83	0.76	0.23

Figure A.7: Stress-strain relation and determined parameters of quasi-static compression tests of pure matrix at $v_{\mathrm{T}} = 0.25\,\mathrm{mm/min}$.

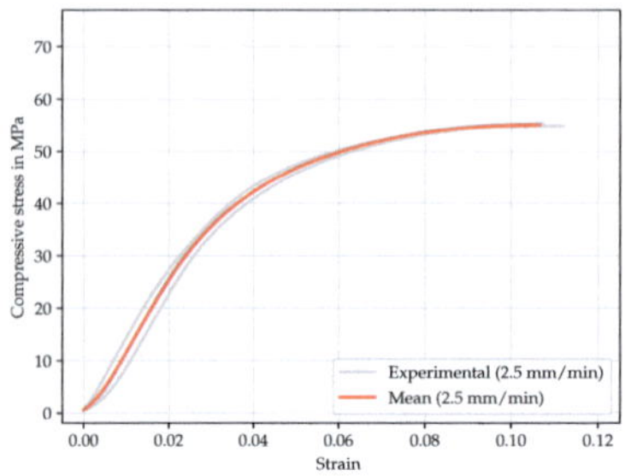

	E_c (MPa)	R^c_m (MPa)	ε^c_F (%)
	1199.55	54.83	10.65
	1385.08	55.37	10.73
	1405.78	54.89	11.19
μ	1330.14	55.03	10.86
s	92.72	0.24	0.24

Figure A.8: Stress-strain relation and determined parameters of quasi-static compression tests of pure matrix at v_T = 2.5 mm/min.

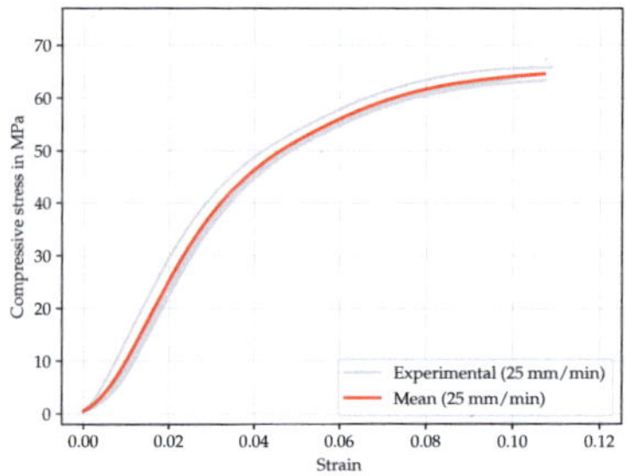

	E_c (MPa)	R^c_m (MPa)	ε^c_F (%)
	1428.26	63.38	10.73
	1486.83	64.53	10.7
	1344.13	65.81	10.87
μ	1419.74	64.57	10.77
s	58.57	0.99	0.07

Figure A.9: Stress-strain relation and determined parameters of quasi-static compression tests of pure matrix at v_T = 25 mm/min.

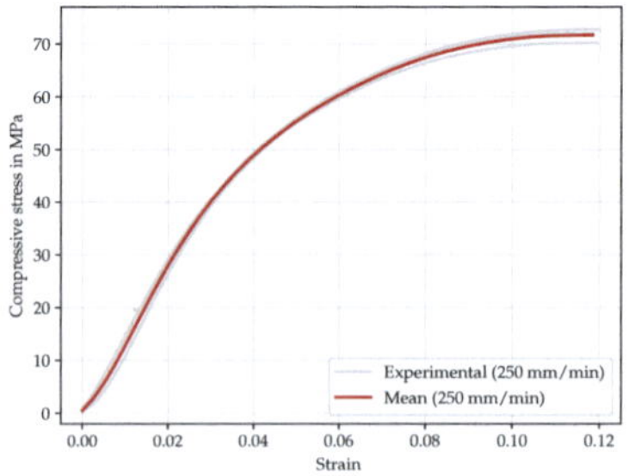

	E_c (MPa)	R^c_m (MPa)	ε^c_F (%)
	1533.29	72.76	12.01
	1543.41	71.99	11.83
	1500.26	70.22	11.93
μ	1525.65	71.66	11.93
s	18.43	1.06	0.07

Figure A.10: Stress-strain relation and determined paramters of quasi-static compression tests of pure matrix at v_T = 250 mm/min.

A.2 Unidirectionally glass fiber-reinforced composite

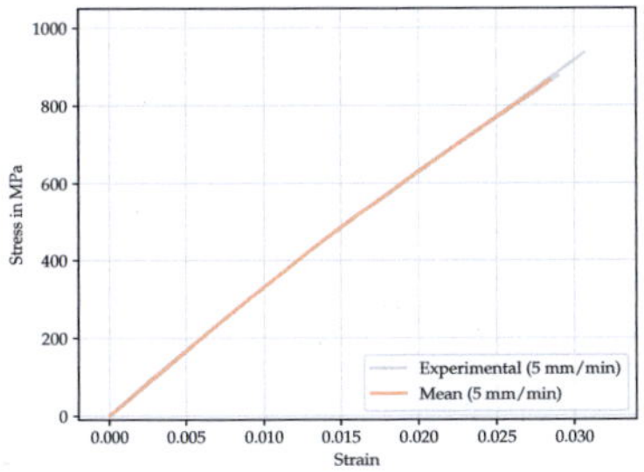

	$E_{\parallel}^{t}$ (GPa)	$R_{\parallel}^{t}$ (MPa)	$\varepsilon_{\parallel}^{t}$ (%)
	33.49	934.25	3.06
	32.91	874.95	2.89
	34.65	859.60	2.84
μ	33.68	889.59	2.93
s	0.89	32.19	0.09

Figure A.11: Stress-strain relation and determined parameters of quasi-static tension tests along the fiber direction of UD specimens at $v_T = 5$ mm/min.

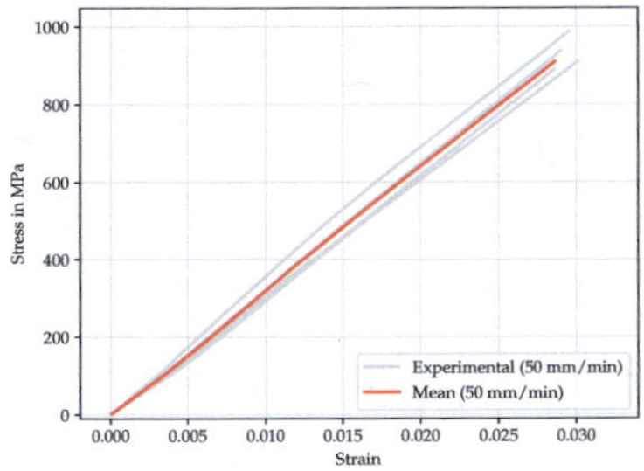

	$E_{\parallel}^{t}$ (GPa)	$R_{\parallel}^{t}$ (MPa)	$\varepsilon_{\parallel}^{t}$ (%)
	26.21	937.43	2.90
	24.94	910.15	3.01
	28.58	889.19	2.86
	28.25	986.99	2.95
μ	27.00	930.94	2.93
s	1.72	36.62	0.06

Figure A.12: Stress-strain relation and determined parameters of quasi-static tension tests along the fiber direction of UD specimens at $v_T = 50$ mm/min.

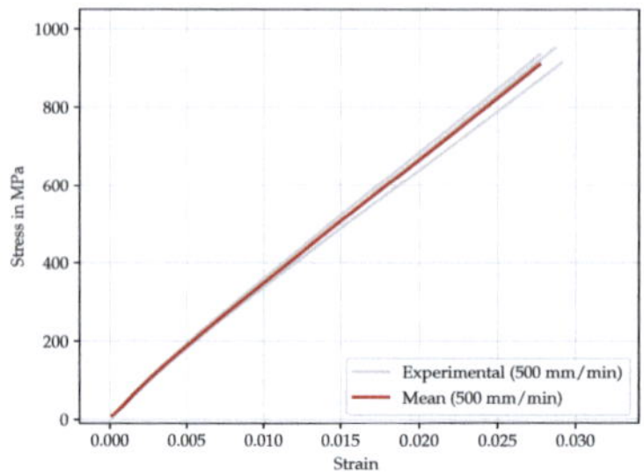

	$E^t_\parallel$ (GPa)	$R^t_\parallel$ (MPa)	$\varepsilon^t_\parallel$ (%)
	39.79	934.49	2.77
	37.24	951.82	2.87
	39.57	914.04	2.91
μ	38.87	933.45	2.85
s	1.42	15.44	0.06

Figure A.13: Stress-strain relation and determined parameters of quasi-static tension tests along the fiber direction of UD specimens at v_T = 500 mm/min.

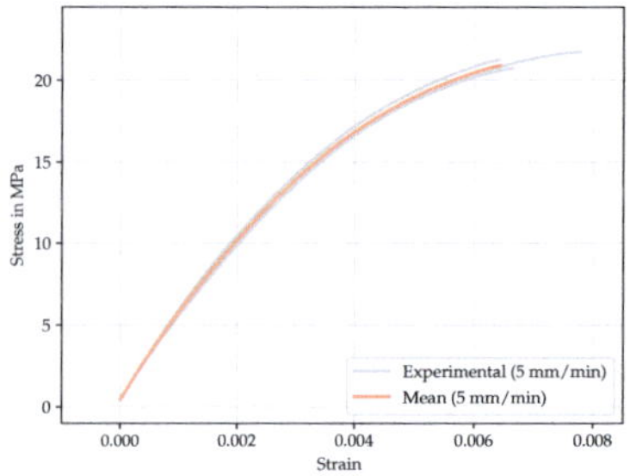

	$E^t_\perp$ (GPa)	$R^t_\perp$ (MPa)	$\varepsilon^t_\perp$ (%)
	4.32	21.73	0.78
	4.40	20.72	0.66
	4.55	21.25	0.64
μ	4.42	21.23	0.69
s	0.11	0.41	0.06

Figure A.14: Stress-strain relation and determined parameters of quasi-static tension tests in transverse direction of UD specimens at v_T = 5 mm/min.

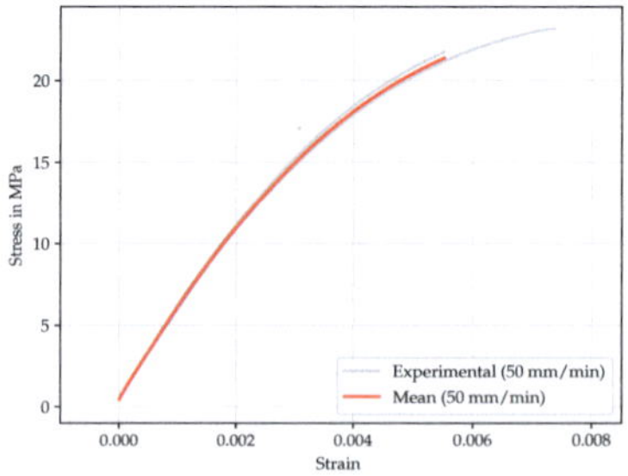

	$E^t_\perp$ (GPa)	$R^t_\perp$ (MPa)	$\varepsilon^t_\perp$ (%)
	4.82	22.73	0.68
	4.77	23.16	0.74
	4.92	21.72	0.55
μ	4.84	22.54	0.65
s	0.09	0.61	0.08

Figure A.15: Stress-strain relation and determined parameters of quasi-static tension tests in transverse direction of UD specimens at v_T = 50 mm/min.

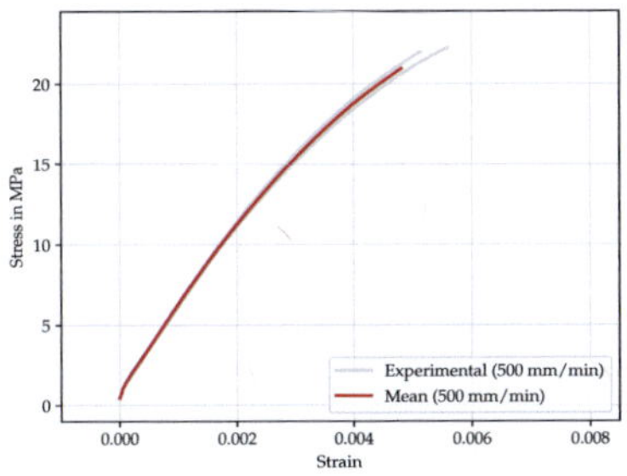

	$E^{t}_{\perp}$ (GPa)	$R^{t}_{\perp}$ (MPa)	$\varepsilon^{t}_{\perp}$ (%)
	4.91	20.94	0.48
	4.97	22.24	0.56
	5.10	21.96	0.51
μ	5.00	21.71	0.52
s	0.11	0.57	0.03

Figure A.16: Stress-strain relation and determined parameters of quasi-static tension tests in transverse direction of UD specimens at $v_T = 500$ mm/min.

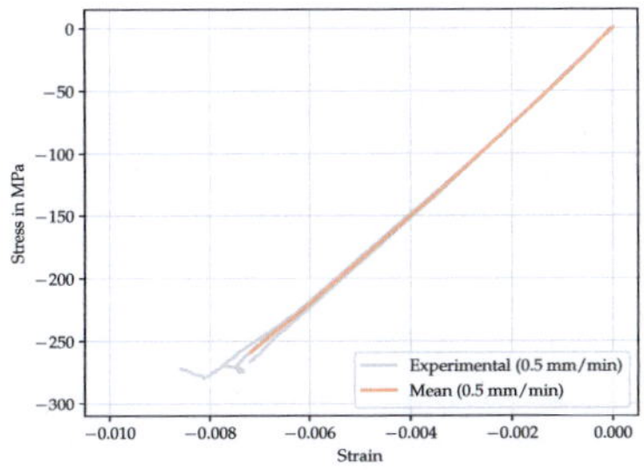

	$E^{c}_{\parallel}$ (GPa)	$R^{c}_{\parallel}$ (MPa)	$\varepsilon^{c}_{\parallel}$ (%)
	36.41	272.04	0.86
	35.50	269.62	0.77
	37.26	266.32	0.72
μ	36.39	269.32	0.78
s	0.88	2.35	0.06

Figure A.17: Stress-strain relation and determined parameters of quasi-static compression tests in longitudinal direction of UD specimens at $v_T = 0.5$ mm/min.

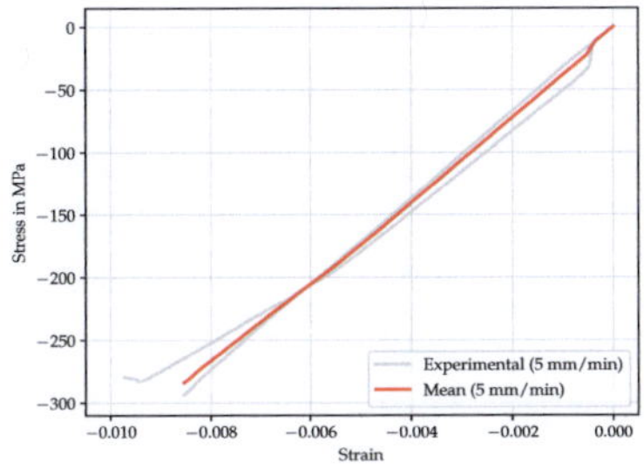

	$E^{c}_{\parallel}$ (GPa)	$R^{c}_{\parallel}$ (MPa)	$\varepsilon^{c}_{\parallel}$ (%)
	34.97	293.58	0.85
	35.16	293.18	0.85
	35.12	279.26	0.97
μ	35.08	288.80	0.89
s	0.08	6.75	0.06

Figure A.18: Stress-strain relation and determined parameters of quasi-static compression tests in longitudinal direction of UD specimens at $v_T = 5$ mm/min.

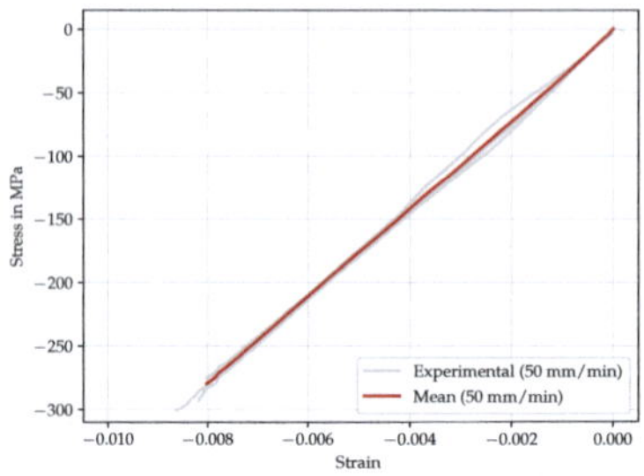

	$E_{\parallel}^{c}$ (GPa)	$R_{\parallel}^{c}$ (MPa)	$\varepsilon_{\parallel}^{c}$ (%)
	32.15	292.92	0.82
	35.41	300.42	0.86
	32.52	275.01	0.80
μ	33.36	289.45	0.83
s	1.79	10.66	0.03

Figure A.19: Stress-strain relation and determined parameters of quasi-static compression tests in longitudinal direction of UD specimens at v_T = 50 mm/min.

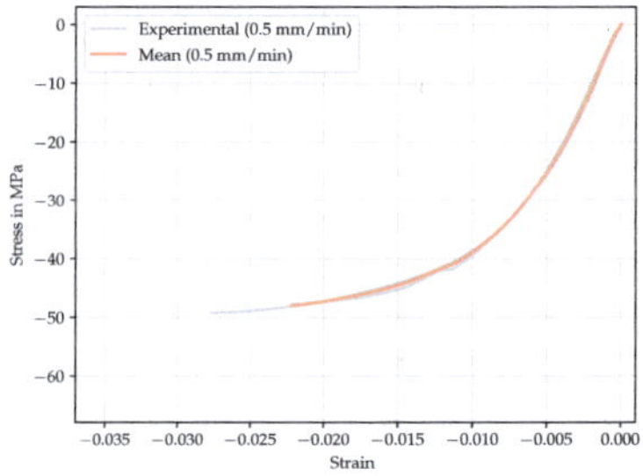

	$E_{\perp}^{c}$ (GPa)	$R_{\perp}^{c}$ (MPa)	$\varepsilon_{\perp}^{c}$ (%)
	5.14	47.83	2.21
	4.87	49.25	2.77
	4.88	49.24	2.76
μ	4.96	48.78	2.59
s	0.13	0.67	0.26

Figure A.20: Stress-strain relation and determined parameters of quasi-static compression tests in transversal direction of UD specimens at v_T = 0.5 mm/min.

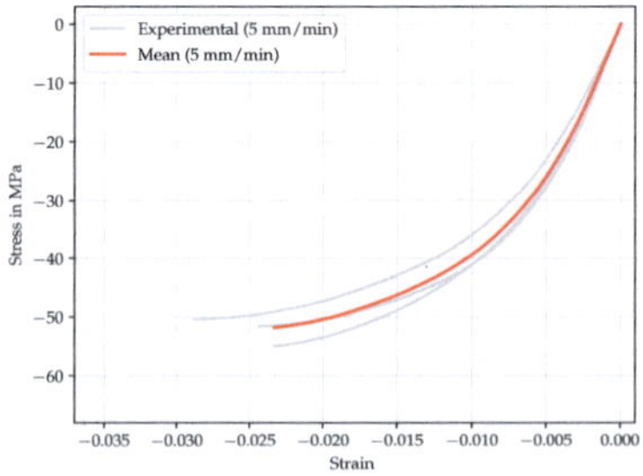

	$E_{\perp}^{c}$ (GPa)	$R_{\perp}^{c}$ (MPa)	$\varepsilon_{\perp}^{c}$ (%)
	4.28	50.42	2.88
	5.15	54.93	2.33
	4.41	51.57	2.44
μ	4.61	52.31	2.55
s	0.47	1.91	0.24

Figure A.21: Stress-strain relation and determined parameters of quasi-static compression tests in transversal direction of UD specimens at v_T = 5 mm/min.

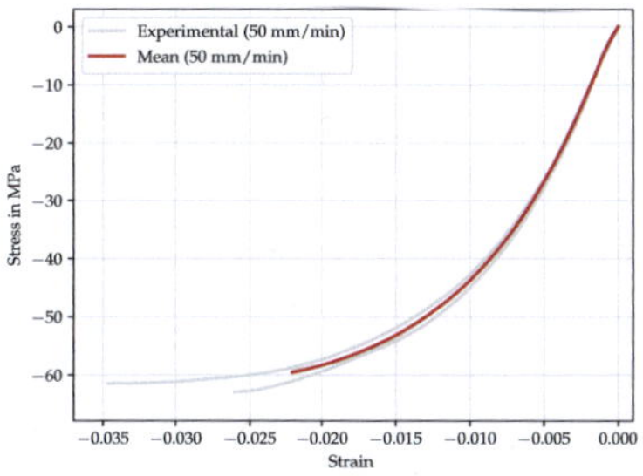

	$E_\perp^c$ (GPa)	$R_\perp^c$ (MPa)	$\varepsilon_\perp^c$ (%)
	5.71	59.01	2.21
	5.38	61.47	3.47
	4.54	62.92	2.60
μ	5.21	61.14	2.76
s	0.60	1.61	0.53

Figure A.22: Stress-strain relation and determined parameters of quasi-static compression tests in transversal direction of UD specimens at v_T = 50 mm/min.

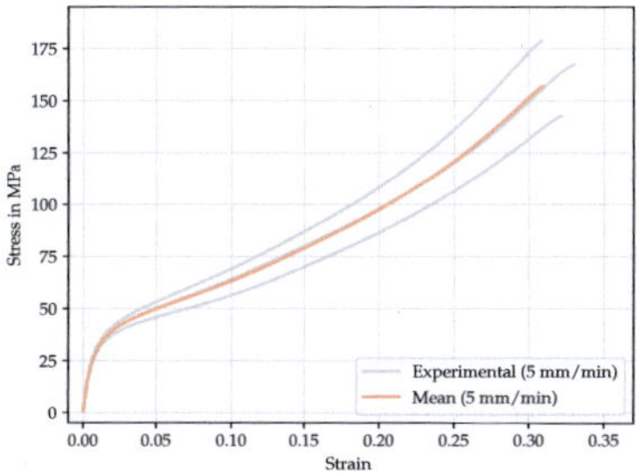

	E_s (GPa)	R_s (MPa)	ε_s (%)
	4.96	167.29	33.00
	5.18	178.86	30.86
	4.83	142.78	32.17
μ	4.99	162.98	32.01
s	0.18	15.04	0.88

Figure A.23: Stress-strain relation and determined parameters of quasi-static shear tests of UD specimens at v_T = 5 mm/min.

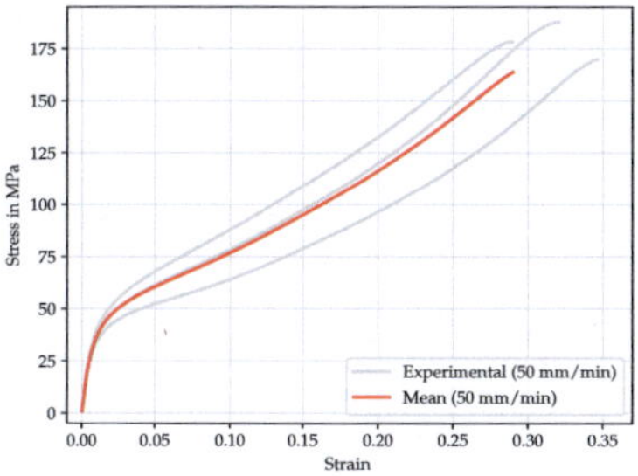

	E_s (GPa)	R_s (MPa)	ε_s (%)
	5.24	169.89	34.64
	5.81	187.76	32.04
	6.10	178.26	28.96
μ	5.72	178.64	31.88
s	0.44	7.30	2.32

Figure A.24: Stress-strain relation and determined parameters of quasi-static shear tests of UD specimens at v_T = 50 mm/min.

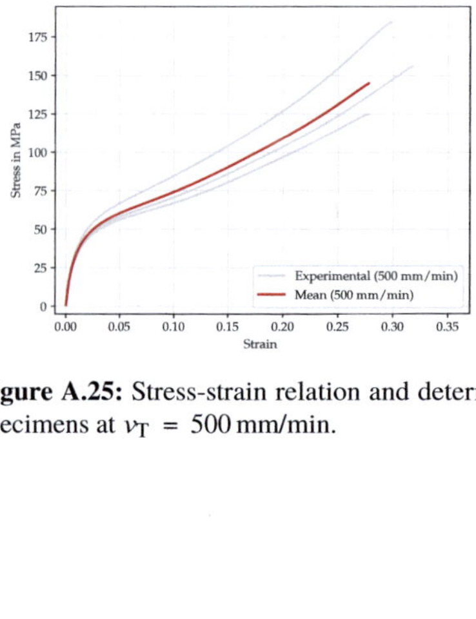

	E_s (GPa)	R_s (MPa)	ε_s (%)
	4.66	124.89	27.78
	5.26	155.89	31.77
	4.55	184.85	29.79
μ	4.82	155.21	29.78
s	0.38	24.48	1.63

Figure A.25: Stress-strain relation and determined parameters of quasi-static shear tests of UD specimens at v_T = 500 mm/min.

A.3 Weave reinforced polypropylene

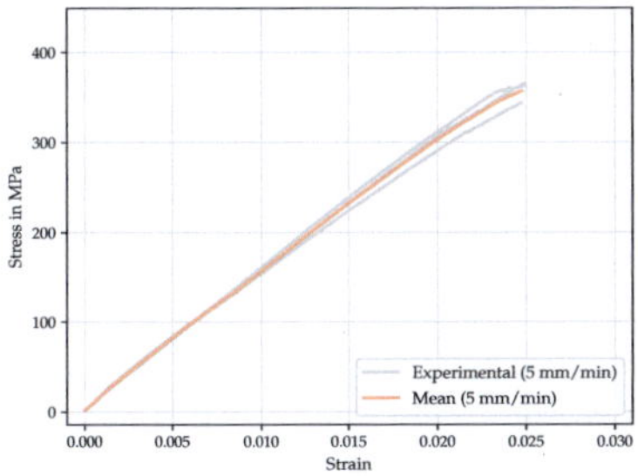

	$E^{t}_{\parallel}$ (GPa)	$R^{t}_{\parallel}$ (MPa)	$\varepsilon^{t}_{\parallel}$ (%)
	16.14	365.12	2.49
	16.60	362.14	2.50
	17.61	343.55	2.47
μ	16.78	356.93	2.49
s	0.75	9.55	0.01

Figure A.26: Stress-strain relation and determined parameters of quasi-static tension tests along the fiber direction of woven specimens at v_T = 5 mm/min.

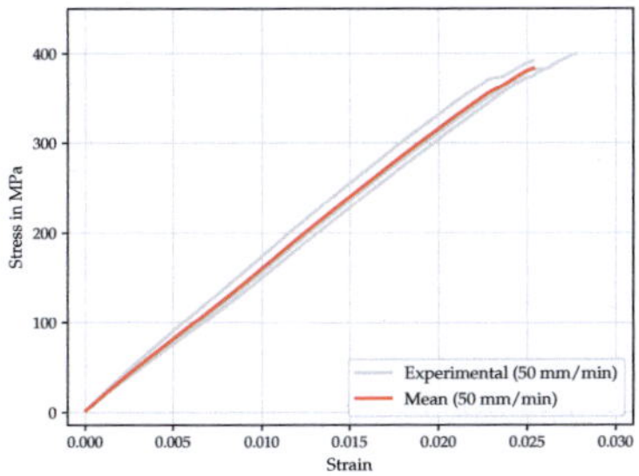

	$E^{t}_{\parallel}$ (GPa)	$R^{t}_{\parallel}$ (MPa)	$\varepsilon^{t}_{\parallel}$ (%)
	15.14	391.40	2.68
	18.57	391.58	2.53
	15.63	399.58	2.77
μ	16.45	394.19	2.66
s	1.86	3.82	0.09

Figure A.27: Stress-strain relation and determined parameters of quasi-static tension tests along the fiber direction of woven specimens at v_T = 50 mm/min.

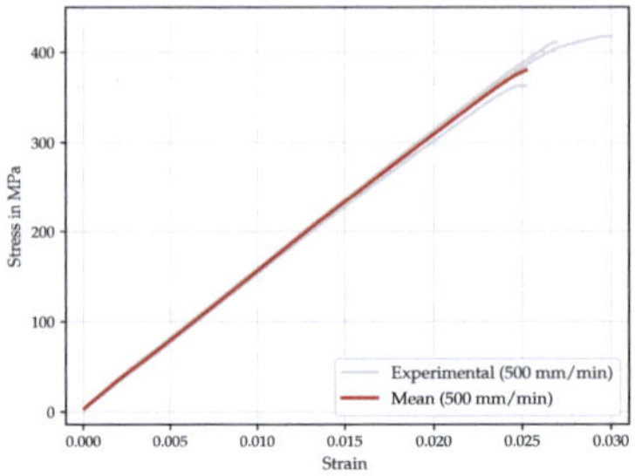

	$E_{\parallel}^{t}$ (GPa)	$R_{\parallel}^{t}$ (MPa)	$\varepsilon_{\parallel}^{t}$ (%)
	16.78	411.36	2.69
	15.37	417.56	2.99
	15.26	362.80	2.52
μ	15.80	397.24	2.74
s	0.86	24.48	0.19

Figure A.28: Stress-strain relation and determined parameters of quasi-static tension tests along the fiber direction of woven specimens at v_T = 500 mm/min.

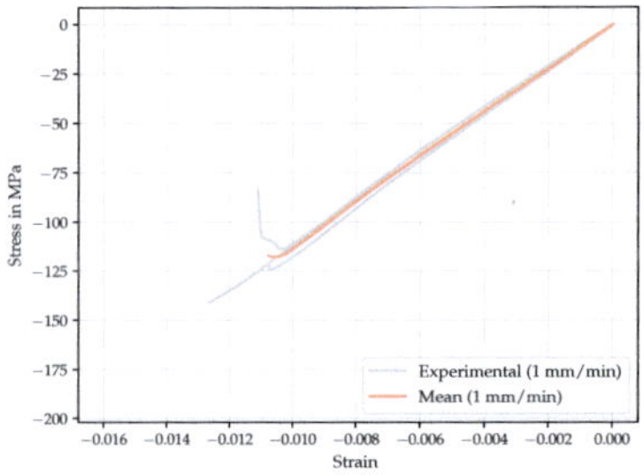

	$E_{\parallel}^{c}$ (GPa)	$R_{\parallel}^{c}$ (MPa)	$\varepsilon_{\parallel}^{c}$ (%)
	10.39	141.10	1.26
	11.88	121.44	1.08
	11.35	83.90	1.11
μ	11.21	115.48	1.15
s	0.75	23.73	0.08

Figure A.29: Stress-strain relation and determined parameters of quasi-static compression tests along the fiber direction of woven specimens at v_T = 1 mm/min.

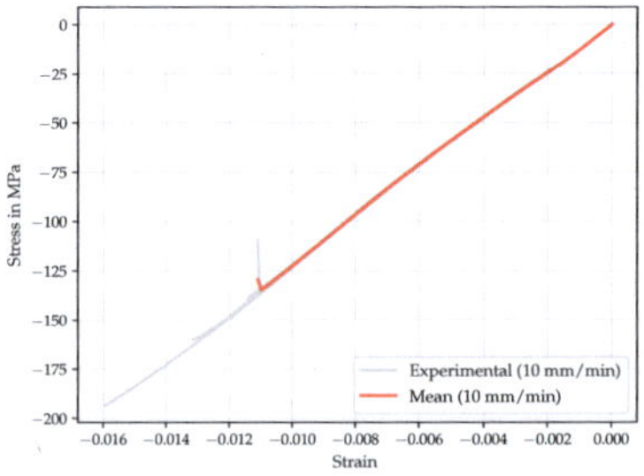

	$E_{\parallel}^{c}$ (GPa)	$R_{\parallel}^{c}$ (MPa)	$\varepsilon_{\parallel}^{c}$ (%)
	12.18	109.23	1.11
	11.56	138.88	1.14
	11.83	193.26	1.59
	12.33	160.10	1.31
μ	11.98	150.37	1.29
s	0.35	30.65	0.19

Figure A.30: Stress-strain relation and determined parameters of quasi-static compression tests along the fiber direction of woven specimens at v_T = 10 mm/min.

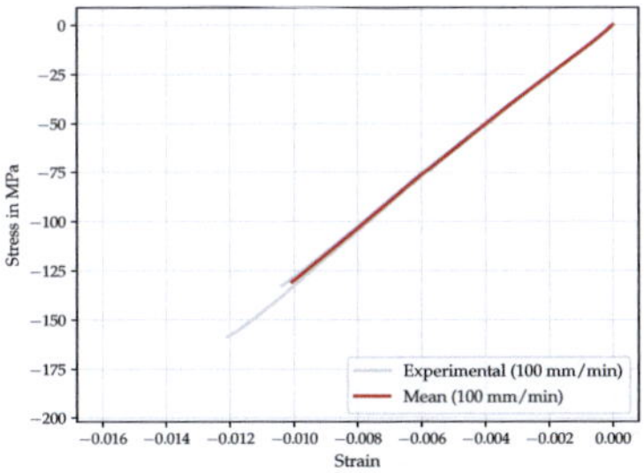

	$E_{\parallel}^{c}$ (GPa)	$R_{\parallel}^{c}$ (MPa)	$\varepsilon_{\parallel}^{c}$ (%)
	12.45	130.04	1.01
	12.18	132.62	1.04
	12.46	158.55	1.21
μ	12.36	140.41	1.09
s	0.16	12.88	0.09

Figure A.31: Stress-strain relation and determined parameters of quasi-static compression tests along the fiber direction of woven specimens at v_T = 100 mm/min.

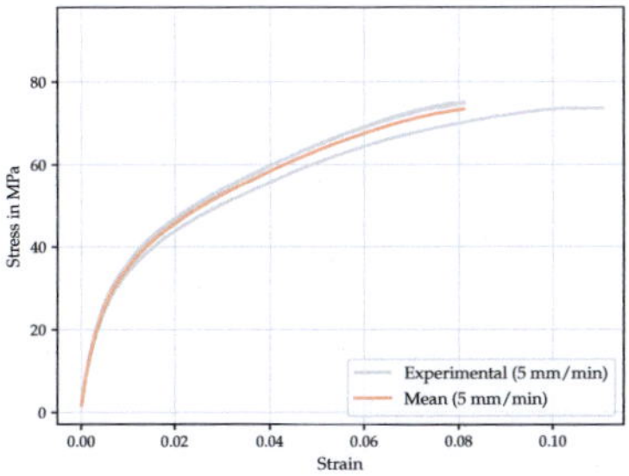

	E_s (GPa)	R_s (MPa)	ε_s (%)
	5.23	73.61	11.04
	5.94	74.79	8.11
	5.53	75.08	8.09
μ	5.56	74.49	9.08
s	0.37	0.64	1.39

Figure A.32: Stress-strain relation and determined parameters of quasi-static ±45° off-axis tension tests of woven specimens at v_T = 5 mm/min.

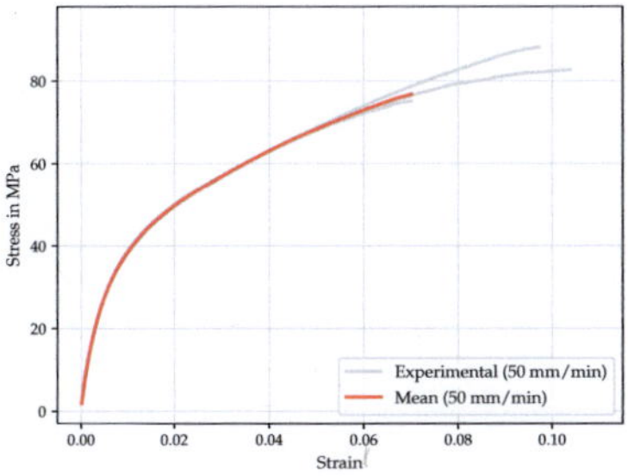

	E_s (GPa)	R_s (MPa)	ε_s (%)
	6.06	88.12	9.72
	5.95	75.11	6.99
	5.88	82.63	10.38
μ	5.96	81.96	9.03
s	0.09	5.33	1.46

Figure A.33: Stress-strain relation and determined parameters of quasi-static ±45° off-axis tension tests of woven specimens at v_T = 50 mm/min.

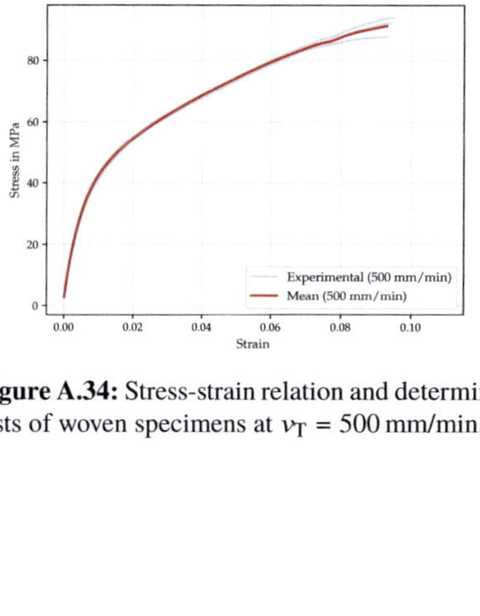

	E_s (GPa)	R_s (MPa)	ε_s (%)
	6.01	87.75	9.29
	6.61	94.04	9.49
	6.60	92.16	9.40
μ	6.40	91.32	9.39
s	0.28	2.64	0.08

Figure A.34: Stress-strain relation and determined parameters of quasi-static ±45° off-axis tension tests of woven specimens at v_T = 500 mm/min.

B

Data sheets

B.1 Data sheet of PP/GF UD material

Celanese
The chemistry inside innovation

CELSTRAN® CFR-TP PP GF70-13 - PP

Description

Celstran® CFR-TP PP GF70-13 is a 70% E-glass fiber by weight PP (polypropylene) continuous fiber (uni-directional) reinforced thermoplastic composite tape. This material exhibits a high strength-to-weight ratio, excellent toughness and chemical resistance. It is well suited for industrial, automotive and sporting goods applications where cost and process ability are critical. The material is available in natural and black colors. Alternate tape widths and thicknesses may be available.

Physical properties	Value	Unit	Test Standard
Density	**1660**	kg/m³	ISO 1183
Fiber Content	**70**	% by wt.	-
Fiber Volume	**45.3**	% by vol.	-
Tape Thickness	**0.25**	mm	-
Tape Width	**305**	mm	-
Tape Areal Weight	**439**	g/m²	-
Fiber Areal Weight	**307**	g/m²	-

Mechanical properties (Tape)	Value	Unit	Test Standard
Tensile Strength, 0°	**931**	MPa	ASTM D 3039M
Tensile Modulus, 0°	**33.9**	GPa	ASTM D 3039M
Tensile Strain at Failure, 0°	**2.99**	%	ASTM D 3039M
Flexural Strength, 0°	**606**	MPa	ASTM D 790/Tape
Flexural Modulus, 0°	**33.2**	GPa	ASTM D 790/Tape
Flexural Strain at Failure, 0°	**2.01**	%	ASTM D 790/Tape

Thermal properties	Value	Unit	Test Standard
Melting temperature, 10°C/min	**173**	°C	ISO 11357-1/-3
Glass transition temperature, 10°C/min	**-10**	°C	ISO 11357-1,-2,-3
CLTE above Tg, parallel	**0.1**	E-4/°C	ISO 11359-2
Start Temp	**23**	°C	ISO 11359-2
End Temp	**50**	°C	ISO 11359-2
CLTE above Tg, normal	**0.66**	E-4/°C	ISO 11359-2
Start Temp	**23**	°C	ISO 11359-2
End Temp	**50**	°C	ISO 11359-2
Thermal conductivity, flow	**0.69**	W/(m K)	ASTM E 1461
Thermal conductivity, crossflow	**0.5**	W/(m K)	ASTM E 1461

Other text information

Compression molding

Celstran® CFR-TP Tape Laminate Processing Guidelines

Celstran® CFR-TP can be molded using a heated platen compression molding press. A hardened steel, aluminum or flexible tooling can be used depending on the application. The tool should be treated with a mold release prior to molding.

The molding cycle consists of the following steps:

1. The platens should be heated above the polymer matrix melt temperature.
2. The individual lamina should be constructed and placed in the tool to achieve the desired laminate reinforcement orientation.
3. The tool is placed between the platens and the platens are closed to achieve a contact pressure on the tool less than 30 psi (2 bar).
4. The tool is allowed to rise in temperature until stabilizing at the initial temperature the platens were set to.
5. The pressure is increased to the desired amount and held for a recommended time.
6. Air and/or water cooling is initiated until the material reaches a temperature sufficiently below the melt and peak crystallization temperatures wherein the pressure is reduced to a contact pressure less than 15 psi (1 bar).
7. The tool is continually cooled until reaching a temperature, typically at or below the glass transition point, at which the pressure is completely removed and the part de-molded from the tool. It should be noted that the choice of tooling, geometry and heating/cooling mechanisms will greatly dictate processing conditions, and thus, optimization specific to the individual molders' capabilities is necessary. Additionally, the resin is what dictates the molding temperatures, whereas the sample thickness is what determines the time. As the thickness increases, the time at melt should also increase to account for the time for heat to conduct to the center of the laminate.

Created: 11.May.2016 Source: www.materialdatacenter.com Page: 1/2

B.2 Data sheet of PP/GF Weave material

TEPEX®
superior in motion

BOND LAMINATES

Material Data Sheet

TEPEX® dynalite 104-RG600(x)/47%
Roving Glass – PP Consolidated Composite Laminate

	Property		*Method ISO*	*Units*	*Longitudinal*	*Transverse*
Material	Reinforcement		Fibres		roving glass	
			Fabric		twill 2/2	
			Area weight	g/m^2	600	
			Yarn	tex	1200	1200
			Weight rate	%	50	50
	Polymer		Polymer		PP	
	Laminate		Density	g/cm^3	1,68	
			Fibre content	% vol.	47	
			Thickness per layer	mm	0,5	
Mechanical	Tensile	Modulus	527-4/5	GPa	20,5	20,1
		Strength	527-4/5	MPa	400	390
		Elongation	527-4/5	%	-	-
		Poisson's ratio	527-4/5		-	-
	Flexural	Modulus	178	GPa	17,5	17,0
		Ultimate stress*	178	MPa	370	365
	Charpy impact strength unnotched	23°C	179/1eU	kJ/m^2	-	-
		-30°C		kJ/m^2	-	-
Thermal	Melting Temperature	per DSC	3146	°C	163	
	Glass transition temperature	per DSC	3146	°C	-	
	Heat deflection temperature	1,80 MPa	75-1/2	°C	158	
	Coefficient of thermal expansion	-30°C to 23°C	ASTM E831	E-6 1/K	-	-
		23°C to 80°C			-	-
	Relative temperature index	20.000 h	IEC 216/1	°C	90	

** 3-Point loading, span-to-depth ratio 16 to 1*

These values are for this specific composition only, the characteristics of composites depend on the reinforcement level and the fibre orientation. Non-standard thickness may also alter some or all of these properties. The data listed here fall within the normal range of product properties, but they should not be used to establish specification limits nor used alone as basis of design.

This information corresponds to our current knowledge on subject. It is offered solely to provide possible suggestions for your own experimentations. It is not intended, however; to substitute for any testing you may need to conduct to determine for yourself the suitability of our products for your particular purposes. This information may be subject to revision as new knowledge and experience becomes available. Since we cannot anticipate all variations in actual end-use conditions. Bond-Laminates makes no warranties and assumes no liability in connection with any use of this information. Nothing in this publication is to be considered as a licence to operate under or a recommendation to infringe any patent right.
Caution: *Do not use this product in medical applications involving permanent implantation in human body.*

Version: 07-01-2009

® Bond-Laminates registered trademark

C

Determination of the adjustment parameter

In order to find the suitable value for A_m that returns a macro damage value d such that Eq. 4.47 is fulfilled, a virtual thought experiment in form of a uniaxial tension test has to be conducted locally (i.e. in each integration point of the element). Therefore, a uniaxial (trial) stress state is assumed (displayed here in Voigt notation)

$$\boldsymbol{\sigma}_0^{\mathrm{UN}} = \begin{pmatrix} \sigma_{0,11}^{\mathrm{UN}} \\ 0 \\ 0 \\ 0 \\ 0 \\ 0 \end{pmatrix}, \tag{C.1}$$

yielding

$$\sigma_{11}^{\mathrm{UN}} = (1-d)\sigma_{0,11}^{\mathrm{UN}} \tag{C.2}$$

and all other entries of $\boldsymbol{\sigma}^{\mathrm{UN}}$ zero. Hence, we can rewrite the rate of energy dissipation per unit volume

$$\frac{\partial \Lambda(\boldsymbol{\sigma}^{\mathrm{UN}})}{\partial d_\sigma} = \Lambda_0(\boldsymbol{\sigma}^{\mathrm{UN}}) \tag{C.3}$$

and

$$\begin{aligned} \frac{\partial d_\sigma}{\partial r} &= \frac{\partial d_\sigma}{\partial d}\frac{\partial d}{\partial r} \\ &= \frac{1}{(1-d)^2}\frac{A_\mathrm{m} r}{r^2}\exp\left(A_\mathrm{m}(1-r)\right). \end{aligned} \tag{C.4}$$

Finally, the damage criterion in case of just reached uniaxial failure reads then as follows

$$F^{\mathrm{d,UN}} = \frac{\left(\sigma_{0,11}\right)^2}{X_\mathrm{c} X_\mathrm{t}} + \sigma_{0,11} \frac{(X_\mathrm{c} - X_\mathrm{t})}{X_\mathrm{c} X_\mathrm{t}} - r = 0 \tag{C.5}$$

and therefore, solving for the critical trial stress $\sigma_{0,11}$

$$\sigma_{0,11}^{\mathrm{UN,crit}} = \frac{(X_\mathrm{t} - X_\mathrm{c}) + \sqrt{(X_\mathrm{c} - X_\mathrm{t})^2 + 4 X_\mathrm{c} X_\mathrm{t} r}}{2}. \tag{C.6}$$

Note that the criterion for damage onset has been defined regarding the trial stresses $\boldsymbol{\sigma}_0$. With the above given information we can rewrite the integral for the dissipated energy per unit volume (dissipated energy density) ψ^D for the uniaxial case

$$\begin{aligned} \psi^{\mathrm{D,UN}} &= \int_1^\infty \frac{\partial \Lambda\left(\sigma_{11}^{\mathrm{UN,crit}}\right)}{\partial d_\sigma} \frac{\partial d_\sigma}{\partial d} \frac{\partial d}{\partial r} \mathrm{d}r = \frac{G_\mathrm{f}}{l^\mathrm{e}} \\ &= \int_1^\infty \left(\Lambda_0 \left(\sigma_{0,11}^{\mathrm{UN,crit}}\right) \frac{\partial d_\sigma}{\partial d} \frac{\partial d}{\partial r} \right) \mathrm{d}r \end{aligned} \tag{C.7}$$

Equation C.7 is solved numerically using the secant method and the definition of damage Eq. 4.46 in order to find the adjustment parameter A_m. This procedure is followed according to MAIMÌ et al. [245]. For geometrically linear calculations A_m has to be computed only once (at the first appearance of damage for each integration point), since the volume of the element is not changing. Non-linear calculations requires the computation in each time increment as the volume does not stay constant and hence l^e differs.

D

Formulation of periodic boundary conditions

Periodic boundary conditions are a frequently used measure when it comes to modeling of repetitive or periodic microstructures. Hereby, representative volume elements are used primarily. In order to allow a straightforward definition within the commercial FE-tool Abaqus (cf. [355]) the corresponding equations are given here. These equations are additional boundary conditions that are considered during calculation of the deformation. The exact formulations can be found by constructing the kinematical relations occurring during the classical six cases of deformation of a representative volume (cf. Fig. D.1) which are divided in three normal and three shear cases. The resulting set of fully three-dimensional equations for cubical RVEs as well as for RVEs with a in-plane periodicity are given in Appendices D.1 and D.2 using the notation of a Abaqus input-file. The used nomenclature for vertices, edges and faces is demonstrated in Fig. D.2. The used vertices, edges and faces must be defined as element sets, since the corresponding equations are evaluated node-wise. Since the introduced equations correspond to a relative connection between opposing sides of the RVE, one point must be encastred. By definition the vertex with the coordinate *XPYPZP* is chosen for that purpose. The user-defined deformation

Table D.1: Assignment of controlling degree of freedoms of reference points.

DOF	RPT	RPS_1	RPS_2
1	XX	YX	XY
2	YY	ZY	YZ
3	ZZ	XZ	ZX

can be controlled by the insertion of additional degree of freedoms (*DOF*)

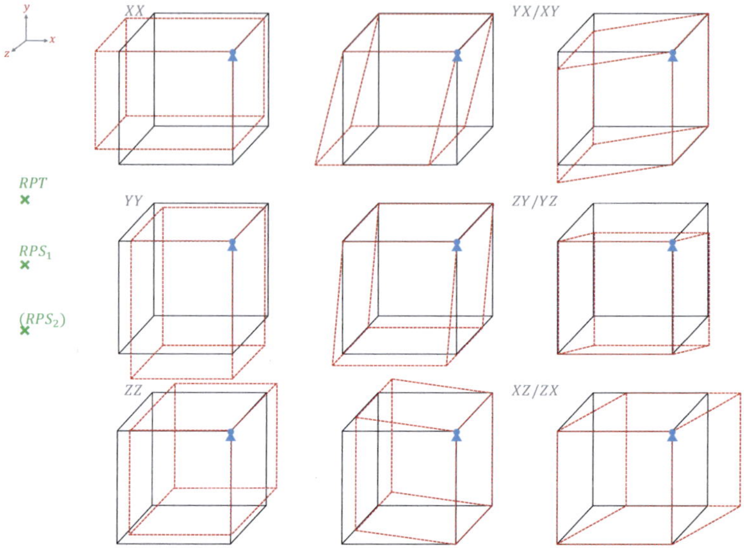

Figure D.1: Depiction of the six classical load cases (three normal tension/compression (*XX*, *YY*, *ZZ*) and three shear load cases (*YX/XY*, *ZY/YZ*, *XZ/ZX*)) that build the basis for the derivation of the periodic boundary conditions. The three displayed reference points provide six additional degrees of freedom that allow the kinematic control of the RVE. Depending on the internal microstructure of the RVE, fluctuating boundaries will form.

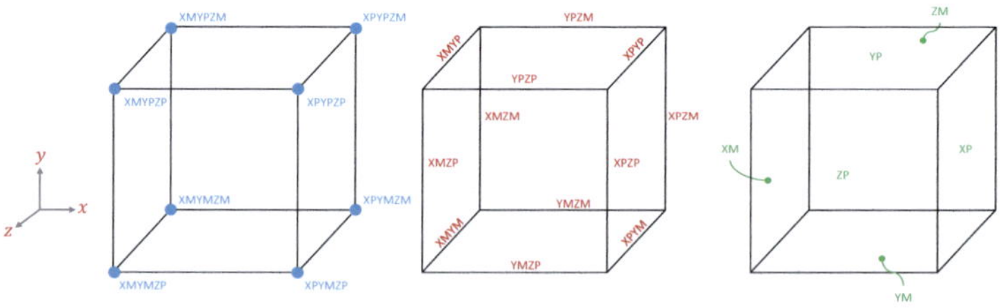

Figure D.2: Nomenclature of vertices, edges and faces of a cubical RVE. P= *plus* and M= *minus* indicate together with the coordinate $X/Y/Z$ the exact location of the respective entity.

which is done by inserting three reference points, RPT, RPS_1, and RPS_2. Each reference points has three DOFs. An overview of the assigned load cases is given in Table D.1. In such a way nine additional degrees of freedom are defined which yields a statically indeterminate system and the number of DOFs has to be reduced. This is done by constraining the double assigned shear degrees. By setting $RPS_2^i = 0 \, \forall \, i \in [1, 3]$, the simple shear cases are obtained. $RPS_2^i = RPS_1^i$ yields the diagonal shear cases. Carefully note, that for the latter case the specified deformation is uniformly distributed in both shear directions. By constraining several DOFs of the reference points (by a certain value or as free), a superposition of deformation boundary conditions is defined and the RVE moves accordingly. Hereby, the inserted equations cause the opposite entities to stay parallel. This yields a deformation that is equal to a behavior

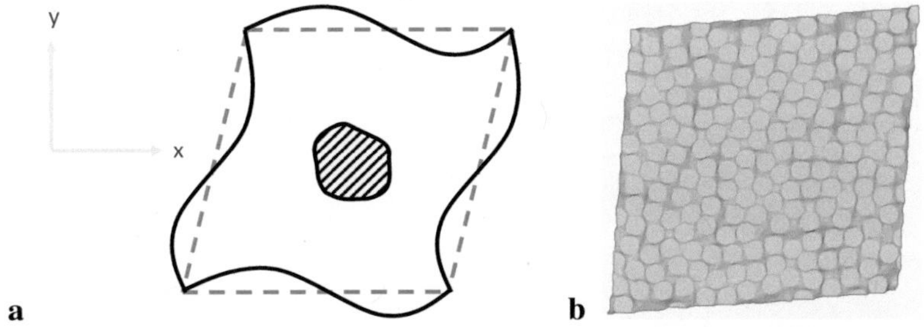

Figure D.3: (a) Two-dimensional RVE deformed under simple YX-shear. The dashed line corresponds to the deformed configuration if no inclusions are present in the volume that could cause perturbations on the boundaries. The full lines show the deformation including the perturbation due to inclusions, hence perturbing the deformation field, finally yielding fluctuations on the boundaries. (b) Example of the deformation of a UD-SRVE ($v_f = 63\%$) under YX/XY-shear.

of an infinite medium, hence forming fluctuating boundaries, depending on the inhomogeneities in the inside of the RVE. Figure D.3a shows two possible configurations of the deformation of a RVE under simple YX-shear. The dashed line corresponds to the deformation that would occur if no inhomogeneities are present in the RVE. The boundaries marked by the full line show fluctuations caused by inclusions in the inner of the RVE. Figure D.3b shows an example of a unidirectionally reinforced $SRVE$ under YX/XY-shear using the below-mentioned periodic boundary conditions.

D.1 Three-dimensional periodic boundary conditions

```
** Vertices
**
*Equation
    5
    SET_XMYMZM, 1, 1.,
    SET_XPYPZP, 1, -1.,
    SET_RPT, 1, -1.,
    SET_RPS_1, 1, -1.,
    SET_RPS_2, 3, -1.
*Equation
    5
    SET_XMYMZM, 2, 1.,
    SET_XPYPZP, 2, -1.,
    SET_RPS_2, 1, -1.,
    SET_RPT, 2, -1.,
    SET_RPS_1, 2, -1.
*Equation
    5
    SET_XMYMZM, 3, 1.,
    SET_XPYPZP, 3, -1.,
    SET_RPS_1, 3, -1.,
    SET_RPS_2, 2, -1.,
    SET_RPT, 3, -1.
*Equation
    5
    SET_XMYPZP, 1, 1.,
    SET_XPYMZM, 1, -1.,
    SET_RPT, 1, -1.,
    SET_RPS_1, 1, 1.,
    SET_RPS_2, 3, 1.
*Equation
    5
    SET_XMYPZP, 2, -1.,
    SET_XPYMZM, 2, 1.,
    SET_RPS_2, 1, 1.,
    SET_RPT, 2, -1.,
    SET_RPS_1, 2, -1.
*Equation
    5
    SET_XMYPZP, 3, -1.,
    SET_XPYMZM, 3, 1.,
    SET_RPS_1, 3, 1.,
    SET_RPS_2, 2, -1.,
    SET_RPT, 3, -1.
*Equation
    5
    SET_XMYPZM, 1, 1.,
    SET_XPYMZP, 1, -1.,
    SET_RPT, 1, -1.,
    SET_RPS_1, 1, 1.,
    SET_RPS_2, 3, -1.
```

```
*Equation
    5
    SET_XMYPZM, 2, 1.,
    SET_XPYMZP, 2, -1.,
    SET_RPS_2, 1, -1.,
    SET_RPT, 2, 1.,
    SET_RPS_1, 2, -1.
*Equation
    5
    SET_XMYPZM, 3, 1.,
    SET_XPYMZP, 3, -1.,
    SET_RPS_1, 3, -1.,
    SET_RPS_2, 2, 1.,
    SET_RPT, 3, -1.
*Equation
    5
    SET_XMYMZP, 1, 1.,
    SET_XPYPZM, 1, -1.,
    SET_RPT, 1, -1.,
    SET_RPS_1, 1, -1.,
    SET_RPS_2, 3, 1.
*Equation
    5
    SET_XMYMZP, 2, 1.,
    SET_XPYPZM, 2, -1.,
    SET_RPS_2, 1, -1.,
    SET_RPT, 2, -1.,
    SET_RPS_1, 2, 1.
*Equation
    5
    SET_XMYMZP, 3, 1.,
    SET_XPYPZM, 3, -1.,
    SET_RPS_1, 3, -1.,
    SET_RPS_2, 2, -1.,
    SET_RPT, 3, 1.
*Equation
    3
    SET_XPYMZP, 1, 1.,
    SET_XPYMZM, 1, -1.,
    SET_RPS_2, 3, 1.
*Equation
    3
    SET_XPYMZP, 2, 1.,
    SET_XPYMZM, 2, -1.,
    SET_RPS_1, 2, 1.
*Equation
    3
    SET_XPYMZP, 3, 1.,
    SET_XPYMZM, 3, -1.,
    SET_RPT, 3, 1.
*Equation
    3
    SET_XPYMZM, 1, 1.,
    SET_XPYPZM, 1, -1,
    SET_RPS_1, 1, -1.
*Equation
```

```
    3
    SET_XPYMZM, 2, 1.,
    SET_XPYPZM, 2, -1,
    SET_RPT, 2, -1.
*Equation
    3
    SET_XPYMZM, 3, 1.,
    SET_XPYPZM, 3, -1,
    SET_RPS_2, 2, -1.
*Equation
    3
    SET_XPYPZM, 1, 1.,
    SET_XPYPZP, 1, -1.,
    SET_RPS_2, 3, -1.
*Equation
    3
    SET_XPYPZM, 2, 1.,
    SET_XPYPZP, 2, -1.,
    SET_RPS_1, 2, -1.
*Equation
    3
    SET_XPYPZM, 3, 1.,
    SET_XPYPZP, 3, -1.,
    SET_RPT, 3, -1.
***********************************
** Edges
**
*Equation
    4
    SET_XMYM, 1, 1.,
    SET_XPYP, 1, -1.,
    SET_RPT, 1, -1.,
    SET_RPS_1, 1, -1.
*Equation
    4
    SET_XMYM, 2, 1.,
    SET_XPYP, 2, -1.,
    SET_RPT, 2, -1.,
    SET_RPS_2, 1, -1.
*Equation
    4
    SET_XMYM, 3, 1.,
    SET_XPYP, 3, -1.,
    SET_RPS_1, 3, -1.,
    SET_RPS_2, 2, -1.
*Equation
    3
    SET_XPYP, 1, 1.,
    SET_XPYM, 1, -1.,
    SET_RPS_1, 1, 1.
*Equation
    3
    SET_XPYP, 2, 1.,
    SET_XPYM, 2, -1.,
    SET_RPT, 2, 1.
*Equation
```

```
    3
    SET_XPYP, 3, 1.,
    SET_XPYM, 3, -1.,
    SET_RPS_2, 2, 1.
*Equation
    4
    SET_XMZM, 1, 1.,
    SET_XPZP, 1, -1.,
    SET_RPT, 1, -1.,
    SET_RPS_2, 3, -1.
*Equation
    4
    SET_XMZM, 2, 1.,
    SET_XPZP, 2, -1.,
    SET_RPS_1, 2, -1.,
    SET_RPS_2, 1, -1.
*Equation
    4
    SET_XMZM, 3, 1.,
    SET_XPZP, 3, -1.,
    SET_RPT, 3, -1.,
    SET_RPS_1, 3, -1.
*Equation
    3
    SET_XPZP, 1, 1.,
    SET_XMZP, 1, -1.,
    SET_RPT, 1, 1.
*Equation
    3
    SET_XPZP, 2, 1.,
    SET_XMZP, 2, -1.,
    SET_RPS_2, 1, 1.
*Equation
    3
    SET_XPZP, 3, 1.,
    SET_XMZP, 3, -1.,
    SET_RPS_1, 3, 1.
*Equation
    4
    SET_YMZM, 1, 1.,
    SET_YPZP, 1, -1.,
    SET_RPS_1, 1, -1.,
    SET_RPS_2, 3, -1.
*Equation
    4
    SET_YMZM, 2, 1.,
    SET_YPZP, 2, -1.,
    SET_RPT, 2, -1.,
    SET_RPS_1, 2, -1.
*Equation
    4
    SET_YMZM, 3, 1.,
    SET_YPZP, 3, -1.,
    SET_RPT, 3, -1.,
    SET_RPS_2, 2, -1.
*Equation
```

```
    3
    SET_YPZP, 1, 1.,
    SET_YPZM, 1, -1.,
    SET_RPS_2, 3, 1.
*Equation
    3
    SET_YPZP, 2, 1.,
    SET_YPZM, 2, -1.,
    SET_RPS_1, 2, 1.
*Equation
    3
    SET_YPZP, 3, 1.,
    SET_YPZM, 3, -1.,
    SET_RPT, 3, 1.
*Equation
    4
    SET_XMYP, 1, 1.,
    SET_XPYM, 1, -1.,
    SET_RPT, 1, -1.,
    SET_RPS_1, 1, 1.
*Equation
    4
    SET_XMYP, 2, 1.,
    SET_XPYM, 2, -1.,
    SET_RPS_2, 1, -1.,
    SET_RPT, 2, 1.
*Equation
    4
    SET_XMYP, 3, 1.,
    SET_XPYM, 3, -1.,
    SET_RPS_1, 3, -1.,
    SET_RPS_2, 2, 1.
*Equation
    4
    SET_YMZP, 1, 1.,
    SET_YPZM, 1, -1.,
    SET_RPS_1, 1, -1.,
    SET_RPS_2, 3, 1.
*Equation
    4
    SET_YMZP, 2, 1.,
    SET_YPZM, 2, -1.,
    SET_RPT, 2, -1.,
    SET_RPS_1, 2, 1.
*Equation
    4
    SET_YMZP, 3, 1.,
    SET_YPZM, 3, -1.,
    SET_RPS_2, 2, -1.,
    SET_RPT, 3, 1.
*Equation
    4
    SET_XMZP, 1, 1.,
    SET_XPZM, 1, -1.,
    SET_RPT, 1, -1.,
    SET_RPS_2, 3, 1.
```

```
*Equation
    4
    SET_XMZP, 2, 1.,
    SET_XPZM, 2, -1.,
    SET_RPS_2, 1, -1.,
    SET_RPS_1, 2, 1.
*Equation
    4
    SET_XMZP, 3, 1.,
    SET_XPZM, 3, -1.,
    SET_RPS_1, 3, -1.,
    SET_RPT, 3, 1.
**********************************
** Faces
**
*Equation
    3
    SET_XM, 1, 1.,
    SET_XP, 1, -1.,
    SET_RPT, 1, -1.
*Equation
    3
    SET_XM, 2, 1.,
    SET_XP, 2, -1.,
    SET_RPS_2, 1, -1.
*Equation
    3
    SET_XM, 3, 1.,
    SET_XP, 3, -1.,
    SET_RPS_1, 3, -1.
*Equation
    3
    SET_YM, 1, 1.,
    SET_YP, 1, -1.,
    SET_RPS_1, 1, -1.
*Equation
    3
    SET_YM, 2, 1.,
    SET_YP, 2, -1.,
    SET_RPT, 2, -1.
*Equation
    3
    SET_YM, 3, 1.,
    SET_YP, 3, -1.,
    SET_RPS_2, 2, -1.
*Equation
    3
    SET_ZM, 1, 1.,
    SET_ZP, 1, -1.,
    SET_RPS_2, 3, -1.
*Equation
    3
    SET_ZM, 2, 1.,
    SET_ZP, 2, -1.,
    SET_RPS_1, 2, -1.
*Equation
```

```
    3
    SET_ZM, 3, 1.,
    SET_ZP, 3, -1.,
    SET_RPT, 3, -1.
```

D.2 In-plane periodic boundary conditions

```
** Vertices
**
*Equation
    4
    SET_XMYMZP, 1, 1.,
    SET_XPYPZP, 1, -1.,
    SET_RPT, 1, -1.,
    SET_RPS_1, 1, -1.
*Equation
    4
    SET_XMYMZP, 2, 1.,
    SET_XPYPZP, 2, -1.,
    SET_RPT, 2, -1.,
    SET_RPS_2, 1, -1.
*Equation
    4
    SET_XMYPZP, 1, 1.,
    SET_XPYMZP, 1, -1.,
    SET_RPT, 1, -1.,
    SET_RPS_1, 1, 1.
*Equation
    4
    SET_XMYPZP, 2, 1.,
    SET_XPYMZP, 2, -1.,
    SET_RPT, 2, 1.,
    SET_RPS_2, 1, -1.
*Equation
    4
    SET_XMYMZM, 1, 1.,
    SET_XPYPZM, 1, -1.,
    SET_RPT, 1, -1.,
    SET_RPS_1, 1, -1.
*Equation
    4
    SET_XMYMZM, 2, 1.,
    SET_XPYPZM, 2, -1.,
    SET_RPT, 2, -1.,
    SET_RPS_2, 1, -1.
*Equation
    4
    SET_XMYPZM, 1, 1.,
    SET_XPYMZM, 1, -1.,
    SET_RPT, 1, -1.,
    SET_RPS_1, 1, 1.
*Equation
    4
    SET_XMYPZM, 2, 1.,
```

```
    SET_XPYMZM, 2, -1.,
    SET_RPT, 2, 1.,
    SET_RPS_2, 1, -1.
*Equation
    3
    SET_XPYMZP, 1, 1.,
    SET_XPYPZP, 1, -1.,
    SET_RPS_1, 1, -1.
*Equation
    3
    SET_XPYMZP, 2, 1.,
    SET_XPYPZP, 2, -1.,
    SET_RPT, 2, -1.
*Equation
    3
    SET_XPYMZM, 1, 1.,
    SET_XPYPZM, 1, -1.,
    SET_RPS_1, 1, -1.
*Equation
    3
    SET_XPYMZM, 2, 1.,
    SET_XPYPZM, 2, -1.,
    SET_RPT, 2, -1.
**********************************
** Edges
**
*Equation
    3
    SET_XMZP, 1, 1.,
    SET_XPZP, 1, -1.,
    SET_RPT, 1, -1.
*Equation
    3
    SET_XMZP, 2, 1.,
    SET_XPZP, 2, -1.,
    SET_RPS_2, 1, -1.
*Equation
    3
    SET_XMZM, 1, 1.,
    SET_XPZM, 1, -1.,
    SET_RPT, 1, -1.
*Equation
    3
    SET_XMZM, 2, 1.,
    SET_XPZM, 2, -1.,
    SET_RPS_2, 1, -1.
*Equation
    4
    SET_XMYM, 1, 1.,
    SET_XPYP, 1, -1.,
    SET_RPT, 1, -1.,
    SET_RPS_1, 1, -1.
*Equation
    4
    SET_XMYM, 2, 1.,
    SET_XPYP, 2, -1.,
```

```
    SET_RPT, 2, -1.,
    SET_RPS_2, 1, -1.
*Equation
    4
    SET_XMYP, 1, 1.,
    SET_XPYM, 1, -1.,
    SET_RPT, 1, -1.,
    SET_RPS_1, 1, 1.
*Equation
    4
    SET_XMYP, 2, 1.,
    SET_XPYM, 2, -1.,
    SET_RPT, 2, 1.,
    SET_RPS_2, 1, -1.
*Equation
    3
    SET_YMZP, 1, 1.,
    SET_YPZP, 1, -1.,
    SET_RPS_1, 1, -1.
*Equation
    3
    SET_YMZP, 2, 1.,
    SET_YPZP, 2, -1.,
    SET_RPT, 2, -1.
*Equation
    3
    SET_YMZM, 1, 1.,
    SET_YPZM, 1, -1.,
    SET_RPS_1, 1, -1.
*Equation
    3
    SET_YMZM, 2, 1.,
    SET_YPZM, 2, -1.,
    SET_RPT, 2, -1.
*Equation
    3
    SET_XPYM, 1, 1.,
    SET_XPYP, 1, -1.,
    SET_RPS_1, 1, -1.,
*Equation
    3
    SET_XPYM, 2, 1.,
    SET_XPYP, 2, -1.,
    SET_RPT, 2, -1.
***********************************
** Faces
**
*Equation
    3
    SET_XM, 1, 1.,
    SET_XP, 1, -1.,
    SET_RPT, 1, -1.
*Equation
    3
    SET_XM, 2, 1.,
    SET_XP, 2, -1.,
```

```
    SET_RPS_2, 1, -1.
*Equation
    3
    SET_YM, 1, 1.,
    SET_YP, 1, -1.,
    SET_RPS_1, 1, -1.
*Equation
    3
    SET_YM, 2, 1.,
    SET_YP, 2, -1.,
    SET_RPT, 2, -1.
**
*Equation
    2
    SET_XMYPZP, 3, 1.,
    SET_XMYMZP, 3, -1.
*Equation
    2
    SET_XMYMZP, 3, 1.,
    SET_XPYMZP, 3, -1.
*Equation
    2
    SET_XPYMZP, 3, 1.,
    SET_XPYPZP, 3, -1.
*Equation
    2
    SET_XMYPZM, 3, 1.,
    SET_XMYMZM, 3, -1.
*Equation
    2
    SET_XMYMZM, 3, 1.,
    SET_XPYMZM, 3, -1.
*Equation
    2
    SET_XPYMZM, 3, 1.,
    SET_XPYPZM, 3, -1.
*Equation
    2
    SET_XMZP, 3, 1.,
    SET_XPZP, 3, -1.
*Equation
    2
    SET_YMZP, 3, 1.,
    SET_YPZP, 3, -1.
*Equation
    2
    SET_XMZM, 3, 1.,
    SET_XPZM, 3, -1.
*Equation
    2
    SET_YMZM, 3, 1.,
    SET_YPZM, 3, -1.
*Equation
    2
    SET_XMYM, 3, 1.,
    SET_XPYM, 3, -1.
```

```
*Equation
    2
    SET_XPYM, 3, 1.,
    SET_XPYP, 3, -1.
*Equation
    2
    SET_XPYP, 3, 1.,
    SET_XMYP, 3, -1.
*Equation
    2
    SET_XM, 3, 1.,
    SET_XP, 3, -1.
*Equation
    2
    SET_YM, 3, 1.,
    SET_YP, 3, -1.
```

Bibliography

[1] Alfano, G. and Crisfield, M. A. Finite element interface models for the delamination analysis of laminated composites: Mechanical and computational issues. *International Journal for Numerical Methods in Engineering*, 50(7):1701–1736, March 2001.

[2] Allix, O. and Corigliano, A. Geometrical and interfacial non-linearities in the analysis of delamination in composites. *International Journal of Solids and Structures*, 36(15):2189–2216, May 1999.

[3] Altenbach, H., Bolchoun, A., and Kolupaev, V. A. Phenomenological yield and failure criteria. In *Plasticity of Pressure-Sensitive Materials*, pages 49–152. Springer Berlin Heidelberg, Berlin, Heidelberg, 2014.

[4] Anand, L. and Gurtin, M. E. A theory of amorphous solids undergoing large deformations, with application to polymeric glasses. *International Journal of Solids and Structures*, 40(6):1465–1487, March 2003.

[5] Arencón, D., Velasco, J. I., Realinho, V., Sánchez-Soto, M., and Gordillo, A. Fracture toughness of glass microsphere-filled polypropylene and polypropylene/poly (ethylene terephthalate-co-isophthalate) blend-matrix composites. *Journal of Materials Science*, 42(1):19–29, January 2007.

[6] Arencón, D., Velasco, J. I., Rodríguez-Pérez, M. A., and de Saja, J. A. Poly(propylene)/poly(ethylene terephthalate-co-isophthalate) blends and glass bead filled composites: Microstructure and thermomechanical properties: PP/co-PET Blends and Glass Bead Filled Composites. *Journal of Applied Polymer Science*, 94(4):1841–1852, November 2004.

[7] Arencón, D., Maspoch, M. L., and Velasco, J. I. The effect of compatibilizing and coupling agents on the mechanical properties of glass bead filled PP/PET blends. *Macromolecular Symposia*, 194(1):225–232, April 2003.

[8] Arencón, D. and Velasco, J. I. Fracture toughness of polypropylene-based particulate composites. *Materials*, 2(4):2046–2094, November 2009.

[9] Arendts, F., Drechsler, K., and Brandt, J. Manufacturing and mechanical performance of composites with 3D woven fiber reinforcement. In *Proceedings of the 38th International SAMPE Symposium*, page 2118, 1989.

[10] Argon, A. S. A theory for the low-temperature plastic deformation of glassy polymers. *Philosophical Magazine*, 28(4):839–865, October 1973.

[11] Arruda, E. M. and Boyce, M. C. Evolution of plastic anisotropy in amorphous polymers during finite straining. *International Journal of Plasticity*, 9(6):697–720, January 1993.

[12] Arruda, E. M. and Boyce, M. C. A three-dimensional constitutive model for the large stretch behavior of rubber elastic materials. *Journal of the Mechanics and Physics of Solids*, 41(2):389–412, February 1993.

[13] Arruda, E. M., Boyce, M. C., and Jayachandran, R. Effects of strain rate, temperature and thermomechanical coupling on the finite strain deformation of glassy polymers. *Mechanics of Materials*, 19(2-3):193–212, January 1995.

[14] Asaro, R. J. Micromechanics of crystals and polycrystals. In *Advances in Applied Mechanics*, volume 23, pages 1–115. Elsevier, 1983.

[15] Asp, L. E., Berglund, L. A., and Gudmundson, P. Effects of a composite-like stress state on the fracture of epoxies. *Composites Science and Technology*, 53(1):27–37, January 1995.

[16] Asp, L. E., Edgren, F., and Sjögren, A. Effects of stitch pattern on the mechanical properties of non-crimp fabric composites. *Proceeding of the 11 ECCM*, pages 31–05, 2004.

[17] Aubourg, P.-A. *Modélisation du comportement jusqu'à rupture de composites stratifies tissés*. PhD thesis, L.M.A. Marseille, 1998.

[18] Ayranci, C. and Carey, J. 2D braided composites: A review for stiffness critical applications. *Composite Structures*, 85(1):43–58, September 2008.

[19] Bažant, Z. P. and Oh, B. H. Crack band theory for fracture of concrete. *Matériaux et Constructions*, 16(3):155–177, May 1983.

[20] Baaser, H., Martin, R. J., and Neff, P. Inconsistency of uhyper and umat in Abaqus for compressible hyperelastic materials. 2017.

[21] Backhaus, G. *Deformationsgesetze: mathematische und physikalische Grundlagen, elastisches und inelastisches Materialverhalten, große Verformungen, Fließvorgänge, Elastizität und Viskoelastizität, Plastizität und Viskoplastizität*. Akad.-Verlag, 1983.

[22] Bahei-El-Din, Y., Rajendran, A., and Zikry, M. A micromechanical model for damage progression in woven composite systems. *International Journal of Solids and Structures*, 41(9-10):2307–2330, May 2004.

[23] Baker, C., Morscher, G. N., Pujar, V. V., and Lemanski, J. R. Transverse cracking in carbon fiber reinforced polymer composites: Modal acoustic emission and peak frequency analysis. *Composites Science and Technology*, 116:26–32, September 2015.

[24] Barbero, E., Lonetti, P., and Sikkil, K. Finite element continuum damage modeling of plain weave reinforced composites. *Composites Part B: Engineering*, 37(2-3):137–147, April 2005.

[25] Barbero, E. J. *Finite element analysis of composite materials using Abaqus*. CRC press, 2013.

[26] Barbero, E. J. *Finite element analysis of composite materials using ANSYS*. Composite materials: Analysis and design. CRC Press, Taylor & Francis Group, Boca Raton, second edition edition, 2014.

[27] Bardenhagen, S., Stout, M., and Gray, G. Three-dimensional, finite deformation, viscoplastic constitutive models for polymeric materials. *Mechanics of Materials*, 25(4):235–253, May 1997.

[28] Basista, M., Krajcinovic, D., and Sumarac, D. Micromechanics, phenomenology and statistics of brittle deformation. In *Computational plasticity: Fundamentals and Applications - Proceedings of the Third International Conference held in Barcelona, 6-10 April, 1992*, pages 1479–1490. Swansea: Pineridge Press, 1992.

[29] Becker, F. *Entwicklung einer Beschreibungsmethodik für das mechanische Verhalten unverstärkter Thermoplaste bei hohen Deformationsgeschwindigkeiten*. PhD thesis, Martin-Luther-Universität Halle-Wittenberg, Halle (Saale), 2009.

[30] Benmedakhene, S., Kenane, M., and Benzeggagh, M. Initiation and growth of delamination in glass/epoxy composites subjected to static and dynamic loading by acoustic emission monitoring. *Composites Science and Technology*, 59(2):201–208, February 1999.

[31] Benveniste, Y. A new approach to the application of Mori-Tanaka's theory in composite materials. *Mechanics of Materials*, 6(2):147–157, June 1987.

[32] Benveniste, Y., Dvorak, G., and Chen, T. Stress fields in composites with coated inclusions. *Mechanics of Materials*, 7(4):305–317, June 1989.

[33] Bergstrom, J. *Mechanics of solid polymers: theory and computational modeling*. Elsevier, Boston, MA, 2015.

[34] Betten, J. Damage tensors in continuum mechanics. *Journal de Mécanique théorique et appliquée*, 2(1):13–32, 1983.

[35] Bigaud, D. and Hamelin, P. From geometrical description to mechanical prediction - Application to woven fabric composites. *Science and Engineering of Composite Materials*, 7(4), January 1998.

[36] Bigaud, D. and Hamelin, P. Stiffness and failure modelling of 2D and 3D textile-reinforced composites by means of imbricate-type elements approaches. *Computers & Structures*, 80(27-30):2253–2264, November 2002.

[37] Bogdanovich, A. E. Multi-scale modeling, stress and failure analyses of 3-D woven composites. *Journal of Materials Science*, 41(20):6547–6590, November 2006.

[38] Bogdanovich, A. and Pastore, C. M. *Mechanics of textile and laminated composites: with applications to structural analysis*. Chapman & Hall, London ; New York, 1996.

[39] Bonnet, B. *Comportement au choc de matreriax composites pour applications automobiles*. PhD thesis, l'Ecole Nationale Supérieure des Mines de Paris, 2005.

[40] Bowden, P. B. The yield behaviour of glassy polymers. In *The Physics of Glassy Polymers*, pages 279–339. Springer Netherlands, Dordrecht, 1973.

[41] Boyce, M. C., Parks, D. M., and Argon, A. S. Large inelastic deformation of glassy polymers. part I: rate dependent constitutive model. *Mechanics of Materials*, 7(1):15–33, September 1988.

[42] Boyce, M., Weber, G., and Parks, D. On the kinematics of finite strain plasticity. *Journal of the Mechanics and Physics of Solids*, 37(5):647–665, January 1989.

[43] Brockenbrough, J., Suresh, S., and Wienecke, H. Deformation of metal-matrix composites with continuous fibers: geometrical effects of fiber distribution and shape. *Acta Metallurgica et Materialia*, 39(5):735–752, May 1991.

[44] Brusselle-Dupend, N., Lai, D., Feaugas, X., Guigon, M., and Clavel, M. Experimental characterization and modeling of the uniaxial mechanical behavior of polypropylene before necking. *Oil & Gas Science and Technology*, 57(2):155–168, March 2002.

[45] Byun, J.-H. and Chou, T.-W. Elastic properties of three-dimensional angle-interlock fabric preforms. *The Journal of The Textile Institute*, 81(4):538–548, January 1990.

[46] Byun, J.-H., Whitney, T. J., Du, G.-W., and Chou, T.-W. Analytical characterization of two-step braided composites. *Journal of Composite Materials*, 25(12):1599–1618, December 1991.

[47] CADFEM. *ANSYS suite of engineering simulation software: Users's manual*. ANSYS, Inc., 2016.

[48] Camanho, P. P., Davila, C. G., and de Moura, M. F. Numerical simulation of mixed-mode progressive delamination in composite materials. *Journal of Composite Materials*, 37(16):1415–1438, August 2003.

[49] Camanho, P. P., Dávila, C. G., Pinho, S. T., Iannucci, L., and Robinson, P. Prediction of in situ strengths and matrix cracking in composites under transverse tension and in-plane shear. *Composites Part A: Applied Science and Manufacturing*, 37(2):165–176, February 2006.

[50] Car, E., Zalamea, F., Oller, S., Miquel, J., and Oñate, E. Numerical simulation of fiber reinforced composite materials—two procedures. *International Journal of Solids and Structures*, 39(7):1967–1986, April 2002.

[51] Carol, I., Rizzi, E., and Willam, K. A unified theory of elastic degradation and damage based on a loading surface. *International Journal of Solids and Structures*, 31(20):2835–2865, October 1994.

[52] Carol, I., Rizzi, E., and Willam, K. On the formulation of anisotropic elastic degradation. I. Theory based on a pseudo-logarithmic damage tensor rate. *International Journal of Solids and Structures*, 38(4):491–518, January 2001.

[53] Carol, I., Rizzi, E., and Willam, K. On the formulation of anisotropic elastic degradation.: II. Generalized pseudo-Rankine model for tensile

damage. *International Journal of Solids and Structures*, 38(4):519–546, January 2001.

[54] Carvelli, V. and Poggi, C. A homogenization procedure for the numerical analysis of woven fabric composites. *Composites Part A: Applied Science and Manufacturing*, 32(10):1425–1432, October 2001.

[55] Carvelli, V. and Taliercio, A. A micromechanical model for the analysis of unidirectional elastoplastic composites subjected to 3D stresses. *Mechanics Research Communications*, 26(5):547–553, 1999.

[56] Carvelli, V., D'Ettorre, A., and Lomov, S. V. Acoustic emission and damage mode correlation in textile reinforced PPS composites. *Composite Structures*, 163:399–409, March 2017.

[57] Carvelli, V., Pazmino, J., Lomov, S. V., Bogdanovich, A. E., Mungalov, D. D., and Verpoest, I. Quasi-static and fatigue tensile behavior of a 3D rotary braided carbon/epoxy composite. *Journal of Composite Materials*, 47(25):3195–3209, November 2013.

[58] Carvelli, V. and Poggi, C. A numerical approach for the failure analysis of textile composites. In *14th International Conference on Composite Materials ICCM-14*, pages 1–8, 2003.

[59] Catalanotti, G., Camanho, P., and Marques, A. Three-dimensional failure criteria for fiber-reinforced laminates. *Composite Structures*, 95:63–79, January 2013.

[60] Chaboche, J. L. Continuum damage mechanics: Part I—General concepts. *Journal of Applied Mechanics*, 55(1):59, 1988.

[61] Chaboche, J. L. Continuum damage mechanics: Part II—Damage growth, crack initiation, and crack growth. *Journal of Applied Mechanics*, 55(1):65, 1988.

[62] Chaboche, J.-L. *Description thermodinamique et phenomenologique de la viscoplasticite cyclique avec endommagement*. Univ. Pierre et Marie Curie, 1978.

[63] Chaboche, J.-L. Continuous damage mechanics — A tool to describe phenomena before crack initiation. *Nuclear Engineering and Design*, 64(2):233–247, April 1981.

[64] Chaboche, J.-L. Thermodynamic formulation of constitutive equations and application to the viscoplasticity and viscoelasticity of metals and polymers. *International Journal of Solids and Structures*, 34(18):2239–2254, June 1997.

[65] Chaboche, J. Anisotropic creep damage in the framework of continuum damage mechanics. *Nuclear Engineering and Design*, 79(3):309–319, June 1984.

[66] Chaboche, J. A review of some plasticity and viscoplasticity constitutive theories. *International Journal of Plasticity*, 24(10):1642–1693, October 2008.

[67] Chaboche, J., Kruch, S., Maire, J., and Pottier, T. Towards a micromechanics based inelastic and damage modeling of composites. *International Journal of Plasticity*, 17(4):411–439, January 2001.

[68] Chamis, C., Wang, S., Steven Johnson, W., Stinchcomb, W., and Sendeckyj, G. Mechanics of composite materials: Past, present, and future. *Journal of Composites Technology and Research*, 11(1):3, 1989.

[69] Chang, Y., Jiao, G., Wang, B., and Liu, W. Elastic behavior analysis of 3D angle-interlock woven ceramic composites. *Acta Mechanica Solida Sinica*, 19(2):152–159, June 2006.

[70] Chen, B., Lang, E. J., and Chou, T.-W. Experimental and theoretical studies of fabric compaction behavior in resin transfer molding. *Materials Science and Engineering: A*, 317(1-2):188–196, October 2001.

[71] Chen, J., Morozov, E., and Shankar, K. A combined elastoplastic damage model for progressive failure analysis of composite materials and structures. *Composite Structures*, 94(12):3478–3489, December 2012.

[72] Chow, C. L. and Wang, J. An anisotropic theory of elasticity for continuum damage mechanics. *International Journal of Fracture*, 33(1):3–16, January 1987.

[73] Christensen, R. M. *The theory of materials failure*. Oxford University Press, 2013.

[74] Clarke, S. and Morales, A. A comparative assessment of textile preforming techniques. *FIBER-TEX*, 90:125–134, 1990.

[75] Coleman, B. D. and Gurtin, M. E. Thermodynamics with internal state variables. *The Journal of Chemical Physics*, 47(2):597–613, July 1967.

[76] Coleman, B. D. and Noll, W. The thermodynamics of elastic materials with heat conduction and viscosity. *Archive for Rational Mechanics and Analysis*, 13(1):167–178, December 1963.

[77] Cordebois, J. P. and Sidoroff, F. Endommagement anisotrope en élasticité et plasticité. *JMTA, Numéro spécial*, pages 45–60, 1982.

[78] Cousigné, O., Moncayo, D., Coutellier, D., Camanho, P., Naceur, H., and Hampel, S. Development of a new nonlinear numerical material model for woven composite materials accounting for permanent deformation and damage. *Composite Structures*, 106:601–614, December 2013.

[79] Cox, B. N. and Dadkhah, M. S. The macroscopic elasticity of 3D woven composites. *Journal of Composite Materials*, 29(6):785–819, April 1995.

[80] Crist, B. Yield processes in glassy polymers. In *The Physics of Glassy Polymers*, pages 155–212. Springer Netherlands, Dordrecht, 1997.

[81] Crookston, J. J., Long, A. C., and Jones, I. A. A summary review of mechanical properties prediction methods for textile reinforced polymer composites. *Proceedings of the Institution of Mechanical Engineers, Part L: Journal of Materials: Design and Applications*, 219(2):91–109, April 2005.

[82] Crookston, J., Robitaille, F., and Long, A. *A systematic study of the mechanical properties of textile composite unit cells based on geometric modelling*. Society of Manufacturing Engineers, 2000.

[83] Cuitino, A. M. and Ortiz, M. Computational modelling of single crystals. *Modelling and Simulation in Materials Science and Engineering*, 1(3):225–263, April 1993.

[84] Daggumati, S., De Baere, I., Van Paepegem, W., Degrieck, J., Xu, J., Lomov, S., and Verpoest, I. Local damage in a 5-harness satin weave composite under static tension: Part I – Experimental analysis. *Composites Science and Technology*, 70(13):1926–1933, November 2010.

[85] Daggumati, S., Van Paepegem, W., Degrieck, J., Xu, J., Lomov, S., and Verpoest, I. Local damage in a 5-harness satin weave composite under static tension: Part II – Meso-FE modelling. *Composites Science and Technology*, 70(13):1934–1941, November 2010.

[86] D'Amato, E. Analysis of nonlinearities in mechanical behavior of textile composites. *WIT Transactions on The Built Environment*, 59, 2002.

[87] Dassow, J. W. A. *Messung und Beschreibung der Querkontraktionszahl von teilkristallinen Thermoplasten als nichtlinear viskoelastischer Materialkennwert*. Verlag der Augustinus-Buchh., Aachen, 1996.

[88] Davila, C. G., Camanho, P. P., and Rose, C. A. Failure criteria for FRP laminates. *Journal of Composite Materials*, 39(4):323–345, February 2005.

[89] de Borst, R. Theoretical and computational issues in localisation and failure. In *European Congress on Computational Methods in Applied Sciences and Engineering*, pages 1–15, 2000.

[90] Deuschle, H. M. *3D failure analysis of UD fibre reinforced composites Puck's theory within FEA*. Inst. für Statik und Dynamik der Luft- und Raumfahrtkonstruktionen, Stuttgart, 2010.

[91] DIN EN ISO 1183-1:2012. Plastics - Methods for determining the density of non-cellular plastics – Part 1: Immersion method, liquid pyknometer method and titration method, 2013.

[92] DIN EN ISO 14126. Fibre-reinforced plastic composites- Determination of compressive properties in the in-plane direction, 2000.

[93] DIN EN ISO 3167. Plastics - Multipurpose test specimens, 2014.

[94] DIN EN ISO 527-1. Plastics - Determination of tensile properties - Part 1: General principles, June 2012.

[95] DIN EN ISO 527-4. Plastics - Determination of tensile properties - Part 4: Test conditions for isotropic and anisotropic fibre-reinforced plastic composites, 1997.

[96] DIN EN ISO 604. Plastics - Determination of compressive properties, December 2003.

[97] Ding, Y., Wenger, W., and McIlhagger, R. Structural characterization and mechanical properties of 3-D woven composites. *Broadening horizons with advanced materials and processes*, pages 1–9, 1993.

[98] Dixit, A. and Mali, H. S. Modeling techniques for predicting the mechanical properties of woven-fabric textile composites: A review. *Mechanics of Composite Materials*, 49(1):1–20, March 2013.

[99] Doitrand, A., Fagiano, C., Chiaruttini, V., Leroy, F., Mavel, A., and Hirsekorn, M. Experimental characterization and numerical modeling of damage at the mesoscopic scale of woven polymer matrix composites under quasi-static tensile loading. *Composites Science and Technology*, 119:1–11, November 2015.

[100] Doyle, T. and Ericksen, J. Nonlinear elasticity. In *Advances in Applied Mechanics*, volume 4, pages 53–115. Elsevier, 1956.

[101] Dragon, A. and Mróz, Z. A continuum model for plastic-brittle behaviour of rock and concrete. *International Journal of Engineering Science*, 17(2):121–137, January 1979.

[102] Drucker, D. C. and Prager, W. Soil mechanics and plastic analysis or limit design. *Quarterly of applied mathematics*, 10(2):157–165, 1952.

[103] Edgren, F., Mattsson, D., Asp, L. E., and Varna, J. Formation of damage and its effects on non-crimp fabric reinforced composites loaded in tension. *Composites Science and Technology*, 64(5):675–692, April 2004.

[104] El-Hage, C., Younès, R., Aboura, Z., Benzeggagh, M., and Zoaeter, M. Analytical and numerical modeling of mechanical properties of orthogonal 3D CFRP. *Composites Science and Technology*, 69(1):111–116, January 2009.

[105] El-Hage, C. *Modélisation du comportement élastique endommageable de matériaux composites à renfort tridimensionnel.* PhD thesis, Université de Technologie de Compiègne, 2006.

[106] Eringen, A. C., editor. *Continuum mechanics of single-substance bodies.* Number 2 in Continuum physics / ed. by A. Cemal Eringen. Acad. Press, New York, NY, 1975.

[107] Eshelby, J. D. The determination of the elastic field of an ellipsoidal inclusion, and related problems. *Proceedings of the Royal Society A: Mathematical, Physical and Engineering Sciences*, 241(1226):376–396, August 1957.

[108] Eshelby, J. D. The elastic field outside an ellipsoidal inclusion. *Proceedings of the Royal Society A: Mathematical, Physical and Engineering Sciences*, 252(1271):561–569, October 1959.

[109] Eshelby, J. Elastic inclusions and inhomogeneities. *Progress in solid mechanics*, 2(1):89–140, 1961.

[110] Eyring, H. Viscosity, plasticity, and diffusion as examples of absolute reaction rates. *The Journal of Chemical Physics*, 4(4):283–291, April 1936.

[111] Fang, G., Liang, J., Lu, Q., Wang, B., and Wang, Y. Investigation on the compressive properties of the three dimensional four-directional braided composites. *Composite Structures*, 93(2):392–405, January 2011.

[112] Farley, G. L., Smith, B. T., and Maiden, J. Compression response of thick layer composite laminates with through-the-thickness reinforcement. *Journal of Reinforced Plastics and Composites*, 11(7):787–810, July 1992.

[113] Fish, J. *Practical multiscaling*. John Wiley & Sons Inc, Chichester, West Sussex, United Kingdom, 2014.

[114] Fish, J. and Yu, Q. Multiscale damage modelling for composite materials: Theory and computational framework. *International Journal for Numerical Methods in Engineering*, 52(12):161–191, September 2001.

[115] Fish, J. and Yu, Q. Two-scale damage modeling of brittle composites. *Composites Science and Technology*, 61(15):2215–2222, November 2001.

[116] Fish, J., Yu, Q., and Shek, K. Computational damage mechanics for composite materials based on mathematical homogenization. *International Journal for Numerical Methods in Engineering*, 45(11):1657–1679, August 1999.

[117] Fonseka, G. U. and Krajcinovic, D. The continuous damage theory of brittle materials, Part 2: Uniaxial and plane response modes. *Journal of Applied Mechanics*, 48(4):816, 1981.

[118] Ganesh, V. K. and Naik, N. K. Failure behavior of plain weave fabric laminates under on-axis uniaxial tensile loading: III—Effect of fabric geometry. *Journal of Composite Materials*, 30(16):1823–1856, November 1996.

[119] Ganesh, V. K. and Naik, N. K. Failure behavior of plain weave fabric laminates under on-axis uniaxial tensile loading: I—Laminate geometry. *Journal of Composite Materials*, 30(16):1748–1778, November 1996.

[120] Garnich, M. R. and Karami, G. Finite element micromechanics for stiffness and strength of wavy fiber composites. *Journal of Composite Materials*, 38(4):273–292, February 2004.

[121] Geers, M. G., Kouznetsova, V., and Brekelmans, W. A. Gradient-enhanced computational homogenization for the micro-macro scale transition. *Le Journal de Physique IV*, 11(PR5):Pr5–145–Pr5–152, September 2001.

[122] Geers, M., Kouznetsova, V. G., and Brekelmans, W. A. M. Multi-scale first-order and second-order computational homogenization of

microstructures towards continua. *International Journal for Multiscale Computational Engineering*, 1(4):371–386, 2003.

[123] Ghosh, S. *Micromechanical analysis and multi-scale modeling using the Voronoi cell finite element method.* CRC Press, 2011.

[124] Ghosh, S., Lee, K., and Moorthy, S. Multiple scale analysis of heterogeneous elastic structures using homogenization theory and Voronoi cell finite element method. *International Journal of Solids and Structures*, 32(1):27–62, January 1995.

[125] Ghosh, S., Lee, K., and Moorthy, S. Two scale analysis of heterogeneous elastic-plastic materials with asymptotic homogenization and Voronoi cell finite element model. *Computer Methods in Applied Mechanics and Engineering*, 132(1-2):63–116, May 1996.

[126] Göktepe, S. Micro-macro approaches to rubbery and glassy polymers: Predictive micromechanically-based models and simulations, November 2007.

[127] Gommers, B., Verpoest, I., and Van Houtte, P. The Mori–Tanaka method applied to textile composite materials. *Acta Materialia*, 46(6):2223–2235, March 1998.

[128] Govaert, L. E., Timmermans, P. H. M., and Brekelmans, W. A. M. The influence of intrinsic strain softening on strain localization in polycarbonate: Modeling and experimental validation. *Journal of Engineering Materials and Technology*, 122(2):177, 2000.

[129] Govindjee, S., Kay, G. J., and Simo, J. C. Anisotropic modelling and numerical simulation of brittle damage in concrete. *International Journal for Numerical Methods in Engineering*, 38(21):3611–3633, November 1995.

[130] Green, A. E. and Adkins, J. E. *Large elastic deformations*. Clarendon Press, Oxford, 2nd ed edition, 1970.

[131] Green, A. E. and Naghdi, P. M. A general theory of an elastic-plastic continuum. *Archive for Rational Mechanics and Analysis*, 18(4):251–281, January 1965.

[132] Green, S., Matveev, M., Long, A., Ivanov, D., and Hallett, S. Mechanical modelling of 3d woven composites considering realistic unit cell geometry. *Composite Structures*, 118:284–293, December 2014.

[133] Grellmann, W. and Seidler, S. *Kunststoffprüfung*. Carl Hanser Verlag GmbH Co KG, 2015.

[134] Greve, L. and Pickett, A. Modelling damage and failure in carbon/epoxy non-crimp fabric composites including effects of fabric pre-shear. *Composites Part A: Applied Science and Manufacturing*, 37(11):1983–2001, November 2006.

[135] Groves, S., Sanchez, R., Lyon, R., and Brown, A. High strain rate effects for composite materials. In *Eleventh Volume: Composite Materials—Testing and Design*, pages 162–162–15. ASTM International, 100 Barr Harbor Drive, PO Box C700, West Conshohocken, PA 19428-2959, January 1993.

[136] Guagliano, M. and Riva, E. Mechanical behaviour prediction in plain weave composites. *The Journal of Strain Analysis for Engineering Design*, 36(2):153–162, February 2001.

[137] Guedes, J. M. and Kikuchi, N. Preprocessing and postprocessing for materials based on the homogenization method with adaptive finite element methods. *Computer methods in applied mechanics and engineering*, 83(2):143–198, 1990.

[138] Guess, T., Sendeckyj, G., Wang, S., Chaio, T., Feng, W., Steven Johnson, W., Rodericks, G., Stinchcomb, W., Reedy, E., and Reifsnider, K. Comparison of interlocked fabric and laminated fabric kevlar 49/epoxy composites. *Journal of Composites Technology and Research*, 7(4):136, 1985.

[139] Gurson, A. L. Continuum theory of ductile rupture by void nucleation and growth: Part I—Yield criteria and flow rules for porous ductile media. *Journal of Engineering Materials and Technology*, 99(1):2, 1977.

[140] Gutkin, R., Green, C., Vangrattanachai, S., Pinho, S., Robinson, P., and Curtis, P. On acoustic emission for failure investigation in CFRP: Pattern recognition and peak frequency analyses. *Mechanical Systems and Signal Processing*, 25(4):1393–1407, May 2011.

[141] Hahn, H. T. and Pandey, R. A micromechanics model for thermoelastic properties of plain weave fabric composites. *Journal of Engineering Materials and Technology*, 116(4):517, 1994.

[142] Hale, R. D. and Villa, M. Influence of opposing wave nesting in compression-loaded composites. *Journal of Composite Materials*, 37(13):1149–1166, July 2003.

[143] Hallal, A. *Modeling of elastic properties of textile composites. PhD thesis*. PhD thesis, University of Versailles-Saint-Quentin en Yvelines, 2013.

[144] Hallal, A., Younes, R., and Fardoun, F. Review and comparative study of analytical modeling for the elastic properties of textile composites. *Composites Part B: Engineering*, 50:22–31, July 2013.

[145] Hallal, A., Younes, R., Fardoun, F., and Nehme, S. Improved analytical model to predict the effective elastic properties of 2.5D interlock woven fabrics composite. *Composite Structures*, 94(10):3009–3028, October 2012.

[146] Halpin Affdl, J. C. and Kardos, J. L. The Halpin-Tsai equations: A review. *Polymer Engineering and Science*, 16(5):344–352, May 1976.

[147] Hamelin, P. and Bigaud, D. A numerical procedure for elasticity and failure behavior prediction of textile-reinforced composite materials. *Journal of Thermoplastic Composite Materials*, 12(3):201–213, May 1999.

[148] Hansen, N. and Schreyer, H. Thermodynamically consistent theories for elastoplasticity coupled with damage. Technical report, Sandia National Labs., Albuquerque, NM (United States), 1992.

[149] Harding, J. Effect of strain rate and specimen geometry on the compressive strength of woven glass-reinforced epoxy laminates. *Composites*, 24(4):323–332, June 1993.

[150] Hasan, O. A. and Boyce, M. C. A constitutive model for the nonlinear viscoelastic viscoplastic behavior of glassy polymers. *Polymer Engineering and Science*, 35(4):331–344, February 1995.

[151] Hashin, Z. Fatigue failure criteria for unidirectional fiber composites. *Journal of Applied Mechanics*, 48(4):846, 1981.

[152] Hashin, Z. and Shtrikman, S. On some variational principles in anisotropic and nonhomogeneous elasticity. *Journal of the Mechanics and Physics of Solids*, 10(4):335–342, October 1962.

[153] Hashin, Z. and Shtrikman, S. A variational approach to the theory of the elastic behaviour of multiphase materials. *Journal of the Mechanics and Physics of Solids*, 11(2):127–140, March 1963.

[154] Hashin, Z. Theory of mechanical behavior of heterogeneous media. Technical report, Pennsylvania University of Philadelphia, Towne school of civil and mechanical engineering, 1963.

[155] Haward, R. N. and Thackray, G. The use of a mathematical model to describe isothermal stress-strain curves in glassy thermoplastics. *Proceedings of the Royal Society A: Mathematical, Physical and Engineering Sciences*, 302(1471):453–472, January 1968.

[156] Henning, F. and Moeller, E. *Handbuch Leichtbau—Methoden, Werkstoffe, Fertigung (Engl.: Lightweight construction handbook—methods, materials, engineering).* Carl Hanser Verlag, 2011.

[157] Hill, R. A theory of the yielding and plastic flow of anisotropic metals. *Proceedings of the Royal Society A: Mathematical, Physical and Engineering Sciences*, 193(1033):281–297, May 1948.

[158] Hill, R. The elastic behaviour of a crystalline aggregate. *Proceedings of the Physical Society. Section A*, 65(5):349, 1952.

[159] Hill, R. Elastic properties of reinforced solids: Some theoretical principles. *Journal of the Mechanics and Physics of Solids*, 11(5):357–372, September 1963.

[160] Hill, R. A self-consistent mechanics of composite materials. *Journal of the Mechanics and Physics of Solids*, 13(4):213–222, August 1965.

[161] Hill, R. Theory of mechanical properties of fibre-strengthened materials—III. Self-consistent model. *Journal of the Mechanics and Physics of Solids*, 13(4):189–198, August 1965.

[162] Hill, R. Generalized constitutive relations for incremental deformation of metal crystals by multislip. *Journal of the Mechanics and Physics of Solids*, 14(2):95–102, March 1966.

[163] Hill, R. *The mathematical theory of plasticity*, volume 11. Oxford university press, 1998.

[164] Hinton, M. J., Kaddour, A. S., and Soden, P. D. *Failure criteria in fibre reinforced polymer composites: the world-wide failure exercise*. Elsevier, Amsterdam, 2004.

[165] Hivet, G., Laine, B., and Boisse, P. Consistent preprocessor for the unit woven cell for meso-macro analyses of fabric forming. In *ESAFORM 2005*, 2005.

[166] Hochard, C., Aubourg, P.-A., and Charles, J.-P. Modelling of the mechanical behaviour of woven-fabric CFRP laminates up to failure. *Composites Science and Technology*, 61(2):221–230, February 2001.

[167] Hochard, C., Miot, S., Lahellec, N., Mazerolle, F., Herman, M., and Charles, J. Behaviour up to rupture of woven ply laminate structures under static loading conditions. *Composites Part A: Applied Science and Manufacturing*, 40(8):1017–1023, August 2009.

[168] Hoffman, O. The brittle strength of orthotropic materials. *Journal of Composite Materials*, 1(2):200–206, April 1967.

[169] Holzapfel, G. A. *Nonlinear solid mechanics: A continuum approach for engineering*. Wiley, Chichester ; New York, 2000.

[170] Hu, J. *3-D fibrous assemblies: properties, applications and modelling of three-dimensional textile structures*. Elsevier, 2008.

[171] Ishikawa, T. and Chou. One-dimensional micromechanical analysis of woven fabric composites. *AIAA Journal*, 21(12):1714–1721, December 1983.

[172] Ishikawa, T. and Chou, T. W. Stiffness and strength behaviour of woven fabric composites. *Journal of Materials Science*, 17(11):3211–3220, November 1982.

[173] Ishikawa, T. and Chou, T.-W. Elastic behavior of woven hybrid composites. *Journal of Composite Materials*, 16(1):2–19, January 1982.

[174] Ishikawa, T. and Chou, T.-W. In-plane thermal expansion and thermal bending coefficients of fabric composites. *Journal of Composite Materials*, 17(2):92–104, March 1983.

[175] Ishikawa, T. and Chou, T.-W. Nonlinear behavior of woven fabric composites. *Journal of Composite Materials*, 17(5):399–413, September 1983.

[176] Ivanov, D., Lomov, S. V., Verpoest, I., Baudry, F., and Xie, H. Damage initiation and development in triaxial braid and fine structure of damage. *Proceedings of ECCM-12, Biarritz*, 2006.

[177] Ivanov, D. S., Lomov, S. V., Ivanov, S. G., and Verpoest, I. Stress distribution in outer and inner plies of textile laminates and novel boundary conditions for unit cell analysis. *Composites Part A: Applied Science and Manufacturing*, 41(4):571–580, April 2010.

[178] Ivanov, D., Lomov, S. V., Verpoest, I., and Zisman, A. Noise reduction of strain mapping data and identification of damage initiation of carbon-epoxy triaxial braided composite. *Composites testing and model identification (CompTest-2006)*, 2006.

[179] Ivanov, D., Lomov, S. V., and Verpoest, I. Zero and first order homogenization schemes: comparison with a full FE model. In *Proc. 11th US-Japan Conference on Composite Materials (TexComp 7)*, 2004.

[180] Jansson, N. and Larsson, R. A damage model for simulation of mixed-mode delamination growth. *Composite Structures*, 53(4):409–417, September 2001.

[181] John, S., Herszberg, I., and Coman, F. Longitudinal and transverse damage taxonomy in woven composite components. *Composites Part B: Engineering*, 32(8):659–668, December 2001.

[182] Kachanov, L. M. Creep and rupture under complex loading. *Problemi prochnosti*, (6), 1977.

[183] Kachanov, L. M. *Introduction to continuum damage mechanics*. Springer Netherlands, Dordrecht, 1986.

[184] Kachanov, L. M. Time of the rupture process under creep conditions. *Nank SSR Otd Tech Nauk*, 8:26–31, 1958.

[185] Kachanov, M. Continuum model of medium with cracks. *Journal of the engineering mechanics division*, 106(5):1039–1051, 1980.

[186] Kalidindi, S. R. and Abusafieh, A. Longitudinal and transverse moduli and strengths of low angle 3-D braided composites. *Journal of Composite Materials*, 30(8):885–905, May 1996.

[187] Kalidindi, S. R. and Franco, E. Numerical evaluation of isostrain and weighted-average models for elastic moduli of three-dimensional composites. *Composites Science and Technology*, 57(3):293–305, January 1997.

[188] Karahan, M., Lomov, S. V., Bogdanovich, A. E., Mungalov, D., and Verpoest, I. Internal geometry evaluation of non-crimp 3D orthogonal woven carbon fabric composite. *Composites Part A: Applied Science and Manufacturing*, 41(9):1301–1311, September 2010.

[189] Karami, G. and Garnich, M. Effective moduli and failure considerations for composites with periodic fiber waviness. *Composite Structures*, 67(4):461–475, March 2005.

[190] Kellas, S., Morton, J., and Jackson, K. Damage and failure mechanisms in scaled angle-ply laminates. In *Composite Materials: Fatigue and Fracture, Fourth Volume*, pages 257–257–24. ASTM International, January 1993.

[191] Kim, J. W., Medvedev, G. A., and Caruthers, J. M. Observation of yield in triaxial deformation of glassy polymers. *Polymer*, 54(11):2821–2833, May 2013.

[192] King, M., Jearanaisilawong, P., and Socrate, S. A continuum constitutive model for the mechanical behavior of woven fabrics. *International Journal of Solids and Structures*, 42(13):3867–3896, June 2005.

[193] Kinloch, A. J. *Fracture behaviour of polymers*. Springer Science & Business Media, 2013.

[194] Klusemann, B. and Svendsen, B. Homogenization methods for multiphase elastic composites: Comparisons and benchmarks. *Technische Mechanik*, 30(4):374–386, 2010.

[195] Knops, M. *Analysis of failure in fiber polymer laminates*. Springer Berlin Heidelberg, Berlin, Heidelberg, 2008.

[196] Ko, F. K. and Pastore, C. M. Design of complex shaped structures. In *Textiles: Product Design and Marketing*, pages 123–135. The Textile Institute, 1987.

[197] Kouznetsova, V., Brekelmans, W. A. M., and Baaijens, F. P. T. An approach to micro-macro modeling of heterogeneous materials. *Computational Mechanics*, 27(1):37–48, January 2001.

[198] Kouznetsova, V., Geers, M. G. D., and Brekelmans, W. A. M. Multiscale constitutive modelling of heterogeneous materials with a gradient-enhanced computational homogenization scheme. *International Journal for Numerical Methods in Engineering*, 54(8):1235–1260, July 2002.

[199] Kouznetsova, V., Geers, M., and Brekelmans, W. Multi-scale second-order computational homogenization of multi-phase materials: a nested finite element solution strategy. *Computer Methods in Applied Mechanics and Engineering*, 193(48-51):5525–5550, December 2004.

[200] Kowalewski, T., Gałęski, A., and Kryszewski, M. The structure and tensile properties of cold drawn modified chalk filled polypropylene. In *Polymer Blends*, pages 223–241. Springer US, Boston, MA, 1984.

[201] Krajcinovic, D. Constitutive equations for damaging materials. *Journal of Applied Mechanics*, 50(2):355, 1983.

[202] Krajcinovic, D. Continuous damage mechanics: Basic concepts and definitions. *Journal of Applied Mechanics*, 52(4):829, 1985.

[203] Krajcinovic, D. and Fonseka, G. U. The continuous damage theory of brittle materials - Part 1: General theory. *Journal of Applied Mechanics*, 48(4):809, 1981.

[204] Krajcinovic, D. *Damage mechanics*, volume 41. Elsevier, 1996.

[205] Kregers, A. F. and Melbardis, Y. G. Determination of the deformability of three-dimensionally reinforced composites by the stiffness averaging method. *Polymer Mechanics*, 14(1):1–5, 1978.

[206] Kröner, E. and Teodosiu, C. Lattice defect approach to plasticity and viscoplasticity. *Problems of plasticity.(A 75-13527 03-39)*, Noordhoff International Publishing, Leiden:45–82, 1974.

[207] Kröner, E. Allgemeine Kontinuumstheorie der Versetzungen und Eigenspannungen. *Archive for Rational Mechanics and Analysis*, 4(1):273–334, January 1959.

[208] Krueger, R., Paris, I. L., Kevin O'Brien, T., and Minguet, P. J. Comparison of 2D finite element modeling assumptions with results from 3D analysis for composite skin-stiffener debonding. *Composite Structures*, 57(1-4):161–168, July 2002.

[209] Kurashiki, T., Zako, M., and Verpoest, I. Damage development of woven fabric composites considering an effect of mismatch of lay-up. In *Proc. ECCM-10, 10th European Conference on Composite Materials*, pages 1–6, 2002.

[210] Ladeveze, P. A damage computational approach for composites: Basic aspects and micromechanical relations. *Computational Mechanics*, 17(1-2):142–150, December 1995.

[211] Ladeveze, P. and Ledantec, E. Damage modelling of the elementary ply for laminated composites. *Composites Science and Technology*, 43(3):257–267, 1992.

[212] Le Page, B., Guild, F., Ogin, S., and Smith, P. Finite element simulation of woven fabric composites. *Composites Part A: Applied Science and Manufacturing*, 35(7-8):861–872, July 2004

[213] Leckie, F. A. and Hayhurst, D. R. Creep rupture of structures. *Proceedings of the Royal Society A: Mathematical, Physical and Engineering Sciences*, 340(1622):323–347, September 1974.

[214] Leckie, F. A. and Onat, E. T. Tensorial nature of damage measuring internal variables. In *Physical Non-Linearities in Structural Analysis*, pages 140–155. Springer Berlin Heidelberg, Berlin, Heidelberg, 1981.

[215] Lee, C. S., Chung, S. W., Shin, H., and Kim, S. J. Virtual material characterization of 3D orthogonal woven composite materials by large-scale computing. *Journal of Composite Materials*, 39(10):851–863, May 2005.

[216] Lee, E. H. Elastic-plastic deformation at finite strains. *Journal of Applied Mechanics*, 36(1):1, 1969.

[217] Lee, S.-K., Byun, J.-H., and Hong, S. H. Effect of fiber geometry on the elastic constants of the plain woven fabric reinforced aluminum matrix composites. *Materials Science and Engineering: A*, 347(1-2):346–358, April 2003.

[218] Lemaitre, J. Evaluation of dissipation and damage in metals submitted to dynamic loading. *Mechanical behavior of materials*, pages 540–549, 1972.

[219] Lemaitre, J. A three-dimensional ductile damage model applied to deep-drawing froming limits. In *Mechanical Behaviour of Materials*, pages 1047–1053. Elsevier, 1984.

[220] Lemaitre, J. and Chaboche, J. L. Aspect phénoménologique de la rupture par endommagement. *Journal de mécanique appliquée*, 2(3), 1978.

[221] Lemaitre, J. A continuous damage mechanics model for ductile fracture. *Journal of Engineering Materials and Technology*, 107(1):83, 1985.

[222] Lemaitre, J. Coupled elasto-plasticity and damage constitutive equations. *Computer Methods in Applied Mechanics and Engineering*, 51(1-3):31–49, September 1985.

[223] Lemaitre, J. and Chaboche, J.-L. *Mechanics of solid materials*. Cambridge University Press, Cambridge, 1990.

[224] Lemaitre, J., Chaboche, J.-L., Benallal, A., and Desmorat, R. *Mécanique des matériaux solides*. Dunod, 2009.

[225] Lemaitre, J., Desmorat, R., and Sauzay, M. Anisotropic damage law of evolution. *European Journal of Mechanics - A/Solids*, 19(2):187–208, March 2000.

[226] Lemaître, J. and Desmorat, R. *Engineering damage mechanics: Ductile, creep, fatigue and brittle failures*. Springer, Berlin ; New York, 2005.

[227] Li, D., Lu, Z., Chen, L., and Li, J. Microstructure and mechanical properties of three-dimensional five-directional braided composites. *International Journal of Solids and Structures*, 46(18-19):3422–3432, September 2009.

[228] Li, S. and Wongsto, A. Unit cells for micromechanical analyses of particle-reinforced composites. *Mechanics of Materials*, 36(7):543–572, July 2004.

[229] Li, X., Hallett, S. R., and Wisnom, M. R. Predicting the effect of through-thickness compressive stress on delamination using interface elements. *Composites Part A: Applied Science and Manufacturing*, 39(2):218–230, February 2008.

[230] Lion, A. A constitutive model for carbon black filled rubber: Experimental investigations and mathematical representation. *Continuum Mechanics and Thermodynamics*, 8(3):153–169, June 1996.

[231] Lomov, S., Ivanov, D., Verpoest, I., Zako, M., Kurashiki, T., Nakai, H., and Hirosawa, S. Meso-FE modelling of textile composites: Road map, data flow and algorithms. *Composites Science and Technology*, 67(9):1870–1891, July 2007.

[232] Lomov, S. V., Bernal, E., Ivanov, D. S., Kondratiev, S. V., and Verpoest, I. Homogenisation of a sheared unit cell of textile composites: FEA and approximate inclusion model. *Revue Européenne des Éléments Finis*, 14(6-7):709–728, January 2005.

[233] Lomov, S. V., Bogdanovich, A., Ivanov, D., Hamada, K., Kurashiki, T., Zako, M., and others. Finite element modelling of progressive damage in non-crimp 3D orthogonal weave and plain weave E-glass composites. In *2nd World conference on 3D fabrics, Greenville, South Carolina, USA*, 2009.

[234] Lomov, S. V., Bogdanovich, A. E., Ivanov, D. S., Mungalov, D., Karahan, M., and Verpoest, I. A comparative study of tensile properties of non-crimp 3D orthogonal weave and multi-layer plain weave E-glass composites. Part 1: Materials, methods and principal results. *Composites Part A: Applied Science and Manufacturing*, 40(8):1134–1143, August 2009.

[235] Lomov, S. V., Ding, X., Hirosawa, S., Kondratiev, S., Molimard, J., Nakai, H., Vautrin, A., Verpoest, I., and Zako, M. FE simulations of textile composites on unit cell level: validation with full-field strain measurements. In *Proc. SAMPE Europe International Conference 2005*, pages 28–33, 2005.

[236] Lomov, S. V., Verpoest, I., Kondratiev, S., and Borovkov, A. An integrated model strategy for processing and properties of textile composites: New results. In *Proc. 22nd International SAMPE Europe Conference*, pages 379–389, 2001.

[237] Lomov, S., Ivanov, D., Truong, T., Verpoest, I., Baudry, F., Vanden Bosche, K., and Xie, H. Experimental methodology of study of damage initiation and development in textile composites in uniaxial tensile test. *Composites Science and Technology*, 68(12):2340–2349, September 2008.

[238] Lomov, S., Verpoest, I., Peeters, T., Roose, D., and Zako, M. Nesting in textile laminates: Geometrical modelling of the laminate. *Composites Science and Technology*, 63(7):993–1007, May 2003.

[239] Long, A. C. *Composites forming technologies*. Elsevier, 2014.

[240] Lubarda, V. A. and Krajcinovic, D. Damage tensors and the crack density distribution. *International Journal of Solids and Structures*, 30(20):2859–2877, 1993.

[241] Lubliner, J. A model of rubber viscoelasticity. *Mechanics Research Communications*, 12(2):93–99, March 1985.

[242] Lurie, A. I. *Nonlinear theory of elasticity*. Number 36 in North-Holland series in Applied Mathematics and Mechanics. North-Holland, Elsevier Science Pub. Co, Amsterdam, New York, 1990.

[243] Maillet, E., Baker, C., Morscher, G. N., Pujar, V. V., and Lemanski, J. R. Feasibility and limitations of damage identification in composite materials using acoustic emission. *Composites Part A: Applied Science and Manufacturing*, 75:77–83, August 2015.

[244] Maimí, P., Camanho, P. P., Mayugo, J. A., and Dávila, C. G. A continuum damage model for composite laminates: Part I – Constitutive model. *Mechanics of Materials*, 39(10):897–908, October 2007.

[245] Maimí, P., Camanho, P. P., Mayugo, J. A., and Dávila, C. G. A continuum damage model for composite laminates: Part II – Computational implementation and validation. *Mechanics of Materials*, 39(10):909–919, October 2007.

[246] Maimí, P. V., Camanho, P. M. P. R. d., Mayugo Majó, J. A., Dávila, C. G., and others. A thermodynamically consistent damage model for advanced composites. *NASA TM Technical Reports, 2006, num. 214282*, 2006.

[247] Maimi, P., Mayugo, J. A., and Camanho, P. P. A three-dimensional damage model for transversely isotropic composite laminates. *Journal of Composite Materials*, 42(25):2717–2745, September 2008.

[248] Maire, J. and Chaboche, J. A new formulation of continuum damage mechanics (CDM) for composite materials. *Aerospace Science and Technology*, 1(4):247–257, June 1997.

[249] Malvern, L. E. *Introduction to the mechanics of a continuous medium*. Prentice-Hall series in engineering of the physical sciences. Prentice-Hall, Englewood Cliffs, N.J, 1969.

[250] Mandel, J. *Plasticite classique et viscoplasticite*, volume Int. Centre Mechanics and Science, of *Courses and lectures, 97*. Springer, Wien, New York, Udine, 1972.

[251] Marcin, L. *Behaviour modelling up to rupture of woven fabric composite for a reliable design of structures*. Theses, Université Bordeaux 1, January 2010.

[252] Marshall, I. and Thompson, A. B. The cold drawing of high polymers. *Proceedings of the Royal Society A: Mathematical, Physical and Engineering Sciences*, 221(1147):541–557, February 1954.

[253] Martín-Santos, E., Maimí, P., González, E., and Cruz, P. A continuum constitutive model for the simulation of fabric-reinforced composites. *Composite Structures*, 111:122–129, May 2014.

[254] Masters, J. E. and Ifju, P. G. A phenomenological study of triaxially braided textile composites loaded in tension. *Composites Science and Technology*, 56(3):347–358, January 1996.

[255] Matsui, K., Terada, K., and Yuge, K. Two-scale finite element analysis of heterogeneous solids with periodic microstructures. *Computers & Structures*, 82(7-8):593–606, March 2004.

[256] Mattsson, D., Joffe, R., and Varna, J. Methodology for characterization of internal structure parameters governing performance in NCF composites. *Composites Part B: Engineering*, 38(1):44–57, January 2007.

[257] Matzenmiller, A., Lubliner, J., and Taylor, R. A constitutive model for anisotropic damage in fiber-composites. *Mechanics of Materials*, 20(2):125–152, April 1995.

[258] Meldahl, A. Brown Boveri Review. *Baden*, 31:260, 1944.

[259] Melro, A., Camanho, P., and Pinho, S. Generation of random distribution of fibres in long-fibre reinforced composites. *Composites Science and Technology*, 68(9):2092–2102, July 2008.

[260] Melro, A. Analytical and numerical modelling of damage and fracture of advanced composites. 2013.

[261] Melro, A., Camanho, P., Andrade Pires, F., and Pinho, S. Micromechanical analysis of polymer composites reinforced by unidirectional fibres: Part I – Constitutive modelling. *International Journal of Solids and Structures*, 50(11-12):1897–1905, June 2013.

[262] Menzel, A. and Steinmann, P. A theoretical and computational framework for anisotropic continuum damage mechanics at large strains. *International Journal of Solids and Structures*, 38(52):9505–9523, December 2001.

[263] Michel, J., Moulinec, H., and Suquet, P. Effective properties of composite materials with periodic microstructure: A computational approach. *Computer Methods in Applied Mechanics and Engineering*, 172(1-4):109–143, April 1999.

[264] Michel, J. and Suquet, P. Nonuniform transformation field analysis. *International Journal of Solids and Structures*, 40(25):6937–6955, December 2003.

[265] Miehe, C. Exponential map algorithm for stress updates in anisotropic multiplicative elastoplasticity for single crystals. *International Journal for Numerical Methods in Engineering*, 39(19):3367–3390, October 1996.

[266] Miehe, C. Implementierung eines finiten elasto-viskoelastischen-plastoelastischen Materialmodells mit Schädigung zur Beschreibung gefüllter gummiaritger Polymere in das Programmsystem ABAQUS, 1999.

[267] Miehe, C. A micro-macro approach to rubber-like materials - Part I: the non-affine micro-sphere model of rubber elasticity. *Journal of the Mechanics and Physics of Solids*, 52(11):2617–2660, November 2004.

[268] Miehe, C., Apel, N., and Lambrecht, M. Anisotropic additive plasticity in the logarithmic strain space: modular kinematic formulation and implementation based on incremental minimization principles for standard materials. *Computer Methods in Applied Mechanics and Engineering*, 191(47-48):5383–5425, November 2002.

[269] Miehe, C., Göktepe, S., and Méndez Diez, J. Finite viscoplasticity of amorphous glassy polymers in the logarithmic strain space. *International Journal of Solids and Structures*, 46(1):181–202, January 2009.

[270] Miehe, C. and Keck, J. Superimposed finite elastic–viscoelastic–plastoelastic stress response with damage in filled rubbery polymers. Experiments, modelling and algorithmic implementation. *Journal of the Mechanics and Physics of Solids*, 48(2):323–365, February 2000.

[271] Miehe, C. and Koch, A. Computational micro-to-macro transitions of discretized microstructures undergoing small strains. *Archive of Applied Mechanics*, 72(4):300–317, 2002.

[272] Miehe, C. *Kanonische Modelle multiplikativer Elasto-Plastizität: thermodynamische Formulierung und numerische Implementation*. Inst. f. Baumechanik u. Numerische Mechanik, Univ. Hannover, 1993.

[273] Miehe, C. A constitutive frame of elastoplasticity at large strains based on the notion of a plastic metric. *International Journal of Solids and Structures*, 35(30):3859–3897, October 1998.

[274] Miehe, C. A formulation of finite elastoplasticity based on dual co- and contra-variant eigenvector triads normalized with respect to a plastic metric. *Computer Methods in Applied Mechanics and Engineering*, 159(3-4):223–260, July 1998.

[275] Miehe, C. and Apel, N. Anisotropic elastic-plastic analysis of shells at large strains. A comparison of multiplicative and additive approaches to enhanced finite element design and constitutive modelling. *International Journal for Numerical Methods in Engineering*, 61(12):2067–2113, November 2004.

[276] Miehe, C. and Lambrecht, M. Algorithms for computation of stresses and elasticity moduli in terms of Seth-Hill's family of generalized strain tensors. *Communications in Numerical Methods in Engineering*, 17(5):337–353, April 2001.

[277] Miehe, C., Méndez Diez, J., Göktepe, S., and Schänzel, L.-M. Coupled thermoviscoplasticity of glassy polymers in the logarithmic strain space based on the free volume theory. *International Journal of Solids and Structures*, 48(13):1799–1817, June 2011.

[278] Mikhaluk, D., Truong, T., Borovkov, A., Lomov, S., and Verpoest, I. Experimental observations and finite element modelling of damage initiation and evolution in carbon/epoxy non-crimp fabric composites. *Engineering Fracture Mechanics*, 75(9):2751–2766, June 2008.

[279] Mitchell, G. *Topics in the numerical analysis of inelastic solids*. PhD thesis, Department of Civil Engineering, University College of Swansea, 1990.

[280] Mori, T. and Tanaka, K. Average stress in matrix and average elastic energy of materials with misfitting inclusions. *Acta Metallurgica*, 21(5):571–574, May 1973.

[281] Mouritz, A., Bannister, M., Falzon, P., and Leong, K. Review of applications for advanced three-dimensional fibre textile composites. *Composites Part A: Applied Science and Manufacturing*, 30(12):1445–1461, December 1999.

[282] Mura, T. *Micromechanics of defects in solids*. Springer Science & Business Media, 2013.

[283] Murakami, S. Anisotropic damage theory and its application to creep crack growth analysis. *Constitutive Laws for Engineering Materials: Theory and Applications.*, 1:187–194, 1987.

[284] Murakami, S. Mechanical modeling of material damage. *Journal of Applied Mechanics*, 55(2):280, 1988.

[285] Murakami, S. and Ohno, N. A continuum theory of creep and creep damage. In *Creep in Structures: 3rd Symposium, Leicester, UK, September 8–12, 1980*, pages 422–444. Springer Berlin Heidelberg, Berlin, Heidelberg, 1981.

[286] Naghdi, P. M. A critical review of the state of finite plasticity. *ZAMP Zeitschrift für angewandte Mathematik und Physik*, 41(3):315–394, May 1990.

[287] Naik, N. K., Azad, S. N. M., and Prasad, P. D. Stress and failure analysis of 3D angle interlock woven composites. *Journal of Composite Materials*, 36(1):93–123, January 2002.

[288] Naik, N. K. and Ganesh, V. K. Failure behavior of plain weave fabric laminates under on-axis uniaxial tensile loading: I—Analytical predictions. *Journal of Composite Materials*, 30(16):1779–1822, November 1996.

[289] Naik, N. K. and Sridevi, E. An analytical method for thermoelastic analysis of 3D orthogonal interlock woven composites. *Journal of Reinforced Plastics and Composites*, 21(13):1149–1191, September 2002.

[290] Naik, N. and Ganesh, V. Prediction of on-axes elastic properties of plain weave fabric composites. *Composites Science and Technology*, 45(2):135–152, January 1992.

[291] Naik, N. and Kuchibhotla, R. Analytical study of strength and failure behaviour of plain weave fabric composites made of twisted yarns. *Composites Part A: Applied Science and Manufacturing*, 33(5):697–708, May 2002.

[292] Naik, N. and Shembekar, P. Elastic behavior of woven fabric composites: I - Lamina analysis. *Journal of Composite Materials*, 26(15):2196–2225, December 1992.

[293] Naik, N. and Shembekar, P. Elastic behavior of woven fabric composites: III - Laminate design. *Journal of Composite Materials*, 26(17):2522–2541, December 1992.

[294] Naik, R. A. Failure analysis of woven and braided fabric reinforced composites. *Journal of Composite Materials*, 29(17):2334–2363, November 1995.

[295] Neitzel, M., Mitschang, P., and Breuer, U. *Handbuch Verbundwerkstoffe: Werkstoffe, Verarbeitung, Anwendung*. Carl Hanser Verlag GmbH Co KG, 2014.

[296] Neto, E. A. d. S., Perić, D., and Owen, D. R. J. *Computational methods for plasticity: Theory and applications*. Wiley, Chichester, West Sussex, UK, 2008.

[297] Nicoletto, G. and Riva, E. Failure mechanisms in twill-weave laminates: FEM predictions vs. experiments. *Composites Part A: Applied Science and Manufacturing*, 35(7-8):787–795, July 2004.

[298] Norman, T. L., Allison, P., Baldwin, J. W., Gracias, B. K., and Seesdorf, D. Effect of tow alignment on the mechanical performance of 3D woven textile composites. *Composites Manufacturing*, 4(4):209–215, December 1993.

[299] Oda, M. A method for evaluating the effect of crack geometry on the mechanical behavior of cracked rock masses. *Mechanics of Materials*, 2(2):163–171, August 1983.

[300] Olave, M., Vanaerschot, A., Lomov, S. V., and Vandepitte, D. Internal geometry variability of two woven composites and related variability of the stiffness. *Polymer Composites*, 33(8):1335–1350, August 2012.

[301] Oller, S., Canet, J. M., and Zalamea, F. Composite material behavior using a homogenization double scale method. *Journal of Engineering Mechanics*, 1:65–79, 2005.

[302] Onat, E. Representation of mechanical behavior in the presence of internal damage. *Engineering Fracture Mechanics*, 25(5-6):605–614, January 1986.

[303] Ortiz, M. A constitutive theory for the inelastic behavior of concrete. *Mechanics of Materials*, 4(1):67–93, March 1985.

[304] Osswald, T. A. and Menges, G. *Materials science of polymers for engineers*. Carl Hanser Verlag GmbH Co KG, 2012.

[305] Owens, B. C., Whitcomb, J. D., and Varghese, J. Effect of finite thickness and free edges on stresses in plain weave composites. *Journal of Composite Materials*, 44(6):675–692, March 2010.

[306] Papadopoulos, P. and Lu, J. A general framework for the numerical solution of problems in finite elasto-plasticity. *Computer Methods in Applied Mechanics and Engineering*, 159(1-2):1–18, July 1998.

[307] Papadopoulos, P. and Lu, J. On the formulation and numerical solution of problems in anisotropic finite plasticity. *Computer Methods in Applied Mechanics and Engineering*, 190(37-38):4889–4910, June 2001.

[308] Paris, F. A study of failure criteria of fibrous composite materials. *NASA contractor report 2001-210661*, 2001.

[309] Pastore, C. M. Illustrated glossary of textile terms for composites. 1993.

[310] Peerlings, R. H. J. *Enhanced damage modelling for fracture and fatigue*. 1999.

[311] Perić, D. On a class of constitutive equations in viscoplasticity: Formulation and computational issues. *International Journal for Numerical Methods in Engineering*, 36(8):1365–1393, April 1993.

[312] Perie, G., Lomov, S. V., Verpoest, I., and Marsal, D. Meso-scale modelling and homogenisation of interlock reinforced composite. In *17th International Conference on Composite Materials (ICCM-17)*, 2009.

[313] Perzyna, P. Fundamental problems in viscoplasticity. In *Advances in Applied Mechanics*, volume 9, pages 243–377. Elsevier, 1966.

[314] Perzyna, P. Thermodynamic theory of viscoplasticity. In *Advances in Applied Mechanics*, volume 11, pages 313–354. Elsevier, 1971.

[315] Pinho, S. T., Dàvila, C. G., Camanho, P. P., Iannucci, L., and Robinson, P. Failure models and criteria for FRP under in-plane or three-dimensional stress states including shear non-linearity. 2005.

[316] Pochiraju, K. and Chou, T. Three-dimensionally woven and braided composites. I: A model for anisotropic stiffness prediction. *Polymer Composites*, 20(4):565–580, August 1999.

[317] Praud, F., Chatzigeorgiou, G., Chemisky, Y., and Meraghni, F. Hybrid micromechanical-phenomenological modelling of anisotropic damage and anelasticity induced by micro-cracks in unidirectional composites. *Composite Structures*, 182:223–236, December 2017.

[318] Prodromou, A. G., Lomov, S. V., and Verpoest, I. The method of cells and the mechanical properties of textile composites. *Composite Structures*, 93(4):1290–1299, March 2011.

[319] Puck, A. *Festigkeitsanalyse von Faser-Matrix-Laminaten: Modelle für die Praxis*. Hanser, München, 1996.

[320] Puck, A., Kopp, J., and Knops, M. Guidelines for the determination of the parameters in Puck's action plane strength criterion. *Composites Science and Technology*, 62(3):371–378, February 2002.

[321] Puck, A. and Schürmann, H. Failure analysis of FRP laminates by means of physically based phenomenological models. In *Failure Criteria in Fibre-Reinforced-Polymer Composites*, pages 832–876. Elsevier, 2004.

[322] Pukánszky, B., Van Es, M., Maurer, F. H. J., and Vörös, G. Micromechanical deformations in particulate filled thermoplastics: Volume strain measurements. *Journal of Materials Science*, 29(9):2350–2358, January 1994.

[323] Pyrz, R. Correlation of microstructure variability and local stress field in two-phase materials. *Materials Science and Engineering: A*, 177(1-2):253–259, April 1994.

[324] Pyrz, R. Quantitative description of the microstructure of composites. Part I: Morphology of unidirectional composite systems. *Composites Science and Technology*, 50(2):197–208, January 1994.

[325] Quek, S. C., Waas, A., Shahwan, K. W., and Agaram, V. Compressive response and failure of braided textile composites: Part 2 — Computations. *International Journal of Non-Linear Mechanics*, 39(4):649–663, June 2004.

[326] Quek, S. C., Waas, A. M., Shahwan, K. W., and Agaram, V. Analysis of 2D triaxial flat braided textile composites. *International Journal of Mechanical Sciences*, 45(6-7):1077–1096, June 2003.

[327] Rabotnov, Y. N. On the equation of state of creep. *Proceedings of the Institution of Mechanical Engineers, Conference Proceedings*, 178(1):2–117–2–122, June 1963.

[328] Raghava, R., Caddell, R. M., and Yeh, G. S. The macroscopic yield behaviour of polymers. *Journal of Materials Science*, 8(2):225–232, 1973.

[329] Rakotoarisoa, C., Laurin, F., Hirsekorn, M., Maire, J., and Olivier, L. Development of a fatigue model for 3D woven polymer matrix composites based on a damage model. *Proceedings of ECCM15*, 2011.

[330] Ramakrishna, S. Characterization and modeling of the tensile properties of plain weft-knit fabric-reinforced composites. *Composites Science and Technology*, 57(1):1–22, January 1997.

[331] Reckling, K.-A. *Plastizitätstheorie und ihre Anwendung auf Festigkeitsprobleme*. Springer-Verlag, 2013.

[332] Reese, S., Raible, T., and Wriggers, P. Finite element modelling of orthotropic material behaviour in pneumatic membranes. *International Journal of Solids and Structures*, 38(52):9525–9544, December 2001.

[333] Reese, S. Meso-macro modelling of fibre-reinforced rubber-like composites exhibiting large elastoplastic deformation. *International Journal of Solids and Structures*, 40(4):951–980, February 2003.

[334] Rice, J. Inelastic constitutive relations for solids: An internal-variable theory and its application to metal plasticity. *Journal of the Mechanics and Physics of Solids*, 19(6):433–455, November 1971.

[335] Rider, J. G. and Hargreaves, E. Yielding of oriented poly(vinyl chloride). *Journal of Polymer Science Part A-2: Polymer Physics*, 7(5):829–844, May 1969.

[336] Rivlin, R. S. and Ericksen, J. L. Stress-deformation relations for isotropic materials. In *Collected Papers of R.S. Rivlin: Volume I and II*, pages 911–1013. Springer New York, New York, 1997.

[337] Robertson, R. E. Theory for the plasticity of glassy polymers. *The Journal of Chemical Physics*, 44(10):3950–3956, May 1966.

[338] Rösler, J., Harders, H., and Bäker, M. *Mechanisches Verhalten der Werkstoffe*. Springer-Verlag, 2016.

[339] Rösner, A. *Beschreibung des viskoelastischen und viskoplastischen Deformationsverhaltens von kontinuierlich faserverstärktem Polyamid-6*. Karlsruher Schriftenreihe Fahrzeugsystemtechnik. KIT Scientific Publishing, 2015.

[340] Ruan, X. and Chou, T. W. Experimental and theoretical studies of the elastic behavior of knitted-fabric composites. *Composites Science and Technology*, 56(12):1391–1403, 1996.

[341] Sankar, B. V. and Marrey, R. V. Analytical method for micromechanics of textile composites. *Composites Science and Technology*, 57(6):703–713, January 1997.

[342] Sause, M., Gribov, A., Unwin, A., and Horn, S. Pattern recognition approach to identify natural clusters of acoustic emission signals. *Pattern Recognition Letters*, 33(1):17–23, January 2012.

[343] Schenkel, G. Kunststoffe. (63):49 – 53, 1973.

[344] Schirmaier, F., Weiland, J., Kärger, L., and Henning, F. A new efficient and reliable algorithm to determine the fracture angle for Puck's 3d matrix failure criterion for UD composites. *Composites Science and Technology*, 100:19–25, August 2014.

[345] Schleicher, F. v. Der Spannungszustand an der Fliessgrenze (Plastizitätsbedingung). *ZAMM-Journal of Applied Mathematics and Mechanics/Zeitschrift für Angewandte Mathematik und Mechanik*, 6(3):199–216, 1926.

[346] Schröder, J. *Theoretische und algorithmische Konzepte zur phänomenologischen Beschreibung anisotropen Materialverhaltens*. PhD thesis, Institut für Mechanik und numerische Mechanik, Universität Hannover, Hannover, 1996.

[347] Schreiber, W. *Zur Gestaltung und Dimensionierung von Antriebswellen aus Faser-Kunststoff-Verbunden*. PhD thesis, VDI-Verlag, 1989.

[348] Searles, K., Odegard, G., and Kumosa, M. Micro- and mesomechanics of 8-harness satin woven fabric composites: I - Evaluation of elastic behavior. *Composites Part A: Applied Science and Manufacturing*, 32(11):1627–1655, November 2001.

[349] Shembekar, P. and Naik, N. Elastic behavior of woven fabric composites: II - Laminate analysis. *Journal of Composite Materials*, 26(15):2226–2246, December 1992.

[350] Sherburn, M. TexGen open source project, online at http://texgen.sourceforge.net/, 2018.

[351] Shokrieh, M. M. and Mazloomi, M. S. An analytical method for calculating stiffness of two-dimensional tri-axial braided composites. *Composite Structures*, 92(12):2901–2905, November 2010.

[352] Simo, J. and Hughes, T. *Elastoplasticity and viscoplasticity: Computational aspects*. Stanford Univ., Division of Applied Mechanics, 1988.

[353] Simo, J. and Ju, J. Strain- and stress-based continuum damage models - I. Formulation. *International Journal of Solids and Structures*, 23(7):821–840, January 1987.

[354] Simo, J. and Ju, J. Strain- and stress-based continuum damage models - II. Computational aspects. *International Journal of Solids and Structures*, 23(7):841–869, January 1987.

[355] SIMULIA. *Abaqus online documentation: Version 6.14*. Dassault Systemes Simulia Corp., 2014.

[356] Skrzypek, J. J. *Plasticity and creep. Theory, examples, and problems*. CRC Press, Boca Raton, 1993.

[357] Smit, R., Brekelmans, W., and Meijer, H. Prediction of the mechanical behavior of nonlinear heterogeneous systems by multi-level finite element modeling. *Computer Methods in Applied Mechanics and Engineering*, 155(1-2):181–192, March 1998.

[358] Sánchez-Palencia, E. Non-homogeneous media and vibration theory. *Lecture notes in physics*, 127, 1980.

[359] Soden, P. Lamina properties, lay-up configurations and loading conditions for a range of fibre-reinforced composite laminates. *Composites Science and Technology*, 58(7):1011–1022, July 1998.

[360] Spahn, J. *An efficient multiscale method for modeling progressive damage in composite materials*. Number 9 in Forschungsbericht / Technische Universität Kaiserslautern, Lehrstuhl für Technische Mechanik. Techn. Univ. Kaiserslautern, Lehrstuhl für Techn. Mechanik, Kaiserslautern, 2015.

[361] Sørensen, B. and Talreja, R. Effects of nonuniformity of fiber distribution on thermally-induced residual stresses and cracking in ceramic matrix composites. *Mechanics of Materials*, 16(4):351–363, November 1993.

[362] Stassi-D'Alia, F. Flow and fracture of materials according to a new limiting condition of yelding. *Meccanica*, 2(3):178–195, 1967.

[363] Suaris, W. A damage theory for concrete incorporating crack growth characteristics. *Constitutive Laws for Engineering Materials, Theory and Applications*, 2:931–998, 1987.

[364] Sugimoto, K., Nakai, A., and Hamada, H. Effect of lamination sequence on mechanical behaviour of woven composites. In *Proceedings of the 7th international conference on textile composites (TexComp-7)*, pages 1–4, 2004.

[365] Suhubi, E. S. Thermoelastic solids in continuum pyhsics. In *Thermoelastic solids in continuum pyhsics*. New York, 1975.

[366] Sun, C. Comparative evaluation of failure analysis methods for composite laminates. Technical report, 1996.

[367] Suquet, P. M. Local and global aspects in the mathematical theory of plasticity. *Plasticity Today: Modelling, Methods and Applications*, Elsevier:279–309, 1985.

[368] Tabiei, A. and Ivanov, I. Materially and geometrically non-linear woven composite micro-mechanical model with failure for finite element simulations. *International Journal of Non-Linear Mechanics*, 39(2):175–188, March 2004.

[369] Tan, P., Tong, L., and Steven, G. Behavior of 3D orthogonal woven CFRP composites. Part II. FEA and analytical modeling approaches. *Composites Part A: Applied Science and Manufacturing*, 31(3):273–281, March 2000.

[370] Tan, P., Tong, L., Steven, G., and Ishikawa, T. Behavior of 3D orthogonal woven CFRP composites. Part I. Experimental investigation. *Composites Part A: Applied Science and Manufacturing*, 31(3):259–271, March 2000.

[371] Tan, P., Tong, L., and Steven, G. P. Micromechanics models for mechanical and thermomechanical properties of 3d through-the-thickness angle interlock woven composites. *Composites Part A: Applied Science and Manufacturing*, 30(5):637–648, May 1999.

[372] Tan, P., Tong, L., and Steven, G. P. Mechanical behavior for 3-D orthogonal woven E-glass/epoxy composites. *Journal of Reinforced Plastics and Composites*, 20(4):274–303, March 2001.

[373] Tandon, G. P. and Weng, G. J. The effect of aspect ratio of inclusions on the elastic properties of unidirectionally aligned composites. *Polymer Composites*, 5(4):327–333, October 1984.

[374] Tang, X. and Whitcomb, J. D. General techniques for exploiting periodicity and symmetries in micromechanics analysis of textile composites. *Journal of Composite Materials*, 37(13):1167–1189, July 2003.

[375] Tang, X. and Whitcomb, J. D. Progressive failure behaviors of 2D woven composites. *Journal of Composite Materials*, 37(14):1239–1259, July 2003.

[376] Tarnopol'skii, Y. M., Polyakov, V. A., and Zhigun, I. G. Composite materials reinforced with a system of three straight, mutually orthogonal fibers: I. Calculation of the elastic characteristics. *Polymer Mechanics*, 9(5):754–759, 1975.

[377] Taya, M. and Mura, T. On stiffness and strength of an aligned short-fiber reinforced composite containing fiber-end cracks under uniaxial applied stress. *Journal of Applied Mechanics*, 48(2):361, 1981.

[378] Taya, M. On stiffness and strength of an aligned short-fiber reinforced composite containing penny-shaped cracks in the matrix. *Journal of Composite Materials*, 15(3):198–210, May 1981.

[379] Taya, M. and Arsenault, R. J. *Metal matrix composites: Thermomechanical behavior*. Elsevier, 2016.

[380] Taya, M. and Chou, T.-W. On two kinds of ellipsoidal inhomogeneities in an infinite elastic body: An application to a hybrid composite. *International Journal of Solids and Structures*, 17(6):553–563, 1981.

[381] Ten Hompel, M. *Textile Werkstoffe für den Leichtbau: Techniken-Verfahren-Materialien*. Springer, 2011.

[382] Terada, K. and Kikuchi, N. A class of general algorithms for multiscale analyses of heterogeneous media. *Computer methods in applied mechanics and engineering*, 190(40):5427–5464, 2001.

[383] Thieme, M., Boehm, R., Gude, M., and Hufenbach, W. Probabilistic failure simulation of glass fibre reinforced weft-knitted thermoplastics. *Composites Science and Technology*, 90:25–31, January 2014.

[384] Tomita, Y. and Tanaka, S. Prediction of deformation behavior of glassy polymers based on molecular chain network model. *International Journal of Solids and Structures*, 32(23):3423–3434, December 1995.

[385] Tresca, H. *Mémoires sur l'écoulement des corps solides*. Imprimerie impériale, 1869.

[386] Trias, D., Costa, J., Mayugo, J., and Hurtado, J. Random models versus periodic models for fibre reinforced composites. *Computational Materials Science*, 38(2):316–324, December 2006.

[387] Trias, D., Costa, J., Turon, A., and Hurtado, J. Determination of the critical size of a statistical representative volume element (SRVE) for carbon reinforced polymers. *Acta Materialia*, 54(13):3471–3484, August 2006.

[388] Truesdell, C. *The elements of continuum mechanics*. Springer Berlin Heidelberg, Berlin, Heidelberg, 1984.

[389] Truesdell, C. and Noll, W. *The non-linear field theories of mechanics*. Springer Berlin Heidelberg, Berlin, Heidelberg, 2004.

[390] Truong, C. T., Ho Chiew, J., Stepan Vladimirovitch, L., and Ignace, V. Sheared biaxial multi-ply carbon fabrics reinforced epoxy composites: the mechanical properties and damage initiation. In *Proc. SAMPE Europe International Conference 2005*, pages 252–257, 2005.

[391] Truong, T. C., Vettori, M., Lomov, S., and Verpoest, I. Carbon composites based on multi-axial multi-ply stitched preforms. Part 4. Mechanical properties of composites and damage observation. *Composites Part A: Applied Science and Manufacturing*, 36(9):1207–1221, September 2005.

[392] Tsai, S. W. and Wu, E. M. A general theory of strength for anisotropic materials. *Journal of Composite Materials*, 5(1):58–80, January 1971.

[393] Tschoegl, N. W. Failure surfaces in principal stress space. *Journal of Polymer Science Part C: Polymer Symposia*, 32(1):239–267, 1971.

[394] Tucker III, C. L. and Liang, E. Stiffness predictions for unidirectional short fiber composites: Review and evaluation. *Composites Science and Technology*, 59(5):655–671, April 1999.

[395] Turon, A., Camanho, P., Costa, J., and Dávila, C. A damage model for the simulation of delamination in advanced composites under variable-mode loading. *Mechanics of Materials*, 38(11):1072–1089, November 2006.

[396] Uemura, M., Ataka, N., Fukuda, H., and Ben, G. Practical FRP structural strength calculations. *Japan Society of Reinforced Plastics*, 1984.

[397] Uetsuji, Y., Zako, M., and Nishiyabu, K. Numerical analysis and in-situ SEM observation of damage development for woven fabric composite materials. *Journal of the Society of Materials Science, Japan(Japan)*, 51(10):1147–1153, 2002.

[398] Valanis, K. A theory of damage in brittle materials. *Engineering Fracture Mechanics*, 36(3):403–416, January 1990.

[399] van Hartingsveldt, E. and van Aartsen, J. Interfacial debonding in polyamide-6/glass bead composites. *Polymer*, 30(11):1984–1991, November 1989.

[400] Vandeurzen, P., Ivens, J., and Verpoest, I. A three-dimensional micromechanical analysis of woven-fabric composites: I. Geometric analysis. *Composites Science and Technology*, 56(11):1303–1315, January 1996.

[401] Vandeurzen, P., Ivens, J., and Verpoest, I. A three-dimensional micromechanical analysis of woven-fabric composites: II. Elastic analysis. *Composites Science and Technology*, 56(11):1317–1327, January 1996.

[402] Verpoest, I. and Lomov, S. Virtual textile composites software : Integration with micro-mechanical, permeability and structural analysis. *Composites Science and Technology*, 65(15-16):2563–2574, December 2005.

[403] Vincent, P. The necking and cold-drawing of rigid plastics. *Polymer*, 1:7–19, January 1960.

[404] Voigt, W. Ueber die Beziehung zwischen den beiden Elasticitätsconstanten isotroper Körper. *Annalen der Physik*, 274(12):573–587, 1889.

[405] von Mises, R. Mechanik der festen Körper im plastisch-deformablen Zustand. *Nachrichten von der Gesellschaft der Wissenschaften zu Göttingen, Mathematisch-Physikalische Klasse*, 1913(4):582–592, 1913.

[406] Voyiadjis, G. Z. and Park, T. Anisotropic damage effect tensors for the symmetrization of the effective stress tensor. *Journal of Applied Mechanics*, 64(1):106, 1997.

[407] Voyiadjis, G. Z. and Deliktas, B. A coupled anisotropic damage model for the inelastic response of composite materials. *Computer Methods in Applied Mechanics and Engineering*, 183(3-4):159–199, March 2000.

[408] Wang, L., Wu, J., Chen, C., Zheng, C., Li, B., Joshi, S. C., and Zhou, K. Progressive failure analysis of 2D woven composites at the meso-micro scale. *Composite Structures*, 178:395–405, October 2017.

[409] Wei, S., Li-Hua, P., Yun-Guo, Y., Zeng, S., and Xian-Dong, T. Elastic damage and energy dissipation in anisotropic solid material. *Engineering Fracture Mechanics*, 33(2):273–281, January 1989.

[410] Wevers, M. and Surgeon, M. Acoustic emission and composites. 2000.

[411] Whitcomb, J. and Srirengan, K. Effect of various approximations on predicted progressive failure in plain weave composites. *Composite Structures*, 34(1):13–20, January 1996.

[412] Whitcomb, J., Srirengan, K., and Chapman, C. Evaluation of homogenization for global/local stress analysis of textile composites. *Composite Structures*, 31(2):137–149, January 1995.

[413] Whitcomb, J. and Tang, X. Effective moduli of woven composites. *Journal of Composite Materials*, 35(23):2127–2144, December 2001.

[414] Whitcomb, J. D. Three-dimensional stress analysis of plain weave composites. In *Composite Materials: Fatigue and Fracture (Third Volume)*. ASTM International, 1991.

[415] Whitcomb, J. D. Analysis of new composite architectures. *NASA contractor report 198506*, 1996.

[416] Whitcomb, J. D., Chapman, C. D., and Tang, X. Derivation of boundary conditions for micromechanics analyses of plain and satin weave composites. *Journal of Composite Materials*, 34(9):724–747, May 2000.

[417] Whitney, T. J. and Chou, T.-W. Modeling of 3-D angle-interlock textile structural composites. *Journal of Composite Materials*, 23(9):890–911, September 1989.

[418] Willis, J. Variational and related methods for the overall properties of composites. In *Advances in Applied Mechanics*, volume 21, pages 1–78. Elsevier, 1981.

[419] Wongsto, A. and Li, S. Micromechanical FE analysis of UD fibre-reinforced composites with fibres distributed at random over the transverse cross-section. *Composites Part A: Applied Science and Manufacturing*, 36(9):1246–1266, September 2005.

[420] Woo, K. and Whitcomb, J. D. A post-processor approach for stress analysis of woven textile composites. *Composites Science and Technology*, 60(5):693–704, April 2000.

[421] Wu, C.-T. and McCullough, R. Constitutive relationships for heterogeneous materials(elastic properties of polycrystals and composite materials). *Developments in composite materials-1.(A 78-38276 16-24) London, Applied Science Publishers, Ltd.*, pages 119–187, 1977.

[422] Wu, P. and van der Giessen, E. On improved network models for rubber elasticity and their applications to orientation hardening in glassy polymers. *Journal of the Mechanics and Physics of Solids*, 41(3):427–456, March 1993.

[423] Wu, P. and van der Giessen, E. Analysis of shear band propagation in amorphous glassy polymers. *International Journal of Solids and Structures*, 31(11):1493–1517, June 1994.

[424] Wu, P. and van der Giessen, E. On neck propagation in amorphous glassy polymers under plane strain tension. *International Journal of Plasticity*, 11(3):211–235, January 1995.

[425] Xiao, H., Bruhns, O. T., and Meyers, A. Elastoplasticity beyond small deformations. *Acta Mechanica*, 182(1-2):31–111, March 2006.

[426] Xu, J., Lomov, S. V., Verpoest, I., Daggumati, S., Paepegem, W. V., Degrieck, J., and Olave, M. A progressive damage model of textile composites on meso-scale using finite element method: Static damage analysis. *Journal of Composite Materials*, 48(25):3091–3109, October 2014.

[427] Yang, J.-M., Ma, C.-L., and Chou, T.-W. Fiber inclination model of three-dimensional textile structural composites. *Journal of Composite Materials*, 20(5):472–484, September 1986.

[428] Yazdani, S. and Schreyer, H. An anisotropic damage model with dilatation for concrete. *Mechanics of Materials*, 7(3):231–244, December 1988.

[429] Yuan, Z. and Fish, J. Toward realization of computational homogenization in practice. *International Journal for Numerical Methods in Engineering*, 73(3):361–380, January 2008.

[430] Zako, M., Uetsuji, Y., and Kurashiki, T. Finite element analysis of damaged woven fabric composite materials. *Composites Science and Technology*, 63(3):507–516, 2003.

[431] Zeman, J. and Šejnoha, M. Homogenization of balanced plain weave composites with imperfect microstructure: Part I—Theoretical formulation. *International Journal of Solids and Structures*, 41(22-23):6549–6571, November 2004.

[432] Zhong, S., Guo, L., liu, G., Lu, H., and Zeng, T. A continuum damage model for three-dimensional woven composites and finite element implementation. *Composite Structures*, 128:1–9, September 2015.

[433] Zhou, Y., Lu, Z., and Yang, Z. Progressive damage analysis and strength prediction of 2D plain weave composites. *Composites Part B: Engineering*, 47:220–229, April 2013.

[434] Zhuk, A. V., Knunyants, N. N., Oshmyan, V. G., Topolkaraev, V. A., and Berlin, A. A. Debonding microprocesses and interfacial strength in particle-filled polymer materials. *Journal of Materials Science*, 28(17):4595–4606, 1993.

Karlsruher Schriftenreihe Fahrzeugsystemtechnik (ISSN 1869-6058)

Herausgeber: FAST Institut für Fahrzeugsystemtechnik

Band 1 Urs Wiesel
Hybrides Lenksystem zur Kraftstoffeinsparung im schweren Nutzfahrzeug. 2010
ISBN 978-3-86644-456-0

Band 2 Andreas Huber
Ermittlung von prozessabhängigen Lastkollektiven eines hydrostatischen Fahrantriebsstrangs am Beispiel eines Teleskopladers. 2010
ISBN 978-3-86644-564-2

Band 3 Maurice Bliesener
Optimierung der Betriebsführung mobiler Arbeitsmaschinen. Ansatz für ein Gesamtmaschinenmanagement. 2010
ISBN 978-3-86644-536-9

Band 4 Manuel Boog
Steigerung der Verfügbarkeit mobiler Arbeitsmaschinen durch Betriebslasterfassung und Fehleridentifikation an hydrostatischen Verdrängereinheiten. 2011
ISBN 978-3-86644-600-7

Band 5 Christian Kraft
Gezielte Variation und Analyse des Fahrverhaltens von Kraftfahrzeugen mittels elektrischer Linearaktuatoren im Fahrwerksbereich. 2011
ISBN 978-3-86644-607-6

Band 6 Lars Völker
Untersuchung des Kommunikationsintervalls bei der gekoppelten Simulation. 2011
ISBN 978-3-86644-611-3

Band 7 3. Fachtagung
Hybridantriebe für mobile Arbeitsmaschinen. 17. Februar 2011, Karlsruhe. 2011
ISBN 978-3-86644-599-4

Die Bände sind unter www.ksp.kit.edu als PDF frei verfügbar oder als Druckausgabe bestellbar.

Band 8 Vladimir Iliev
Systemansatz zur anregungsunabhängigen Charakterisierung des Schwingungskomforts eines Fahrzeugs. 2011
ISBN 978-3-86644-681-6

Band 9 Lars Lewandowitz
Markenspezifische Auswahl, Parametrierung und Gestaltung der Produktgruppe Fahrerassistenzsysteme. Ein methodisches Rahmenwerk. 2011
ISBN 978-3-86644-701-1

Band 10 Phillip Thiebes
Hybridantriebe für mobile Arbeitsmaschinen. Grundlegende Erkenntnisse und Zusammenhänge, Vorstellung einer Methodik zur Unterstützung des Entwicklungsprozesses und deren Validierung am Beispiel einer Forstmaschine. 2012
ISBN 978-3-86644-808-7

Band 11 Martin Gießler
Mechanismen der Kraftübertragung des Reifens auf Schnee und Eis. 2012
ISBN 978-3-86644-806-3

Band 12 Daniel Pies
Reifenungleichförmigkeitserregter Schwingungskomfort – Quantifizierung und Bewertung komfortrelevanter Fahrzeugschwingungen. 2012
ISBN 978-3-86644-825-4

Band 13 Daniel Weber
Untersuchung des Potenzials einer Brems-Ausweich-Assistenz. 2012
ISBN 978-3-86644-864-3

Band 14 **7. Kolloquium Mobilhydraulik. 27./28. September 2012 in Karlsruhe.** 2012
ISBN 978-3-86644-881-0

Band 15 4. Fachtagung
Hybridantriebe für mobile Arbeitsmaschinen 20. Februar 2013, Karlsruhe. 2013
ISBN 978-3-86644-970-1

Band 16 Hans-Joachim Unrau
Der Einfluss der Fahrbahnoberflächenkrümmung auf den Rollwiderstand, die Cornering Stiffness und die Aligning Stiffness von Pkw-Reifen. 2013
ISBN 978-3-86644-983-1

Band 17 Xi Zhang
Untersuchung und Entwicklung verschiedener Spurführungsansätze für Offroad-Fahrzeuge mit Deichselverbindung. Noch nicht erschienen
ISBN 978-3-7315-0005-6

Band 18 Stefanie Grollius
Analyse des gekoppelten Systems Reifen-Hohlraum-Rad-Radführung im Rollzustand und Entwicklung eines Rollgeräuschmodells. 2013
ISBN 978-3-7315-0029-2

Band 19 Tobias Radke
Energieoptimale Längsführung von Kraftfahrzeugen durch Einsatz vorausschauender Fahrstrategien. 2013
ISBN 978-3-7315-0069-8

Band 20 David Gutjahr
Objektive Bewertung querdynamischer Reifeneigenschaften im Gesamtfahrzeugversuch. 2014
ISBN 978-3-7315-0153-4

Band 21 Neli Ovcharova
Methodik zur Nutzenanalyse und Optimierung sicherheitsrelevanter Fahrerassistenzsysteme. 2014
ISBN 978-3-7315-0176-3

Band 22 Marcus Geimer, Christian Pohlandt
Grundlagen mobiler Arbeitsmaschinen. 2014
ISBN 978-3-7315-0188-6

Band 23 Timo Kautzmann
Die mobile Arbeitsmaschine als komplexes System. 2014
ISBN 978-3-7315-0187-9

Band 24 Roman Weidemann
Analyse der mechanischen Randbedingungen zur Adaption der oszillierenden Hinterschneidtechnik an einen Mobilbagger. 2014
ISBN 978-3-7315-0193-0

Band 25 Yunfan Wei
Spurführungsregelung eines aktiv gelenkten Radpaars für Straßenbahnen. 2014
ISBN 978-3-7315-0232-6

Band 26 David Schmitz
Entwurf eines fehlertoleranten Lenkventils für Steer-by-Wire Anwendungen bei Traktoren. 2014
ISBN 978-3-7315-0264-7

Band 27 Christian Schwab
Beitrag zu einer universellen Baggerschnittstelle zur Übertragung elektrischer und hydraulischer Leistung sowie elektronischer Signale für komplexe Anbaugeräte. 2014
ISBN 978-3-7315-0281-4

Band 28 Peter Dengler
Untersuchung zum effizienten Betrieb von Hydraulikzylindern in Konstantdrucksystemen unter Verwendung einer Zwischendruckleitung. 2015
ISBN 978-3-7315-0295-1

Band 29 Manuel Bös
Untersuchung und Optimierung der Fahrkomfort- und Fahrdynamikeigenschaften von Radladern unter Berücksichtigung der prozessspezifischen Randbedingungen. 2015
ISBN 978-3-7315-0310-1

Band 30 5. Fachtagung
Hybride und energieeffiziente Antriebe für mobile Arbeitsmaschinen. 25. Februar 2015, Karlsruhe. 2015
ISBN 978-3-7315-0323-1

Band 31 Michael Eckert
Energieoptimale Fahrdynamikregelung mehrmotoriger Elektrofahrzeuge. 2015
ISBN 978-3-7315-0332-3

Band 32 Martin Scherer
Beitrag zur Effizienzsteigerung mobiler Arbeitsmaschinen. Entwicklung einer elektrohydraulischen Bedarfsstromsteuerung mit aufgeprägtem Volumenstrom. 2015
ISBN 978-3-7315-0339-2

Band 33 Rinaldo Arnold
Automatische Abstimmung der Sekundärseite eines dreiphasigen Systems zur berührungslosen induktiven Energieübertragung. 2015
ISBN 978-3-7315-0355-2

Band 34 Johannes Gültlinger
Kraftübertragung und Fahrbahnverschleiß durch Spikereifen. 2015
ISBN 978-3-7315-0358-3

Die Bände sind unter www.ksp.kit.edu als PDF frei verfügbar oder als Druckausgabe bestellbar.

Band 35	Thorsten Dreher **Energieeffizienz von Konstantdrucksystemen mit sekundärgeregelten Antrieben beim Einsatz in mobilen Arbeitsmaschinen.** 2015 ISBN 978-3-7315-0377-4
Band 36	Steffen Kölling **Konzeptionelle Untersuchung zur Neigekompensation von Stromabnehmern.** 2015 ISBN 978-3-7315-0387-3
Band 37	Michael Fritz **Entwicklungswerkzeuge für die Fahrzeugklimatisierung von Nutzfahrzeugen.** 2015 ISBN 978-3-7315-0384-2
Band 38	Ralf Oberfell **Stochastische Simulation von Energieflüssen im Nutzfahrzeug. Ein einsatzorientiertes Bewertungs- und Optimierungsverfahren.** 2015 ISBN 978-3-7315-0403-0
Band 39	Christoph Sturm **Bewertung der Energieeffizienz von Antriebssystemen mobiler Arbeitsmaschinen am Beispiel Bagger.** 2015 ISBN 978-3-7315-0404-7
Band 40	Florian Netter **Komplexitätsadaption integrierter Gesamtfahrzeugsimulationen.** 2016 ISBN 978-3-7315-0414-6
Band 41	Markus Springmann **Auslegung eines asynchronen Langstatorlinearmotors mit großem Luftspalt als Straßenbahnantrieb.** 2015 ISBN 978-3-7315-0418-4
Band 42	Alexander Basler **Eine modulare Funktionsarchitektur zur Umsetzung einer gesamtheitlichen Betriebsstrategie für Elektrofahrzeuge.** 2015 ISBN 978-3-7315-0421-4
Band 43	Hans-Georg Wahl **Optimale Regelung eines prädiktiven Energiemanagements von Hybridfahrzeugen.** 2015 ISBN 978-3-7315-0422-1

Band 44	Jennifer Heck **Zur Simulation des Rad-Schiene-Verschleißes bei Straßenbahnen.** 2016 ISBN 978-3-7315-0443-6
Band 45	Moritz Vaillant **Design Space Exploration zur multikriteriellen Optimierung elektrischer Sportwagenantriebsstränge: Variation von Topologie und Komponenteneigenschaften zur Steigerung von Fahrleistungen und Tank-to-Wheel Wirkungsgrad.** 2016 ISBN 978-3-7315-0452-8
Band 46	Philip Nagel **Entwicklung einer Betriebsstrategie zur Energierückgewinnung in hybriden Mehrverbrauchersystemen.** 2016 ISBN 978-3-7315-0479-5
Band 47	Matthias Pfriem **Analyse der Realnutzung von Elektrofahrzeugen in kommerziellen Flotten zur Definition einer bedarfsgerechten Fahrzeugauslegung.** 2016 ISBN 978-3-7315-0489-4
Band 48	Mohanad El-Haji **Ontologie-basierte Definition von Anforderungen an Validierungswerkzeuge in der Fahrzeugtechnik.** 2016 ISBN 978-3-7315-0496-2
Band 49	**9. Kolloquium Mobilhydraulik 22./23. September 2016 in Karlsruhe.** 2016 ISBN 978-3-7315-0573-0
Band 50	6. Fachtagung **Hybride und energieeffiziente Antriebe für mobile Arbeitsmaschinen. 15. Februar 2017, Karlsruhe.** 2017 ISBN 978-3-7315-0601-0
Band 51	Fabian Schirmaier **Experimentelle Untersuchung und Simulation des Umformverhaltens nähgewirkter unidirektionaler Kohlenstofffasergelege.** 2017 ISBN 978-3-7315-0620-1
Band 52	Mathias Cabrera Cano **Neuronale Netze mit externen Laguerre-Filtern zur automatischen numerischen Vereinfachung von Getriebemodellen.** 2017 ISBN 978-3-7315-0621-8

Die Bände sind unter www.ksp.kit.edu als PDF frei verfügbar oder als Druckausgabe bestellbar.

Band 53 Arwed Schmidt
Flottenbetrieb von elektrischen und autonomen Serviceagenten im städtischen Personennahverkehr. 2017
ISBN 978-3-7315-0633-1

Band 54 Katharina Knaisch
Untersuchung von Spulensystemen zur induktiven Energieübertragung von Elektrofahrzeugen. Vergleich von Topologien und Entwicklung einer Auslegungsmethodik. 2017
ISBN 978-3-7315-0630-0

Band 55 Frank Christof Stalter
Ansätze zur akustischen Optimierung von Reifen und Fahrbahnen für Elektrofahrzeuge unter Antriebsmoment. 2017
ISBN 978-3-7315-0645-4

Band 56 Steffen Rose
Modellbildung und Simulation von mobilen Arbeitsmaschinen. Untersuchungen zu systematischen Modellvereinfachungen in der Simulation von Antriebssystemen am Beispiel Bagger. 2017
ISBN 978-3-7315-0684-3

Band 57 Ulrico Peckelsen
Objective Tyre Development. Definition and Analysis of Tyre Characteristics and Quantification of their Conflicts. 2017
ISBN 978-3-7315-0713-0

Band 58 Stefan Haag
Sequentieller Versuch zur HiL-unterstützten Validierung hybrider Antriebssysteme mit gekoppelten Antriebseinheiten. 2018
ISBN 978-3-7315-0752-9

Band 59 Dirk Hülsebusch
Fahrerassistenzsysteme zur energieeffizienten Längsregelung - Analyse und Optimierung der Fahrsicherheit. 2018
ISBN 978-3-7315-0755-0

Band 60 Christian Pohlandt
Intelligentes Gesamtmaschinenmanagement für elektrische Antriebssysteme. 2018
ISBN 978-3-7315-0774-1

Band 61 Oliver Maier
Modellbasierte Entwicklung eines aktiven Sicherheitssystems für elektrifizierte Fahrräder. 2018
ISBN 978-3-7315-0778-9

Band 62	Stephan Rhode **Robust and Regularized Algorithms for Vehicle Tractive Force Prediction and Mass Estimation.** 2018 ISBN 978-3-7315-0807-6
Band 63	Péter Megyesi **Methode zur Einbindung realer Nutzerprofile in die ökologische und ökonomische Bewertung von Fahrzeugkonzepten.** 2018 ISBN 978-3-7315-0808-3
Band 64	Benedikt Reick **Methode zur Analyse und Bewertung von stufenlosen Traktorgetrieben mit mehreren Schnittstellen.** 2018 ISBN 978-3-7315-0815-1
Band 65	Matthias Graf **Entwicklung und Validierung einer neuen Methode zur Charakterisierung des Scherverhaltens von Carbonfaser-gewebe mit Binderauftrag unter Normalkraft und viskoser Reibung bei hohen Prozesstemperaturen.** 2018 ISBN 978-3-7315-0838-0
Band 66	Christoph Kühn **Simulation von Fahrspielen und Energieflüssen in Nahverkehrssystemen.** 2018 ISBN 978-3-7315-0740-6
Band 67	7. Fachtagung **Hybride und energieeffiziente Antriebe für mobile Arbeitsmaschinen. 20. Februar 2019, Karlsruhe.** 2019 ISBN 978-3-7315-0873-1
Band 68	Oliver Krauss **Experimentelle Untersuchungen zum Innengeräusch von Fahrzeugluftreifen.** 2019 ISBN 978-3-7315-0867-0
Band 69	Tristan Reich **Beurteilung der Prüfprozesseignung bei Fahrzeugversuchen mit mobilen Arbeitsmaschinen.** 2019 ISBN 978-3-7315-0848-9
Band 70	Matthias Bürckert **Realitätsnahe Bewertung des Einflusses der Oberflächen-spannung flüssiger Zwischenmedien auf den maximalen Reibschluss zwischen Reifen und Fahrbahn.** 2019 ISBN 978-3-7315-0891-5

Band 71 Matthias Greiner
Verfahren zur Prädiktion des Rollwiderstands bei variablen Betriebsparametern auf Basis standardisierter Rollwiderstandsmessungen. 2019
ISBN 978-3-7315-0898-4

Band 72 David Hummelberger
Hybride Werkstoffsysteme: Systematische Betrachtung und Bewertung der physikalischen Wirkmechanismen. 2019
ISBN 978-3-7315-0901-1

Band 73 Michael Kerber
Analyse dynamischer Reifeneigenschaften und deren Auswirkung auf den Schwingungskomfort. 2019
ISBN 978-3-7315-0944-8

Band 74 Kilian Berthold
Techno-ökonomische Auslegungsmethodik für die Elektrifizierung urbaner Busnetze. 2019
ISBN 978-3-7315-0953-0

Band 75 Bernd Wassertheurer
Reifenmodellierung für die Fahrdynamiksimulation auf Schnee, Eis und nasser Fahrbahn. 2019
ISBN 978-3-7315-0957-8

Band 76 Kai-Lukas Bauer
Echtzeit-Strategieplanung für vorausschauendes automatisiertes Fahren. 2020
ISBN 978-3-7315-0949-3

Band 77 Thomas Schirle
Systementwurf eines elektromechanischen Fahrwerks für Megacitymobilität. 2020
ISBN 978-3-7315-0995-0

Band 78 Dominik Dörr
Simulation of the thermoforming process of UD fiber-reinforced thermoplastic tape laminates. 2020
ISBN 978-3-7315-0998-1

Band 79 Dominik Robert Naake
Simulation of damage mechanisms in weave reinforced materials based on multiscale modeling. 2020
ISBN 978-3-7315-1005-5